Supramolecular Stereochemistry

NATO ASI Series

Advanced Science Institutes Series

A Series presenting the results of activities sponsored by the NATO Science Committee, which aims at the dissemination of advanced scientific and technological knowledge, with a view to strengthening links between scientific communities.

The Series is published by an international board of publishers in conjunction with the NATO Scientific Affairs Division

A Life Sciences	Plenum Publishing Corporation
B Physics	London and New York
C Mathematical and Physical Sciences	Kluwer Academic Publishers
D Behavioural and Social Sciences	Dordrecht, Boston and London
E Applied Sciences	
F Computer and Systems Sciences	Springer-Verlag
G Ecological Sciences	Berlin, Heidelberg, New York, London,
H Cell Biology	Paris and Tokyo
I Global Environmental Change	

PARTNERSHIP SUB-SERIES

1. Disarmament Technologies	Kluwer Academic Publishers
2. Environment	Springer-Verlag / Kluwer Academic Publishers
3. High Technology	Kluwer Academic Publishers
4. Science and Technology Policy	Kluwer Academic Publishers
5. Computer Networking	Kluwer Academic Publishers

The Partnership Sub-Series incorporates activities undertaken in collaboration with NATO's Cooperation Partners, the countries of the CIS and Central and Eastern Europe, in Priority Areas of concern to those countries.

NATO-PCO-DATA BASE

The electronic index to the NATO ASI Series provides full bibliographical references (with keywords and/or abstracts) to more than 50000 contributions from international scientists published in all sections of the NATO ASI Series.
Access to the NATO-PCO-DATA BASE is possible in two ways:

– via online FILE 128 (NATO-PCO-DATA BASE) hosted by ESRIN, Via Galileo Galilei, I-00044 Frascati, Italy.

– via CD-ROM "NATO-PCO-DATA BASE" with user-friendly retrieval software in English, French and German (© WTV GmbH and DATAWARE Technologies Inc. 1989).

The CD-ROM can be ordered through any member of the Board of Publishers or through NATO-PCO, Overijse, Belgium.

Series C: Mathematical and Physical Sciences – Vol. 473

Supramolecular Stereochemistry

edited by

Jay S. Siegel
Department of Chemistry,
University of California,
San Diego, California, U.S.A.

SPRINGER-SCIENCE+BUSINESS MEDIA, B.V.

Proceedings of the NATO Advanced Research Workshop on
Supramolecular Stereochemistry
Hverageroi, Iceland
September 14–19, 1994

A C.I.P. Catalogue record for this book is available from the Library of Congress.

ISBN 978-94-010-4157-7 ISBN 978-94-011-0353-4 (eBook)
DOI 10.1007/978-94-011-0353-4

Printed on acid-free paper

This book contains the proceedings of a NATO Advanced Research Workshop held within the programme of activities of the NATO Special Programme on Supramolecular Chemistry as part of the activities of the NATO Science Committee.

Other books previously published as a result of the activities of the Special Programme are:

WIPFF, G. (Ed.), *Computational Approaches in Supramolecular Chemistry.* (ASIC 426) 1994. ISBN 0-7923-2767-5

FLEISCHAKER, G.R., COLONNA, S. and LUISI, P.L. (Eds.), *Self-Production of Supramolecular Structures.* From Synthetic Structures to Models of Minimal Living Systems. (ASIC 446) 1994. ISBN 0-7923-3163-X

FABBRIZZI, L., POGGI, A. (Eds.), *Transition Metals in Supramolecular Chemistry.* (ASIC 448) 1994. ISBN 0-7923-3196-6

BECHER, J., SCHAUMBURG, K. (Eds.), *Molecular Engineering for Advanced Materials.* (ASIC 456) 1995. ISBN 0-7923-3347-0

LA MAR, G.N. (Ed.), *Nuclear Magnetic Resonance of Paramagnetic Macromolecules.* (ASIC 457) 1995. ISBN 0-7923-3348-9

to my mother in memory of my father

Table of Contents

Preface

Organizing this meeting on Supramolecular Stereochemistry has been a fulfilling and eye-opening adventure. Good fortune reigned through out the process, from the encouragement I received at the outset, to the generous support from NATO and the US National Science Foundation, to the wonderfully positive response that came from the invitees, to the spectacular facilities and surroundings of the conference site in Iceland to the superior proceedings which led to the creation of this book.

Supramolecular Stereochemistry is a topic with enormous breadth, such that participants from widely diverse backgrounds could be invited. The conference brought together experts in polymer chemistry, bioorganic chemistry, crystallography, material science, dendrimer science, nanochemistry, conformational analysis, molecular recognition chemistry, and topological stereochemistry.

The conference was intensive with 29 lectures and 2 poster sessions in 4 days, but was accented by the beautiful Icelandic environment. The town of Hveragerdi, which hosted the meeting was located in an area 50 Km east of Reykjavik, and provided an excellent isolated spot with a complete supportive infrastructure which allowed us to be about or work without extraneous worries.

It is a special opportunity to be able to document this event with this volume of the NATO-ARW Series. Thanks are due to the publisher for their abundant cooperation in this endeavor.

Jay S. Siegel
La Jolla, California, 1995

Acknowledgments

This workshop would not have been possible without the generous financial support of the NATO Supramolecular Chemistry Program (ARW-931583) and the US National Science Foundation. Publicity materials, programs, and conference flyers were provided by the San Diego Supercomputer Center; design and production by Gail Bamber and Kim Baldridge. Conference organization and reservations were handled by Luana Kaar. Many thanks to all.

Introduction

Jay S. Siegel
Department of Chemistry, University of California, San Diego, La Jolla, California 92093-0358.

The successful connection between molecular structure and energy accounts for the enormous development of chemistry in the 20th century. The concepts of chemical structure and energy were rigidly held apart by those 19th century chemists who believed that seeing molecular structure (if there was such a thing) was a fantasy never to be realized; however, the allure of a tangible model for molecular structure proved irresistible to another group of chemists. These chemists perpetrated the fantasy of molecular structure as a connection between the visual three-dimensional mechanical world and the microscopic world of atoms and molecules, and ultimately founded the field of stereochemistry; the use of physical molecular models as representations of molecules (molecular ensembles) thus became the earmark of this field. One area of modern chemistry that has profited enormously from visual tradition of stereochemical models is supramolecular chemistry. Indeed, a century after the founding of stereochemistry, the scale and complexity of the chemical problems have expanded and the principles of stereochemistry have matured such that any supramolecular process can be well served by a parallel understanding of its stereochemistry.

Fundamentally, the tenets of supramolecular chemistry were first set down by the "grandfather of stereochemistry," Louis Pasteur. The main concepts associated with supramolecular phenomena (e.g. molecular recognition, inclusion, self-assembly, lock-n-key selectivity, extended-range order, and molecular transport) are contained in his address to the French Academy on the principles of enantiomer resolutions and characterizations. Without any specific structural theory, Pasteur realized that symmetry arguments and the then novel physical property of optical activity would allow him to follow a variety of recognition processes and aggregate phenomena. He realized the essential difference between organized systems which were reinforced only by the long range order of the medium (e.g., the chiral nature of quartz) and those with properties inherent to their molecular structure (e.g., the tartaric acids). His discussion of conglomerate vs. racemate formation (e.g., in tartaric acids) presaged the concepts of self-assembly and molecular recognition. He also demonstrated the ability of one molecule to recognize another through the formation of diastereomeric salts and compounds. Even the lock–n–key character of nature did not escape his purview, as he demonstrated the ability of yeast to recognize and selectively degrade only one chiral form of a racemic mixture, leading to the first kinetic resolution. As stereochemistry developed post Pasteur, the connection with what is now supramolecular chemistry continued.

Supramolecular chemistry has been described as the chemistry of the non-covalent bond and this description has motivated much discussion of just what the noncovalent bond is, as well as when should we speak of supramolecules vs. molecules. A similar problem of bonding definitions has arisen in the discussion of diastereomers vs. constitutional isomers and in the area of conformational analysis. The description of the molecule as a complete molecular graph with weighted edges, and the application of fuzzy logic has greatly aided these problems. The definitions of when two things are bonded vs. held together by non-covalent interactions has meaning only within a certain time scale—distance resolution—energy boundary; the same holds for molecular shape or symmetry. These ideas are inherent in Pauling's original definition "that there is a bond between two atoms or groups of atoms in the case that the forces acting between them are such as to lead to the formation of an aggregate with sufficient stability to make it convenient for the chemist to consider it as an independent molecular species." Thus, supramolecular chemistry is not so much a new branch of science as a new orientation within the time scale—distance resolution—energy coordinate system that defines every structural chemistry problem.

Reflecting on this analysis, one might ask whether supramolecular chemistry is simply "old tartrates in new vials." On the contrary, the coincidence between supramolecular chemistry and stereochemistry serves to amplify the synergy between them. For example, the realization that conformational analysis is dominated by intramolecular molecular recognition highlights the inalienable nature of stereochemistry in any discussion of "supramolecularity." Stereochemistry greatly simplifies supramolecular chemistry by dispelling the mystique of supramolecular jargon and sound-bites. Thus, the field becomes open to the development of investigations directed toward specific scientific questions rather than diffuse technical demonstrations. Along with this opening of research avenues emerges what we might call a experimental "uniqueness criterion," the fulfillment of which insures that every experiment is more than a derivative technical display.

The mechanical aspects of stereochemical models motivate chemists to design new structural motifs and mechanical analogies for supramolecular systems. Polymers of novel configuration and complexity, macrocycles and cyclophanes of novel topology, self-assembling monolayers and membranes, natural and unnatural biomolecular superstructures, catalytic antibodies and molecular receptors, and macroscopic and microscopic crystal habits represent a sampling of the targets now graspable to the modern chemist. With the advent of each new form of matter comes the challenge of providing an insightful chemical analysis of its structure and function. Beyond the mainstays of NMR and atomic resolution X-ray diffraction crystallography, modern techniques include scanning tunneling and atomic force microscopy, electron microscopy, laser ultrafast spectroscopy, and low angle X-ray scattering spectroscopy. Each of these structures manifests chemistry in space on the nanometer scale, and thus, we can speak of a "supramolecular stereochemistry" which employs the atomic resolution of molecular models to explain the nanometer/kilodalton metrics of suprachemical structures.

A successful conference on supramolecular stereochemistry should cover a broad range of structural chemical research, but at a deep philosophical level, with the goal of accenting what we can learn from selected labors rather than flaunting the unfiltered mass of labors lost. The conferees should provide a general enough base such that chemists from all fields can take home thought provoking lessons. Indeed, the following contributions were submitted with these goals in mind. The preceding comments are a few of the thoughts I have taken home after exposure to this caucus. The reader is warned that in order to capture the candid nature of the meeting, there has been little editing of these manuscripts. On the whole, this is a significant and provocative anthology of articles, each with its own special fulfillment of the uniqueness criterion as applied to research in supramolecular stereochemistry.

Macromolecular Stereochemistry: The Differing Roles of a Helix Reversal in Dilute Solution, in a Liquid Crystal, and in a Polymer Blend

Mark M. Green
Department of Chemistry, Polytechnic University, Six Metrotech Center, Brooklyn, New York 11201 USA
MGREEN@VM.POLY.EDU

A variant of the Nylons, the polyisocyanates, which were created at the DuPont Corporation in the late 1950's, have proven to be archetypal stiff helical polymers related therefore to the alpha-helical peptides or to DNA. There is though one essential difference, the absence of a predetermined helical sense. In long enough chains this leads to helix reversals (1) which are discovered to enforce a strong cooperativity for chiral effects.

Synthetic work allows the incorporation of minute chiral perturbations which influence the ratio of the right and left handed helical states leading to large optical activities (2.3.4). Statistical thermodynamic methods correlated with chiral optical experiments reveal both the chiral interaction energies which prove to be too small to interpret by structural theory and the energy of the helix reversal (5,6).

The difficulty faced here with the chiral bias energies can be seen as a general consequence of cooperative systems of this class, of which their are several examples including the industrially important blending of polymers, in which accumulation of minute energies control properties. In this situation although appropriate experiments, as above, may yield the values of these energies, their magnitude at this time preclude their interpretation by structural theory.

This energy of the helix reversal is used to add new parameters to a force field to reveal the geometry of the reversal and through this the precise conformational characteristics of the helical state of the polyisocyanates. Longstanding questions could then be answered concerning the unusual affects of solvents on the global dimensions of the polyisocyanates and the assignment of this polymer class as broken rods or worms.

The problem of assigning the helix sense in these optically active polymers could not be solved by existing methods based on circular dichroism theory or by X-ray analysis. This problem was solved by correlating liquid crystal theory with the lyotropic cholesteric properties (7). It was discovered that the enthalpic and entropic terms controlling the cholesteric twisting in a lyotropic liquid crystal could be isolated by studying the temperature dependence. The entropic term could be related to the fitting of screws which allowed therefore a simple connection between macroscopic mechanical relationships and the screws sense of the polymer. This method was found to be applicable to a polysaccharide and to a polypeptide as well as to the polyisocyanate.

The work on liquid crystals brought up the question of the population difference of the kinked helix reversal from dilute solution. This is an issue in liquid crystal theory generally concerned with the affect of the mesogenic organization on conformational states which are not linear. Chiral doping experiments were designed in which polyisocyanates which differed in their helix sense ratios in dilute solution were found to act identically in the liquid crystal formation of cholesteric states. This can only mean that the helix reversals in these polymers are excluded in the liquid crystal. As another test of this idea, the liquid crystal state was formed as aggregates under poor solvent conditions. The optical activity properties of these aggregates showed large increases over dilute solution demonstrating again the exclusion of the reversals. In the course of this work, still quite dilute solutions exhibited the formation of thermally reversible gels in a wide variety of hydrocarbon

J. S. Siegel (ed.), Supramolecular Stereochemistry, 1–2.

solvents. This opened up light scattering experimental opportunities which revealed the source of the gelation. Such phenomena are of wide interest in polymers (8).

Synthetic transformations in the polymer side chains allowed the formation of blends with copolymers of styrene and vinyl phenol. These blends belong to the class of molecular composites (9). The use of optically active polyisocyanates in these blends has opened new possibilities for the study of polymer dynamics and the glass transition state since the amorphous nature of the blend allows the monitoring of the motion of the helix reversals by polarimetry or other chiral optical measurements. This kind of work has unusual opportunities for conformational analysis since the transition from the rubbery to the glassy state on lowering of temperature can be seen as a standoff between the lowered free volume, i.e. the shrinkage of the material and the increased population of low energy conformations in the molecule (10). In addition, the polystyrene matrix in the blend formed here can be varied so that motion is of the Rouse type, involving relatively easy motion of the chains, to motion by reptation (11) where the kinks may be completely constrained in both the glassy and rubbery state. These kinds of experiments with the unusual overlap of the disciplines of stereochemistry and polymer physics will likely have other benefits which are now unforeseen.

REFERENCES

1. Lifson, S., Felder, C.E., Green, M.M. (1992) Helical Conformation, Internal Motion and Helix Sense Reversal in Polyisocyanates and the Preferred Helix Sense of an Optically Active Polyisocyanates. Macromolecules, 25, 4142.

2. Green, M.M., Andreola, C., Munoz, B., Reidy, M.P., Zero, K. (1988) Macromolecular Stereochemistry: A Cooperative Deuterium Isotope Effect Leading to a Large Optical Rotation. J.Am.Chem.Soc., 110, 4063.

3. Green, M.M., Reidy, M.P., Johnson, R.D., Darling, G., O'Leary and Willson, G. (1989) Macromolecular Stereochemistry: The Out-of-Proportion Influence of Optically Active Comonomers on the Conformational Characteristics of Pollyisocyanates. The Sergeants and Soldiers Experiment. J.Am.Chem.Soc., 111, 6452.

4. Green, M.M., Khatri, C., and Peterson, N.C. (1993) A Macromolecular Conformational Change Driven by a Minute Chiral Solvation Energy J.Am.Chem.Soc., 115, 4941.

5. Lifson, S., Andreola, C., Peterson, N.C., and Green, M.M. (1989) A Statistical Thermodynamic Analysis of the Cooperative Source of Helix Sense Preference in Polyisocyanates: The Amplification of a Conformational Equilibrium Isotope Effect. J.Am.Chem.Soc., 111, 8850.

6. Sato, T., Teramoto, A., Lifson, S., Peterson, N.C. and Green M.M. (1994) submitted to Macromolecules.

7. Sato, T., Sato, Y., Umemura, Y., Teramoto, A., Nagamura, Y., Wagner, J., Weng, D., Okamoto, Y., Hatada, K., Green, M.M. (1993) Polyisocyanates and the Interplay of Experiment and Theory in the Formation of Lyotropic Cholesteric States, Macromolecues, 26, 4551.

8. Green, M.M., Khatri, C.A., Reidy, M.P., Levon, K. (1993) Dilute Solution Chiral Optical Changes Signal the Thermally Reversible Gelation of Poly(n-hexyl isocyanate) in Hydrocarbon Solvents, Macromolecules, 26, 4723.

9. Vaidya, M.M., Khatri, C.A., Levon, K. and Green, M.M. (1994) Polymer Preprints (ACS), 35, No. 2.

10. Gibbs, J.H. and DiMarzio, E.A. (1958) The Nature of the Glass Transition and the Glassy State, J.Chem.Phys., 28, 373.

11. Sperling, L.H. (1986) "Introduction to Physical Polymer Science," J. Wiley Publ. N.Y., Chapters 5 and6.

FROM MOLECULAR TO SUPRAMOLECULAR CHIRALITY

J.H. VAN ESCH, N.A.J.M. SOMMERDIJK, R.J.H. HAFKAMP, M.C. FEITERS, AND R.J.M. NOLTE
Department of Organic Chemistry, NSR Center, University of Nijmegen, Toernooiveld, 6525 ED Nijmegen, The Netherlands

ABSTRACT. The formation of supramolecular assemblages from chiral amphiphilic molecules has been studied in monolayers by Brewster angle and epifluorescence microscopy, and in solution by electron microscopy. The expression of the molecular chirality at the supramolecular level depends strongly on the packing properties of the amphiphiles. The morphology of the supramolecular structures can be tuned by changing the pH or by complexation of metal ions.

1. Introduction

Chirality is a characteristic feature of biological systems in which it is one of the keys to highly selective recognition processes and reactions. Expression of chirality takes place at different levels of molecular organization, e.g. in the secundary and tertiary structure of proteins, in protein-DNA complexes, in membranes, and even at the level of cells and tissues. In this regard it is enticing to mimic such supramolecular structures, not only to obtain a better understanding, but also to explore possible applications, e.g. in areas such as catalysis, (bio)sensor research, optical devices, etc.

In our studies we use synthetic amphiphilic molecules as building blocks to construct supramolecular architectures. Amphiphilic molecules contain a hydrophobic as well as a hydrophilic part. Because of this property they form all kind of aggregates in water. In recent years several studies have appeared in the literature, dealing with chiral amphiphilic molecules that self-assemble to give dissymmetric supramolecular aggregates. In the following we report some of our recent results in this area, in particular we will discuss how the expression of chirality can be controlled by small variations in molecular structure, by changing the pH, or by complexation of metal ions. Our particular interest have so-called metallo-vesicles, which are structures formed by amphiphiles having a metal-coordinating ligand.

2. Expression of Chirality in Monolayers

Monolayers at the air-water interface can give valuable information on general properties of amphiphilic molecules [1]. They can also be used to detect chiral interaction and recognition between molecules [2]. In our studies, monolayers were characterized by measuring their π/A isotherms. In addition, we used epifluorescence [3] and Brewster angle microscopy [4] to study their texture.

J. S. Siegel (ed.), Supramolecular Stereochemistry, 3–11.

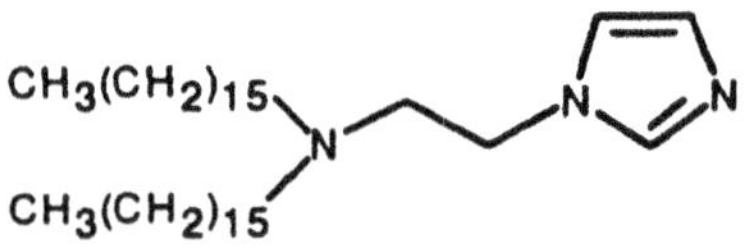

1

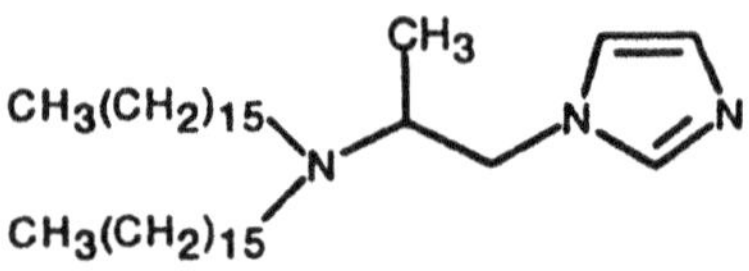

2

Upon dispersion in water the HCl salts of imidazole amphiphiles **1** and **2** form unilamellar vesicles. Addition of copper(II) ions yields complexes of the type copper(imidazole)$_4^{2+}$ [5]. Spreading of pure (S)-**2** or (R)-**2** on an aqueous subphase of pH 5.5 leads to the formation of stable monolayers with a well defined liquid-expanded (LE) to liquid-condensed (LC) phase transition (Figure 1). No effect of the chiral center is detectable when the isotherms of homochiral **2** and its racemate are compared. Such an effect can, however, be seen by fluorescence microscopy. Figure 2 shows fluorescence micrographs of monolayers of compound **2** to which 0.5 mol% of a fluorescent probe was added. Upon compression into the LE - LC phase transition region, the LC phase appears as dark dendritic domains, which originate from a diffusion limited growth process. The picture clearly shows that (S)-**2** forms domains with branches that are bent in a counter-clockwise direction (Figure 2A). On the contrary, amphiphile (R)-**2** yields domains with clockwise turned branches (Figure 2B). The racemic mixture (R,S)-**2** does not show chiral dendritic domains. Instead, more open, fractal-like structures are observed without any specific rotation of the branches (Figure 2C).

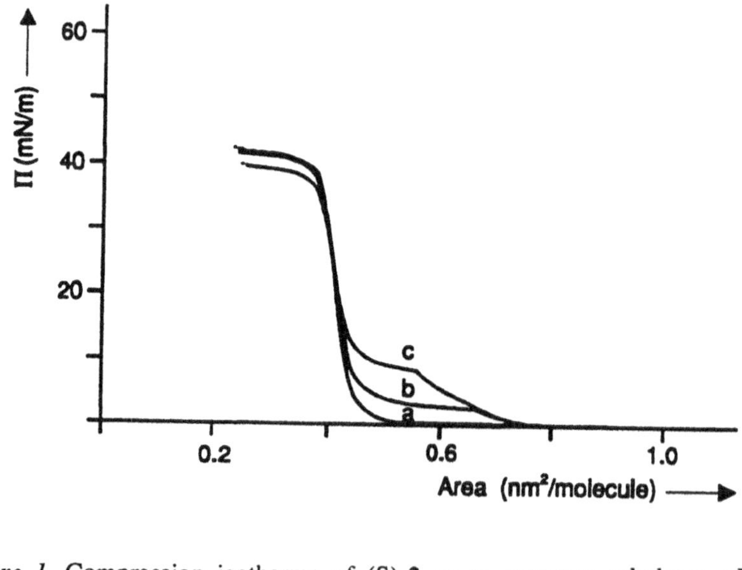

Figure 1. Compression isotherms of (S)-**2** on an aqueous subphase, pH 5.5, at different temperatures: 10 °C (a), 15 °C (b), 20 °C (c).

Figure 2. Fluorescence micrographs of monolayers of (S)-**2** (a), (R)-**2** (b), and of (R,S)-**2** (c) on an aqueous subphase of pH 5.5 (T=20 °C).

Complexation of transition metal ions like copper(II) or zinc(II) to monolayers of **2** changes the isotherms. This suggests that these metals are selectively bound at the interface. The domains formed by the LC phase are, however, not chiral.

The phospholipids **3** and **4** differ in the positional arrangement of the phosphate head group and the amide linked hydrocarbon chain on the glycerol skeleton (Figure 3). The compression isotherms of these compounds show a remarkable difference in the onset of the LE phase, the latter being 350 and 80 Å^2/molecule for **3** and **4**, respectively (Figure 4) [6]. No LC phase is visible upon compression of **3**, probably because the butyrate group is too short to allow a close packing of the hydrocarbon chains, as normally observed in phospholipids. Compression of the isomer **4** to a pressure of 15 mN/m does, however, result in a transition to a LC phase, suggesting a much higher degree of hydrocarbon chain organization in this lipid than in **3**. This may be explained by assuming that the butyrate group of **4** is immersed in the aqueous subphase, resulting in an extended molecular conformation, as depicted in Figure 3.

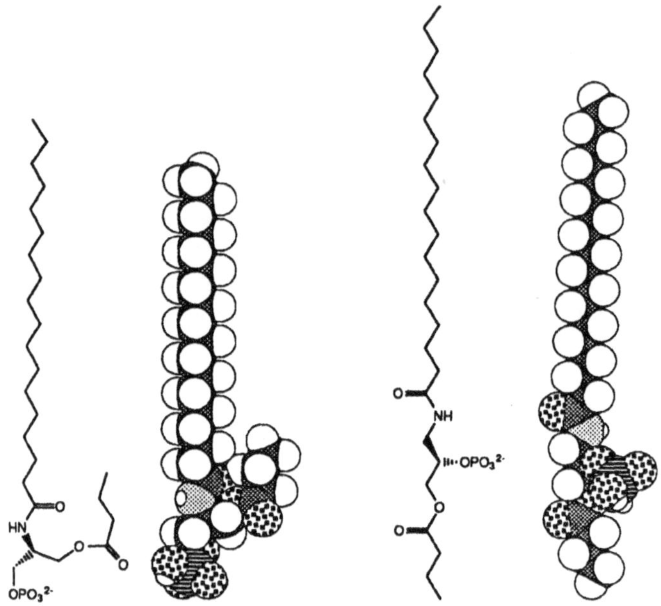

Figure 3. Structures and CPK models of amphiphiles **3** and **4**.

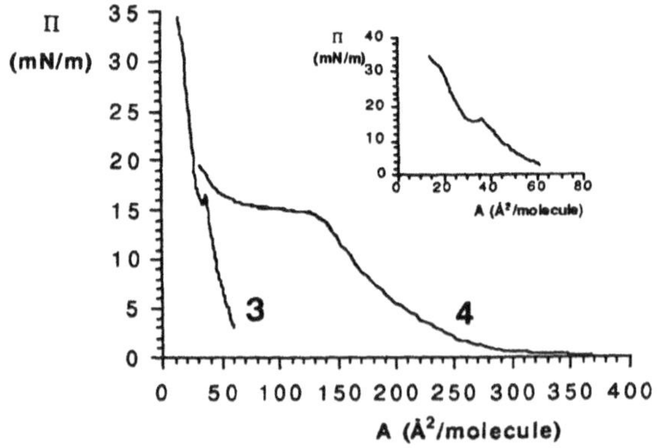

Figure 4. Monolayer isotherms of **3** and **4** at 293 K. Inset: enlargement of isotherm of **3**.

Monolayers of **4** were studied using Brewster angle microscopy. Compression of **4** into the LE - LC phase transition results in the formation of LC domains, which exhibit an overall counter-clockwise pattern (Figure 5). Similar domains are visible by fluorescence microscopy, but the size of the structures is much smaller (Figure 5C). The latter is most likely the result from the fact that the fluorescence probe act as an impurity, which causes an increase in the number of nucleation sites.

Since the size of the head group usually is an important factor in the packing of surfactant molecules, it was investigated whether protonation of the phosphate headgroup could change the supramolecular organization of **4**. Brewster angle microscopy of monolayers of **4** at pH 1.5 visualizes a change in the direction of the patterns (Figure 5D). Apparently, the reduction of the headgroup repulsion by protonation leads to a change in molecular organization, and hence to a different long-range tilt order in the domains.

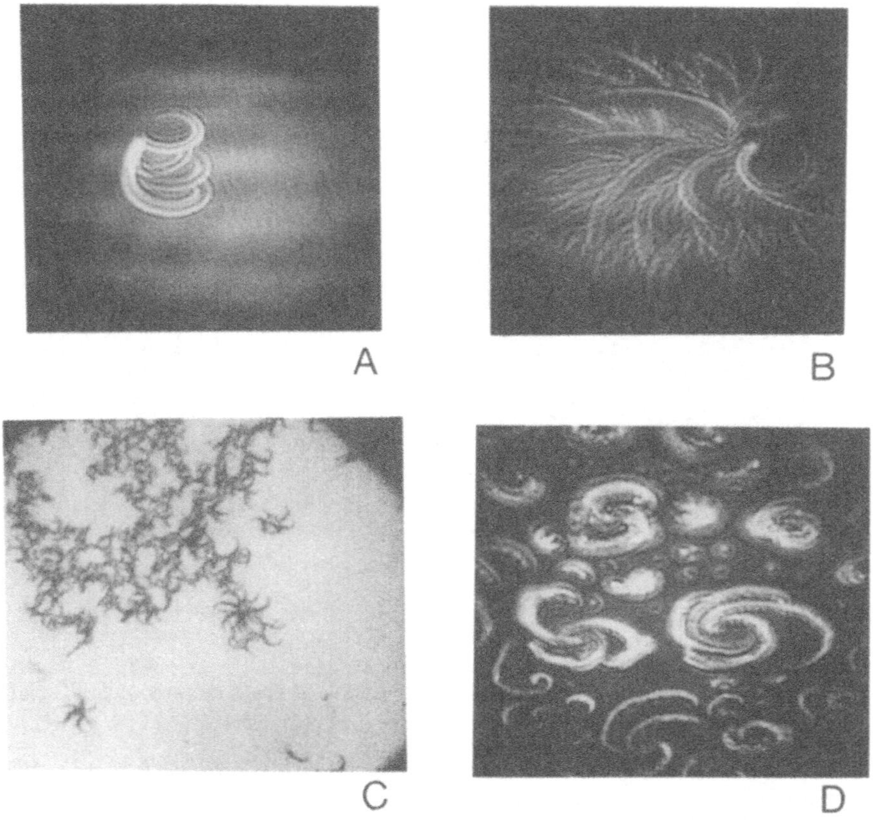

Figure 5. Morphology of monolayers of **4**. (A) Brewster angle micrograph taken at π = 5 mN/m. (B) Idem at π = 16 mN/m (pH = 6.5). (C) Fluorescence micrograph taken at π = 15 mN/m (pH 6.5). (D) Brewster angle micrograph taken at π = 16 mN/m (pH = 1.5).

3. Expression of Chirality in Solution

It is well known that aggregates of amphiphiles in water can form a rich variety of morphologies, which can be visualized by transmission electron microscopy (TEM) on stained or freeze fractured samples. Aqueous dispersions (2% w/w) of the surfactants **3** and **4** were allowed to dry and subsequently shaded with platinum. TEM micrographs of **3** revealed the presence of plate-like structures without any special features (Figure 6, left). Electron micrographs of **4**, however, showed left-handed helical strands with a diameter of 22 nm, which coagulated to form rope-like structures (Figure 6, right) [6]. Apparently, the presence of a chiral center in **3** does not lead to the formation of supramolecular structures with a chiral appearance, whereas in compound **4** it does. A similar behavior was observed for the monolayers of **3** and **4** (see above).

Figure 6. Transmission electron micrographs of dried dispersions of **3** (left) and of **4** (right).

FTIR spectroscopy on oriented films of **3** and **4** confirmed that these compounds are packed in a different way. The spectra showed that both compounds form so-called trans-amide polymers [7], *i.e.* structures in which the amides have a trans conformation and are aligned through intramolecular hydrogen bonding. The ester carbonyl group of **3** was found to participate in a hydrogen bonded network, indicating that it is located in the hydrogen belt area [8]. In a dried film, the ester carbonyl group of **4** displayed a vibration at 1738 cm^{-1}, which is typical for a free ester. After wetting of the film a second peak at 1706 cm^{-1} arose, typical for an ester group bonded to water. This indicates that the butyl group in aggregates of **4** is immersed in the aqueous phase, as depicted in Figure 3. We may conclude that the location and orientation of the butyrate group with respect to the intramolecular axis of **3** and **4** plays an essential role in the ability of these molecules to form chiral supramolecular structures.

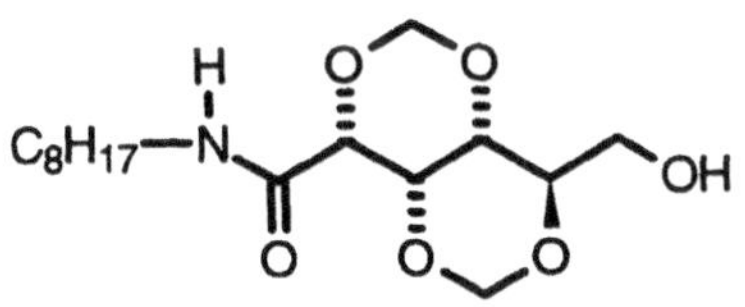

5

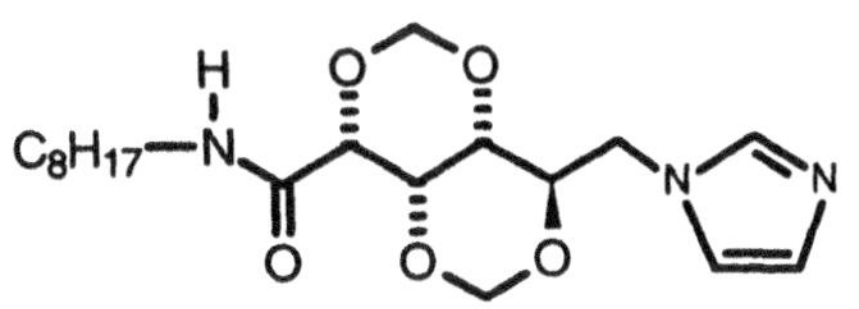

6

Other types of surfactants recently prepared by our group include the sugar derivatives **5** and **6** [9]. Although the polarity of **5** is considerably decreased by the methylene bridges, it is still soluble in water. However, unlike the unprotected derivative [10] it does not form well-defined aggregates, as judged from electron micrographs. Compound **6** is also soluble in water and in aqueous buffered solutions. In a NaOAc/HOAc buffer of pH 4.5, which is well below the pK_a of the imidazole function, protonated **6** self-assembles to form vesicles with diameters ranging from 160 nm to 780 nm (Figure 7A). On the other hand, in a tris(hydroxymethyl)aminomethane (TRIS) buffered solution of pH 8.5, unprotonated **6** forms long fibers (Figures 7B and C) and hollow tubuli (Figure 7C). The fibers have a diameter of 100 nm and a length/width ratio of 500. They are composed of multilayers. The hollow tubuli have diameters of approximately 3 µm.

Addition of $Cu(ClO_4)_2$ to a solution of **6** in water leads to the formation of copper complexes of the type $Cu[\mathbf{6}]_4^{2+}$. These complexes self-assemble to generate a molecular braid composed of 4 bundles of fibers (Figure 7D). The diameter of the braid is approximately 330 nm.

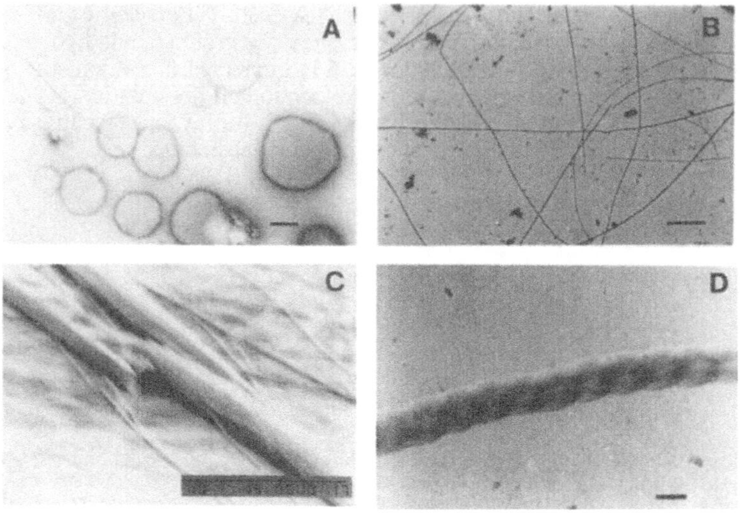

Figure 7. (A) TEM of vesicles of **6** (B) TEM of fibers of **6** (bar is 5 μm) (C) SEM of hollow tubuli and thin fibers of **6** (bar is 10 μm) (D) TEM of a molecular braid formed from the copper complex of **6** (bar is 200 nm).

4. References

1. Gaines, G.L. (1966) *Insoluble Monolayers at Liquid-gas Interfaces,* Wiley Interscience, New York.
2. Arnett, E.M.; Harvey, N.G., and Rose, P.L. (1989) Stereochemistry and molecular recognition in "two dimensions", *Acc. Chem. Res.*, **22**, 131.
3. (a) Weis, R.M. and McConnel, (1984) Two-dimensional chiral crystals of phospholipid, *Nature*, **310**, 47. (b) Knobler, C.M. Seeing phenomena in flatland: studies of monolayers by fluorescence microscopy., *Science*, **249**, 870.
4. Hönig, D. and Möbius, D. (1991) Direct visualization of monolayers at the air-water interface by Brewster angle microscopy, *J. Phys. Chem.*, **95**, 4590.
5. Van Esch, J.H., Damen, M., Feiters, M.C., and Nolte, R.J.M. (1994) Aggregation behaviour and copper-binding properties of surfactants containing imidazole and pyrazole ligands., *Recl. Trav. Chim. Pays-Bas* **113**, 186.
6. Sommerdijk, N.A.J.M., Buynsters, P.J.A.A., Pistorius, A.M.M., Mu Wang, Feiters, M.C., Nolte, R.J.M., and Zwanenburg, B. (1994) Tuning the supramolecular expression of chirality: phospholipid analogues containing amide linkages, *J. Chem. Soc. Chem. Commun.*, in press.
7. Kunitake, T. (1992) Synthetic bilayer membranes: molecular design, self-organization, and applications, *Angew. Chem.*, **104**, 692.
8. Colthup, N.B., Daly, L.H., and Wiberly, S.E. (1964) *Introduction to infrared and Raman Spectroscopy*, Academic press, New York.
9. Hafkamp, R.J.H., Feiters, M.C., and Nolte, R.J.M. (1994) Tunable supramolecular structures from an imidazole-containing gluconamide , *Angew. Chem. Int. Ed. Eng.*, **33**, 986.
10. (a) Fuhrhop, J.-H., Schneider, P., Rosenberg, J., Boekema, E., and Helfrich, W. (1987) The chiral bilayer effect stabilizes micellar fibers, *J. Am. Chem. Soc.*, **109**, 3387. (b) Köning, J., Boetnerr, C., Winkler, H., Zeitler, E., Talmon, Y., and Fuhrhop, J.-H. (1993) Magic angle (54.7°) gradient and minimal surface in quadruple micellar fibers, *J.Am. Chem. Soc.*, **115**, 693.

TEMPLATE INDUCED CONTROL OF STEREOCHEMISTRY FOR THE SYNTHESIS OF POLYMERS

G. WULFF
Heinrich-Heine-University Düsseldorf
Institute of Organic Chemistry and Macromolecular Chemistry
Universitätsstr. 1, 40225 Düsseldorf, Germany

1. Introduction

The general idea of the work presented in this paper focuses on the attempt to control the constitution and the configuration of complex supramolecular arrangements by specific interactions of the constituents with a suitable template molecule. If one uses constituents bearing polymerizable vinyl groups, the whole constituent-template complex can be polymerized and, thus, the original structure will be stabilized and frozen. After removal of the template molecules, polymers with defined structures are obtained by transfer of structural information from template to polymer.

It is well known that a similar transfer of stereochemical information is occurring, during most asymmetric syntheses, from a chiral low molecular-weight auxiliary to the reaction product. Furthermore, in most enzymatic syntheses, an information transfer takes place from a macromolecule to a low molecular-weight substance.

In our case, when we look at the transfer of properties from a low molecular-weight molecule to the resulting macromolecule, the situation becomes much more complicated since a large number of stereogenic centers are formed which are covalently linked. On the other hand, it also provides compared with the before-mentioned examples quite some additional possibilities. With polymers it is not only possible to obtain, by a repetitive property transfer along a long chain, structurally defined linear polymers, but it is also possible to get an information transfer to a plain surface in a two-dimensional manner or, if a crosslinking reaction is performed during the transfer, a real three-dimensional transfer is possible.

2. Molecular Imprinting

Starting with the last example of a three-dimensional information transfer, Scheme 1 shows the principle of molecular imprinting, an approach we have introduced quite some years ago [1,2]. Polymerizable vinyl monomers containing functional groups are attached to suitable template molecules (T). Subsequent copolymerization in the

J. S. Siegel (ed.), Supramolecular Stereochemistry, 13–19.

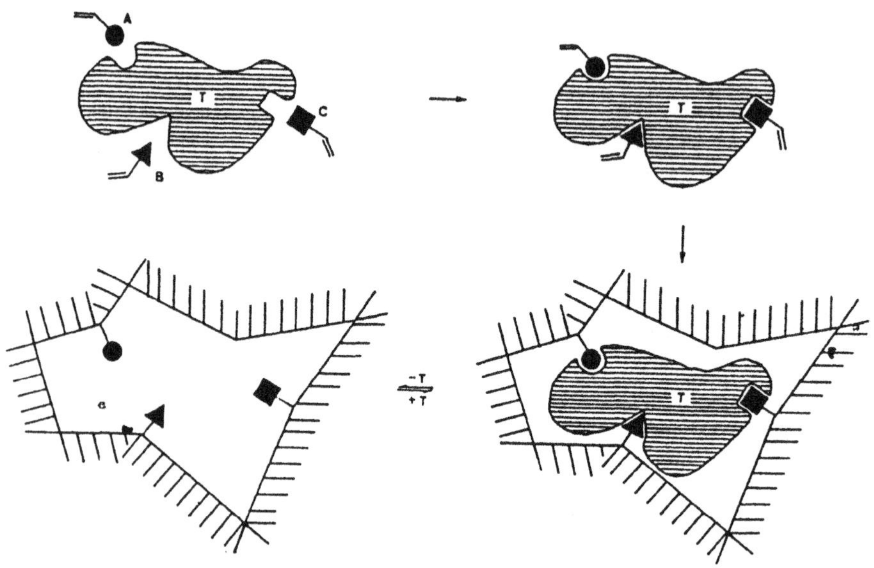

Scheme I Schematic representation of a crosslinking polymerization in the presence of a template to obtain cavities of a specific shape and a defined spatial arrangement of functional groups.

presence of solvent and large concentrations of crosslinking agent produces relatively rigid macroporous polymers. Removal of the template molecules (see Scheme 1) leaves behind cavities in the polymer whose shape and whose arrangement of the functional groups are determined by the template molecules. The relation of template and imprint resembles very much the key-and-lock principle postulated for enzyme catalysis by Emil Fischer exactly 100 years ago [3].

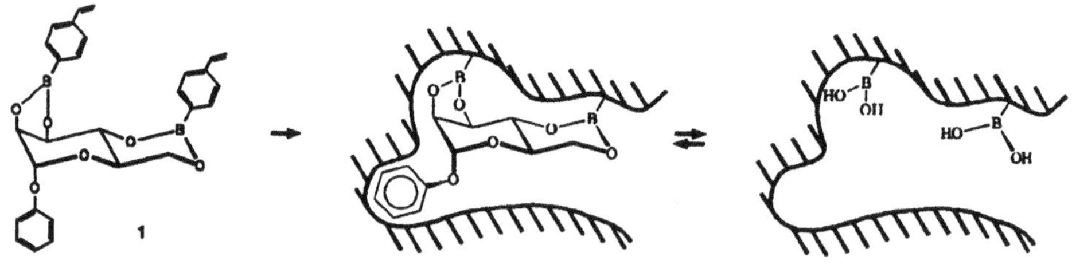

Scheme II Crosslinked polymers prepared from the template monomer 1 and ethylene dimethacrylate with and without the template [2].

Template monomer 1 is shown as an illustration, in which two molecules of 4-vinylphenylboronic acid have been bound to phenyl α-D-mannopyranoside as the template by boronic ester linkages to the free hydroxyl groups. A 1:9 ratio of monomer 1 and ethylene dimethacrylate were copolymerized in an inert solvent to yield a macroporous polymer. From this the template could be split off to the extent of 90-95% providing chiral cavities each bearing a pair of boronic acid groups (Scheme 2). Polymers of this type exhibit an excellent ability for racemic resolution of the racemate of the template. Under batch equilibration the original template enantiomer is preferably incorporated and separation factors α (ratio of K_D/K_L) as high as 4-6 are achieved. These materials can also be used as chromatographic supports in HPLC columns giving resolution higher than $R_S = 4.3$ (see Fig. 1) [4]. Investigations as to the mechanism of this separation show that the orientation of the functional groups within the cavities is the dominating factor for the separation; the shape of the cavity is of secondary importance [5]. Many examples of this procedure have been published in the meanwhile (reviews see [6-8]) by several research groups.

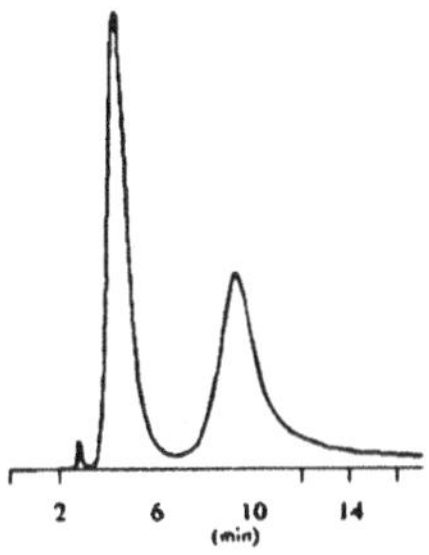

Figure 1 Chromatographic separation of the racemate of the template phenyl α-mannopyranoside in h.p.l.c. on a support of a type shown in Scheme II [4].

Information transfer to a two-dimensional plane has been exemplified by a similar imprinting procedure onto the surface of silica through the formation of siloxane bonds [9]. With this method it was possible to locate two functional groups on a more or less planar silica surface and to investigate the selectivity caused by the distance accuracy of such a pair of functional groups.

3. Preparation of optically active linear vinyl polymers

Of special interest is the transfer of stereochemical information during the preparation of linear vinyl polymers. This procedure enabled us for the first time to obtain optically active vinyl polymers due to the configuration of the main chain (main-chain chirality) [10,11]. For this, two 4-vinylphenylboronic acids were attached to a 3,4-disubstituted D-mannitol derivative **2** and copolymerized with a comonomer. As shown in Fig. 2 the copolymerization involves a cyclopolymerization of monomer **2** since no crosslinking occurred. After splitting off the template, linear copolymers showing strong optical activity were obtained. A wide variety of optically active copolymers have thus been prepared. Also optically active homopolymers have been obtained in a similar manner.

Mechanistic studies show that this polymerization proceeds through an asymmetric cyclopolymerization involving an unusually large 19-membered ring. The optical

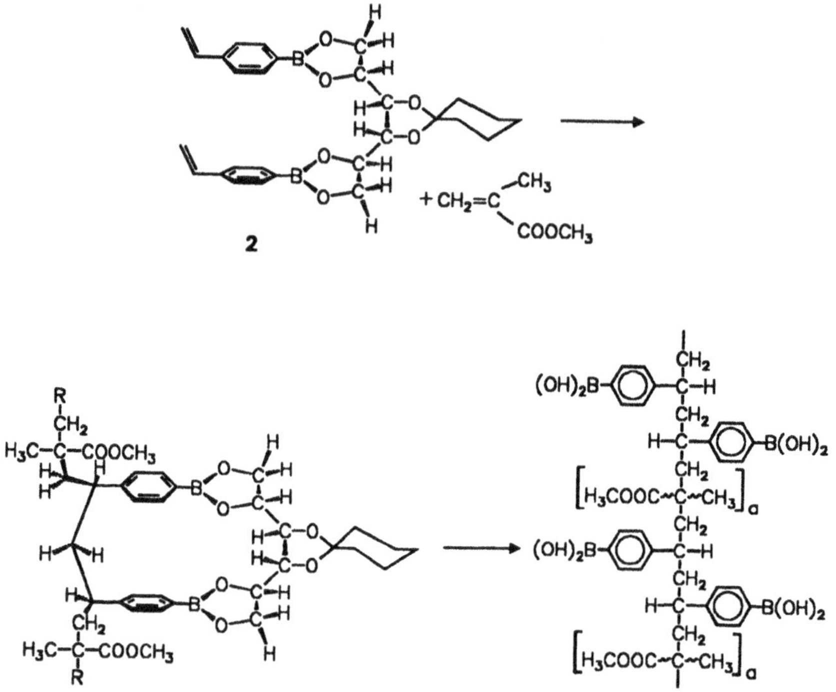

Figure 2 Asymmetric cyclocopolymerization of monomer **2** with methyl methacrylate. The polymer with the template and after removal of the template is shown [11].

activity stems from the formation of chiral distyryl dyads in the main chain. They are separated from one another by the comonomeric unit. The absolute configuration of the prevailing asymmetric dyad building block has been established to be (S,S) as evidenced by the synthesis of oligomeric model compounds. The ratio of meso (R,S):(S,R) and racemic dyads (S,S):(R,R) could be determined by radical cyclization of **2** in the presence of an excess of azobis(isobutyronitrile) (Fig.3) [12]. Investigation of the reaction products showed that the two diastereomeric dyads are produced in a comparable ratio; in both cases the enantiomeric excess is very high.

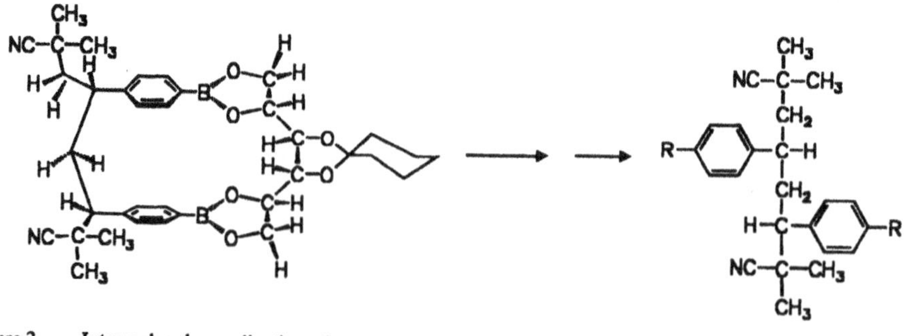

Figure 3 Intramolecular cyclization of monomer **2** in the presence of a high amount of azobis(isobutyronitrile). The main diastereomer after splitting off the template is shown [12].

By changing the reaction conditions it is not only possible to produce a 19-membered ring. Copolymerization of **2** and MMA in the presence of $AlEt_{1.5}Cl_{1.5}$ gives a 21-membered ring owing to insertion of the MMA monomeric unit between the two double bonds of **2** (see Fig. 4) [13]. The reaction, after removal of the template and the boronic acid, results in strictly alternating styrene-MMA copolymers. In this case lower optical activity is expected and also observed.

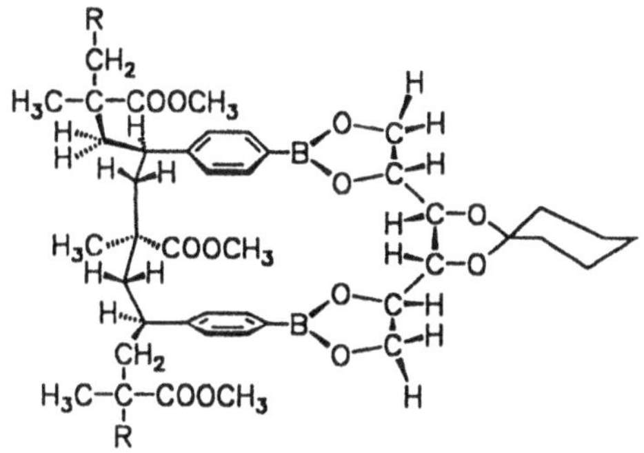

Figure 4 Cyclocopolymerization of monomer **2** with methyl methacrylate by initiation with $AlEt_{1.5}Cl_{1.5}$ [13].

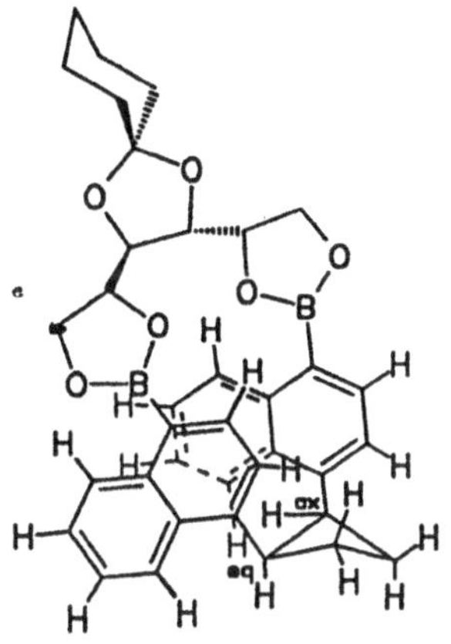

Figure 5 Reaction product of photocyclization of the corresponding naphthyl derivative of **2** [14].

Yet another type of cyclization can be attained with these compounds. Irradiation causes an intramolecular [2+2] cycloaddition of **2** and, even quicker, of the corresponding 4-vinyl-naphthyl derivative (see Fig. 5) [14]. The latter already reacts during measurement of circular dichroism (CD) or UV spectra originating in a strong exciton couplet at 231.8 nm. The pure cyclization product of the naphthyl derivative shows a molar ellipticity [Θ] of $-3 \cdot 10^5$ at 224.3 nm and of $+2.7 \cdot 10^5$ at 240.8 nm. Since the starting compound shows almost no CD, this system might provide a very sensitive method for information storage, in which light of a defined wavelength can be detected very sensitively by analyzing the circular dichroism of the cyclization product.

Being able to prepare an optically active polymer with main chain chirality, we further investigated whether it is possible to extend this procedure to the synthesis of chiral liquid crystalline polymers [15]. For this, monomer **2** was copolymerized with a comonomer carrying a mesogenic group. After splitting off the template, a polymer of structure **3a** was obtained. This polymer proved, after removal of the boron to **3b**, nematic and by means of circular dichroism it was found that the chromophor of the mesogen showed no Cotton effect. Elongation of the chiral dyad in **3a** by the Suzuki reaction led to polymer **3c** carrying terphenyl-groups in the chiral dyad. This polymer was liquid crystalline with cholesteric texture. In solution still no chirality transfer to the mesogenic chromophore was observed according to CD, but polymer **3c** films showed a strong exiton couplet. Therefore, only in the liquid crystalline phase a special chirality transfer is possible through space.

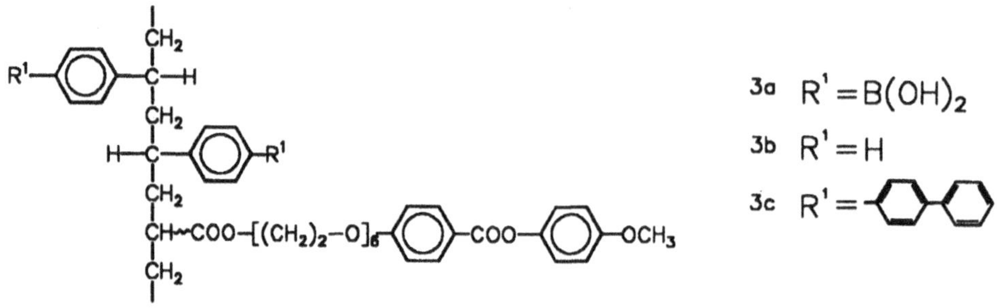

4. Outlook

While in future chiral liquid crystalline phases might be used as ferroelectrics or for nonlinear optics, imprinted polymers are presently investigated in quite a number of directions:

1. Chromatographic supports for racemic separation [6,7],
2. Imprinted cavities as chiral microreactors [16,17],
3. Application as designed catalysts and enzyme mimics [18],
4. Mimics of antibodies for drug assay [19],
5. Imprinted polymers as selective sensoring layer in artificial biosensors [20,21].

5. Acknowledgment

I am indebted to my coworkers who have through their commitment and their ideas contributed to an important share of these results. Their names are given in the references. The financial support of the *Deutsche Forschungsgemeinschaft*, the *Minister für Wissenschaft und Forschung des Landes Nordrhein-Westfalen*, and the *Fonds der Chemischen Industrie* is gratefully acknowledged.

6. References

1. Wulff, G., Sarhan, A., and Zabrocki, K. (1973) Enzyme-analogue built polymers and their use for the resolution of racemates, *Tetrahedron Lett.*, 4329-4332.
2. Wulff, G., Vesper, W., Grobe-Einsler, R., and Sarhan, A. (1977) On the synthesis of polymers containing chiral cavities and their use for the resolution of racemates, *Makromol. Chem.* **178**, 2799-2816.
3. Fischer, E. (1894) Einfluß der Configuration auf die Wirkung der Enzyme, *Ber. Dtsch. Chem. Ges.* **27**, 2985-2993.
4. Wulff, G. and Minarik, M. (1990) Template imprinted polymers for HPLC separation of racemates, *J. Liquid Chromatogr.* **13**, 2987-3000.

5. Wulff, G. and Schauhoff, S. (1991) Racemic resolution of free sugars with macroporous polymers prepared by molecular imprinting. Selectivity dependence on the arrangement of functional groups versus spatial requirements, *J. Org. Chem.* **56**, 395-400.
6. Wulff, G. (1993) Biorecognition in molecularly imprinted polymers: Concept, chemistry, and applications, in T. T. Ngo (ed), *Molecular interactions in bioseparations*, Plenum Press, New York, pp. 363-381.
7. Andersson, L. I., Ekberg, B., and Mosbach, K. (1993) Bioseparation and catalysis in molecularly imprinted polymers, in T. T. Ngo (ed), *Molecular interactions in bioseparations*, Plenum Press, New York, pp. 383-394.
8. Shea, K. J. (1994) Molecular imprinting of synthetic network polymers: The de novo synthesis of macromolecular binding and catalytic sites, *Trends Polymer Science* **2**, 166-173.
9. Wulff, G., Heide, B., and Helfmeier, G. (1986) Molecular recognition through the exact placement of functional groups on rigid matrices via a template approach, *J. Am. Chem. Soc.* **108**, 1089-1091.
10. Wulff, G., Zabrocki, K., and Hohn, J. (1978) Optically active polyvinyl compounds with chirality in the main chain, *Angew. Chem. Int. Ed. Engl.* **17**, 535-537.
11. For a review see: Wulff, G. (1989) Main-chain chirality and optical activity in polymers consisting of C-C chains, *Angew. Chem. Int. Ed. Engl.* **28**, 21-37.
12. Wulff, G. and Kühneweg, B. (1994) Unpublished results.
13. Wulff, G. and Krieger, S. (1994) Investigation on main-chain chiral copolymers with alternating structure, *Macromol. Chem. Phys.* **195** in press.
14. Wulff, G., Krieger, S., Kühneweg, B., and Steigel, A. (1994) Occurence of strong circular dichroism during measurement of CD spectra due to intramolecular cyclization, *J. Am. Chem. Soc.* **116**, 409-410.
15. Wulff, G., Schmidt, H., Witt, H., and Zentel, R. (1994) Cooperativity and transfer of chirality in liquid-crystalline polymers, *Angew. Chem. Int. Ed. Engl.* **33**, 188-191.
16. Wulff, G. and Vietmeier, J. (1989) Enantioselective synthesis of amino acids using polymers possessing chiral cavities obtained by an imprinting procedure with template molecules, *Makromol. Chem.* **190**, 1727-1735.
17. Byström, S. E., Börje, A., and Akermark, B. (1993) Selective reduction of steroid 3- and 17-ketones using $LiAlH_4$ activated template molecules, *J. Am. Chem. Soc.* **115**, 2081-2083.
18. Beach, J. V. and Shea, K. J. (1994) Designed catalysts. A synthetic network polymer that catalyzes the dehydrofluorination of 4-fluoro-4-(p-nitrophenyl)butan-2-one, *J. Am. Chem. Soc.* **116**, 379-380.
19. Vlatakis, G., Andersson, L. I., Müller, R., and Mosbach, K. (1993) Drug assay using antibody mimics made by molecular imprinting, *Nature* **361**, 645-647.
20. Piletsky, S. A., Dubei, I. Y., Fedoryak, D. M., and Kukhar, V. P. (1990) Substrate-selective polymeric membranes. Selective transfer of nucleic acid components, *Biopolimery Kletka* **6**, 55-58.
21. Hedborg, E., Winquist, F., Lundström, I., Andersson, L. I., and Mosbach, K. (1993) Some studies of molecularly-imprinted polymer membranes in combination with field-effect devices, *Sensors Actuators A* **37-38**, 796-799.

DENDRIMERS - NANOSCOPIC SUPERMOLECULES ACCORDING TO DENDRITIC RULES AND PRINCIPLES

DONALD A. TOMALIA
Michigan Molecular Institute
1910 W. St. Andrews Road
Midland, Michigan 48640 USA

1. Introduction

Chemistry is not just a science that involves the study of electron movement to rearrange and connect atoms. It is a philosophy, a way of thinking about the world. It is a way of thinking about the architecture, complexity and function of matter throughout the dimensional hierarchy of the universe, from the simplest atoms to the most complex phenomena--namely life itself. The ideas described here evolved from such a scientific and a philosophical perspective. They deal to a large extent with the analysis and mimicry of simple but pervasive patterns observed in Nature. Furthermore, they focus on the organization of monomer or branch cell units around atomic or molecular reference cores according to dendritic rules and principles. These synthetic constructions have allowed molecular level mimicry of pervasive architectural and mathematical patterns observed at either the macroscopic or at the atomic levels of matter.

2. Mimicry of Macroscopic Branching Patterns Found in Trees and Amplification Patterns Found in Biology

The convergence of three life experiences; namely, a formal education as a synthetic/physical organic chemist, exposure to polymer chemistry in industry and a lifetime passion for tree farming/horticulture, undoubtedly attracted me to the molecular level mimicry problem illustrated in Figure 1. Chemistry used to demonstrate this concept required the stepwise assembly of monomer units into *branch cells* which were then organized into single trunked tree-like ensembles referred to as *dendrons* (Greek word for tree). These dendron constructions allowed us to demonstrate an amplification of matter which has recently been compared to analogous amplification events found in polymerase chain reactions (PCR) and biological cell divisions (see Figure 2)[1]. Growing multiples of these dendrons in an outward fashion from an initiator core(I) produces entities we have coined as *dendrimers* [2].

J. S. Siegel (ed.), Supramolecular Stereochemistry, 21–26.

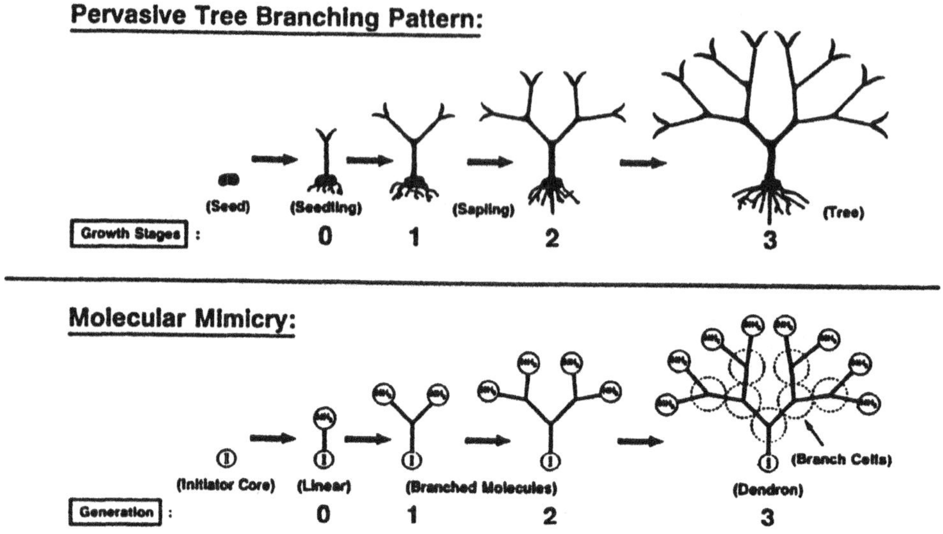

Figure 1. Pervasive tree branching patterns and molecular mimicry with chemical structures.

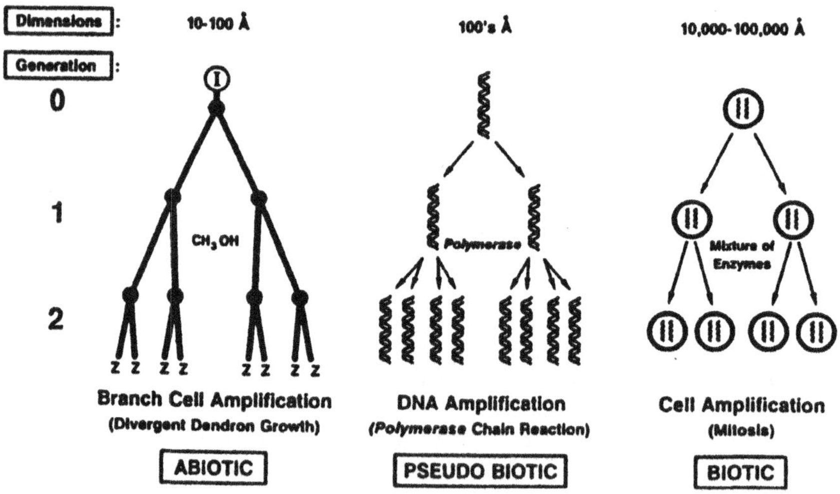

Figure 2. Amplification of matter by biotic, pseudo biotic and abiotic strategies

3. Synthetic Strategies to Dendrimers

Vast arrays of dendrimers differing in composition, size, shape and surface groups can be synthesized by two major synthetic strategies. The first synthetic strategy, namely the "divergent method" was hypothesized by Vögtle[3]; however, these experimental procedures did not produce dendrimer structures. Dendrimers were first synthesized by using experimental methods developed by Denkewalter[4], Newkome[5] and Tomalia[2]. They involved the assembly of multiples of dendrons from an initiator core by growth from the "inside out" (see Figure 3)[6-8]. Conversely, Fréchet[9] assembled dendrons from the "outside in" to produce a chemically reactive dendron trunk which was then anchored in multiples to a common core to produce dendrimers by the "convergent method" shown in Figure 3.

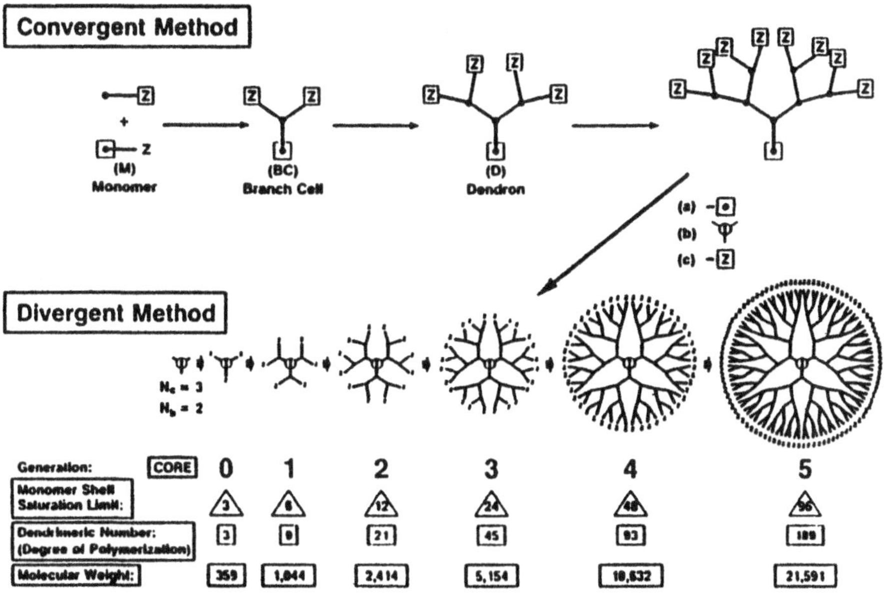

Figure 3. Convergent and divergent synthetic strategies to dendrimer (N_c = 3; N_b = 2), illustrating magic numbers associated with the poly(amidoamine) (PAMAM) dendrimer family.

4. Control of Critical Molecular Design Parameters - Magic Numbers Related to Size, Valency and Molecular Weight

In either case, it is apparent that these constructions follow systematic branching rules (geometric progressions) if the assembly is near ideal. This allows one to predict the number of monomer repeat units assembled per molecule (N_{RU}; degree of polymerization), the number of surface groups (valency, Z) and to calculate dendrimer theoretical molecular weights (MW) as a function of initiator core multiplicity (valency), N_c; branch cell multiplicity (valency) N_b; and generations (G), according to the following relationships:

$$\bullet \quad \boxed{\text{Monomer Shell Saturation Limit}} = \boxed{\text{Valency}}\,(Z) = N_c N_b^{G}$$

$$\bullet\bullet \quad \boxed{\text{Dendrimer Number}} = \boxed{N_{RU}} = N_c\left[\frac{N_b^{G+1}-1}{N_b-1}\right] = \text{(Degree of Polymerization)}$$

$$\bullet\bullet\bullet \quad \boxed{\text{Molecular Weight}} = \boxed{MW} = M_c + N_c\left[M_{RU}\left(\frac{N_b^{G+1}-1}{N_b-1}\right) + M_e N_b^{G+1}\right]$$

For example, Figure 3 illustrates the molecular level amplification of monomer units around an initiator core with multiplicity ($N_c = 3$), designed to produce branch cells with multiplicities of $N_b = 2$. Such is the case for the poly(amidoamine) (PAMAM) dendrimer series. It can be calculated from these geometric progression equations that the characteristic Z values would be 3, 6, 12, 24, 48, etc., the N_{RU} values are 3, 9, 21, 45, 93, etc.; whereas the corresponding molecular weight values are 359, 1044, 2414, 5154, 10632 etc. for generations, G = 0, 1, 2, 3, 4, 5, etc., respectively. These Z values represent the surface valency of a dendrimer. It is a "magic number" or monomer shell saturation limit, beyond which additional monomer or branch cell units cannot be added to a particular generation. The predictability and confirmation of these "magic numbers" has been demonstrated countless times in our group and other laboratories[10].

5. Dendrimers as Fundamental Nanoscopic Building Blocks - "Supermolecular Analogues to Atoms"

The ability to precisely control mass, surface valency and surface functionality in these dendrimer structures clearly validates their proposed roles as fundamental building blocks possessing architecture and function analogous to atoms, albeit, with nanoscopic dimensions (i.e., 1-100nm).

The notion of fixed building blocks of matter that exhibit certain "pervasive patterns" was basic to the atomic theory of John Dalton (1766-1844) as he published his "New System of Chemical Philosophy" in 1808. In fact, without the benefit of understanding quantum mechanics or electronic theory, the operational parameters of our first (atom building block) chemistry set became fairly well defined throughout the 19th century. These parameters, derived from simple logic, were clearly based on experimental observations and generally involved the interpretation of at least the following pervasive patterns.

I. Atoms form chemical bonds

II. Atoms (elements possess well defined masses relative to each other (combining weights)).

III. Atoms (elements) form chemical bonds with well defined valency.

Convinced we have observed these same patterns within dendrimer families, we have been using dendrimers as so called *"supermolecular analogues to atoms"*[10]. Just as most chemical reactions between atoms involve surface electrons to form compounds, so it is also the case with the terminal monomer functionality in dendrimers. Over one hundred different stoichiometric reactions have been documented between dendrimer surfaces and sub-nanoscopic reagents. In general, these reactions merely modify dendrimer surfaces and affect the dimensions of the dendrimers incrementally. However, when nucleophilic dendrimers are allowed to react with other dendrimers possessing electrophilic surfaces, the products obtained are dramatically enhanced in dimensions (i.e., 1-100nm). We call these new structures *nanoscopic compounds, clusters, and polymers*. Figure 4 illustrates this concept. Unlike classical picoscopic/sub-nanoscopic compounds derived from atomic building blocks, these nanoscopic *supermolecular compounds, clusters,* and *assemblies* are large enough to be observed directly by electron microscopy.

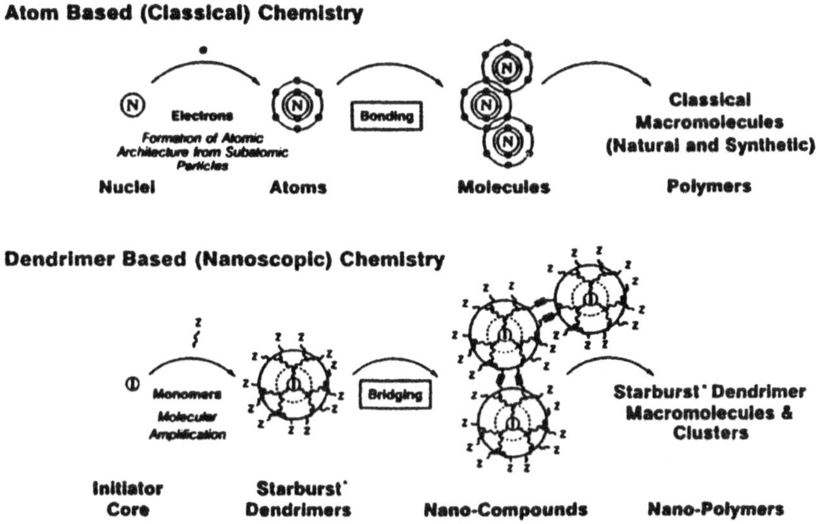

Figure 4. Comparison of classical atom based chemistry to dendrimer based nanoscopic chemistry.

6. Conclusions

Just as the elements have been used as fundamental building blocks to produce literally millions of sub-nanoscopic sized compounds and assemblies, so should it be possible to use dendrimers as fundamental building blocks to produce an infinite variety of supramolecular and supermolecular entities of nanoscopic proportions. As shown in Figure 5, these building blocks may be expected to underpin a myriad of new synthetic strategies to precise structures of designed size and shape transcending the entire range of nanoscopic dimensions.

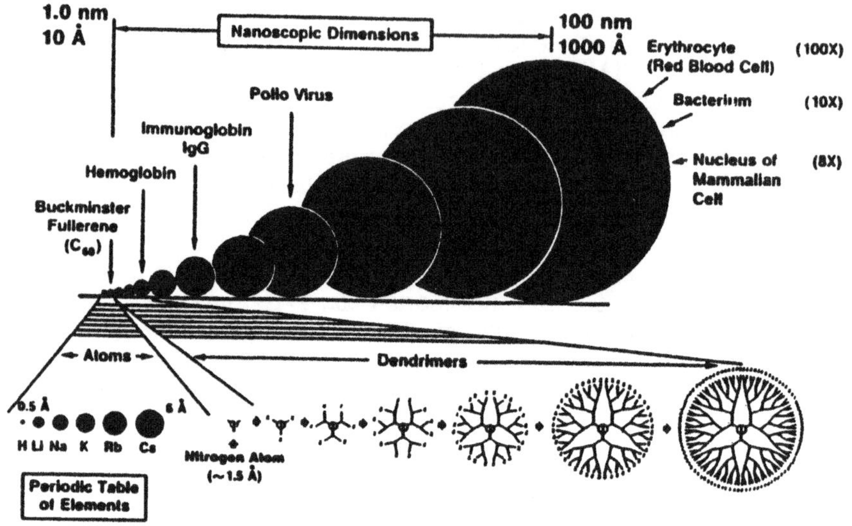

Figure 5. Amplification of matter to produce atomic and dendrimeric building blocks with a comparison to nanoscopic dimensions.

7. References

1. Dvornic, P.R. and Tomalia, D.A. (1994) A family tree for polymers, *Chemistry in Britain*, **30**, 641-645.
2. Tomalia, D.A., Baker, H., Dewald, J., Hall, M., Kallos, G., Martin, S., Roeck, J., Ryder, J., and Smith, P. (1985) A new class of polymers: starburst-dendritic macromolecules, *Polymer J. (Tokyo)*, **17**, 117-132.
3. Buhleier, E., Wehner, W., and Vögtle, F. (1978) "Cascade"- and "nonskid-chain-like" syntheses of molecular cavity topologies, *Synthesis*, 155-158.
4. Denkewalter, R.G., Kolc, J.F., and Lukasavage, W.J. (1983) U.S. Pat. 4410688; (1984) *Chem. Abstr.*, 100, 103907p.
5. Newkome, G.R., Yao, Z.-Q., Baker, G.R., and Gupta, V.K. (1985) Cascade molecules: a new approach to micelles. A [27]-arborol, *J. Org. Chem.*, **50**, 2003-2004.
6. Tomalia, D.A. (1986) Dendritic macromolecules: synthesis of starburst dendrimers, *Macromolecules*, **19**, 2467-2468.
7. Tomalia, D.A., Naylor, A.M., and Goddard III, W.A. (1990) Starburst dendrimers: molecular-level control of size, shape, surface chemistry, topology, and flexibility from atoms to macroscopic matter, *Angew. Chem.*, **102**, 119-157; (1990) *Angew. Chem. Int. Ed. Engl.*, **29**, 138-175.
8. Tomalia, D.A., and Durst, H.D. (1993) Genealogically directed synthesis: starburst*/cascade dendrimers and hyperbranched structures, in E. Weber (ed.), *Topics in Current Chemistry Vol. 165; Supramolecular Chemistry I - Directed Synthesis and Molecular Recognition*, Springer-Verlag, Berlin Heidelberg, pp. 93-313.
9. Hawker, C.J., and Fréchet, J.M.J. (1990) Preparation of polymers with controlled molecular architecture. A new convergent approach to dendritic macromolecules, *J. Am. Chem. Soc.*, **112**, 7638-7647.
10. Tomalia, D.A. (1993) Starburst™/cascade dendrimers: fundamental building blocks for a new nanoscopic chemistry set, *Aldrichimica Acta*, **26**, 91-101.

CONSTRUCTION OF DNA POLYHEDRA AND KNOTS THROUGH SYMMETRY MINIMIZATION

NADRIAN C. SEEMAN, YUWEN ZHANG,
SHOU MING DU AND JUNGHUEI CHEN
Department of Chemistry, New York University
New York, NY 10003 USA

The goals of supramolecular chemistry include structural control on the nanometer scale comparable to that enjoyed by craft workers on the macroscopic scale.[1] The ability to join, couple or weave two molecules together to produce a structure with the same certainty enjoyed by a carpenter, a plumber or a garment worker, would increase greatly the efficiency of materials scientists, chemists, and molecular biologists. Chemists have learned that it is not as simple to create structures from molecules as it is from macroscopic objects: They must rely on intrinsic propensities of precursors, because there are no nails, screws, or threads available to form bonds between atoms. Furthermore, the laws of physics do not permit all conceivable separations of atomic nuclei. In contrast, the components in biological systems often self-assemble spontaneously, by using complementary surfaces to form cohesive structures.

We have exploited the self-assembly properties of DNA. The complementary nature of DNA makes it ideal for use in nanoscale construction. We have used branched DNA molecules to form stick-polyhedra, and we have exploited the self-complementary features of single-stranded molecules to form knotted molecules[2,3]. The DNA self-assembly system provides an excellent demonstration of the principle that the maximization of control is provided by the minimization of symmetry in its broadest sense. In the work described below, we will show three examples of symmetry minimization, and how this leads to molecular control: [1] First, we will show how unusual DNA structures containing branches can be assembled by minimizing the sequence symmetry of the molecule. [2] Next, we will show how the energetic symmetry associated with molecular flexibility can lead to unwanted products, and how it can be eliminated in DNA object synthesis. [3] Finally, we will show how the combination of sequence asymmetry and different conditions can be used to force a single linear DNA molecule to assume a series of different knotted topologies. The systems differ in each case, but the destruction of symmetry always leads to molecular control.

Minimizing Sequence Symmetry to Form Branched DNA Molecules

For the past decade, we have used the self-assembly of DNA molecules as a means to achieve control over the structure of matter on the nanometer scale. DNA offers the most predictable *inter*molecular associations known: The double helix invariably forms in compliance with the complementarity rule that adenine (A) pairs with thymine (T) and

J. S. Siegel (ed.), Supramolecular Stereochemistry, 27–32.

that guanine (G) pairs with cytosine (C). Under conventional conditions, the product of this association is a right-handed double helix approximately 20 Å in diameter with a pitch about 34 Å, containing about 10 nucleotide pairs per turn. Two DNA molecules with single-stranded overhangs, called 'sticky ends' will cohere, and can be linked covalently by DNA ligases. It is possible to design *synthetic* sequences that self-assemble to form branched molecules called 'junctions'. Junctions are designed by minimizing the sequence symmetry and maximizing the base pairing of the target[4].

Figure 1 illustrates a branched junction molecule. The molecule contains 4 single strands (Arabic numbers) arranged into 4 double helical 'arms' (Roman numerals).

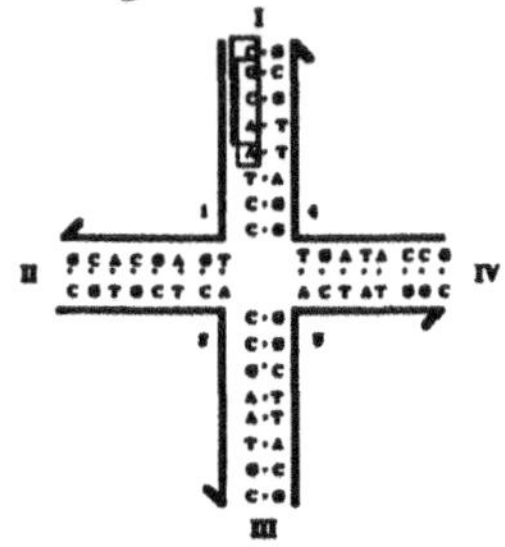

Figure 1. *A Stable DNA Branched Junction.* The junction shown is composed of four strands of DNA; the 3' end of each strand is indicated by the half-arrows.

Nucleic acid engineering involving duplex components is based on directing the tendency of DNA molecules to maximize base pairing. If one wishes to make branched helix axes, the target molecule will contain features that do not correspond to the lowest energy form of DNA. Nevertheless, one can ensure the formation of the 'excited state' molecule, if one precludes more favorable arrangements. Control in this system derives from minimizing DNA sequence symmetry[4].

The minimization of sequence symmetry is illustrated for the junction shown in Figure 1. Each strand contains 16 nucleotides, which have been broken up into 13 overlapping tetramers, indicated by the solid boxes. If we insist that each of the 52 tetramers in the entire molecule be unique, base pairing competitive with the target molecule can come only from redundant trimers. We further insist that the molecule contain no tetramer complementary to a bend. There are 240 asymmetric tetramer elements. To design larger structures, one must divide the strands into longer elements of length **n**, and competition from redundant (**n**-1)-mers can assume greater importance.

Forming DNA Objects By Sticky-Ended Associations

Linear DNA is a construction medium familiar to those involved in biotechnology. The sticky-ended ligation of DNA fragments is the fundamental reaction of biotechnology; it enables genetic engineers to assemble genetic elements with selected functional features. The attachment of sticky ends to a branched nucleic acid structure converts it, in principle, into a nanoscale valence cluster with addressable ends[4]. By treating them as molecular lego, branched DNA molecules become building blocks for assembly into multiply-connected stick-figures and networks. The edges of these figures consist of double helical DNA, and the vertices correspond to the branch points. The basis of self-association is shown in Figure 2, which shows the assembly of a 4-arm junction into a quadrilateral.

When this type of assembly is attempted, one discovers that individual junctions are not rigid[5]. We can expect that the double helical edges of each object are torsionally and flexurally stiff, because the persistence length of DNA is approximately 500 Å[6]: This value leads to a 66.5 Å RMS end-to-end distance for two turns of DNA, nominally 68 Å long. However, the 'valence angles' between the double helical edges of each object are flexible. If a junction is oligomerized to yield oligolaterals, it forms a series of cyclic products: trimers, tetramers, and so on. The flexibility of DNA objects corresponds to non-unique ('symmetric') minima in the potential energy surface that prevent the use of oligomerization to generate either target objects, or the lattice suggested by Figure 2.

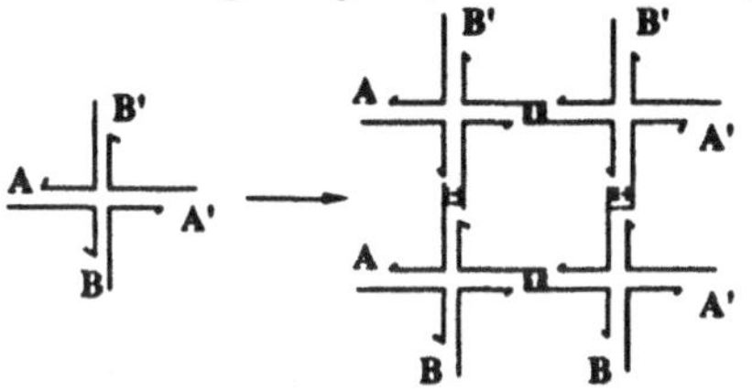

Figure 2. *Formation of a Two-Dimensional Lattice from a Junction with Sticky Ends.* A is a sticky end and A' is its complement. The same relationship exists between B and B'. Four of the monomeric junctions on the left yield the structure on the right. Note that A and B are different from each other. DNA ligase closes the gaps in the complex. Note that the complex has maintained open valences, so that it can be extended to form a 2-D lattice.

Thus, energetic symmetry decreases control on the assembly of target species. So long as one cannot have confidence that a given structural component will be assume a particular structure, one cannot control the products. However, specific sticky-ends can be used to direct the assembly of junctions into target structures. For example, the assembly of a quadrilateral, directed by four different sticky end pairs has been achieved[7].

The most exciting prospects for this system entail the assembly of multiply-connected objects in 3-D. We have constructed from 3-arm junctions a molecule whose helix axes have the connectivity of a cube[8]. A schematic of the 3-dimensional, 3-connected object is shown in Figure 3. It contains 12 edges that consist of double helical DNA. Every edge contains 20 nucleotide pairs of DNA, so we expect that their lengths will be about 68 Å. Each edge contains a unique restriction site, thereby enabling us to validate the synthesis. The synthesis of the cube-like object has been demonstrated by breaking it down to standard catenanes, whose electrophoretic properties are known.

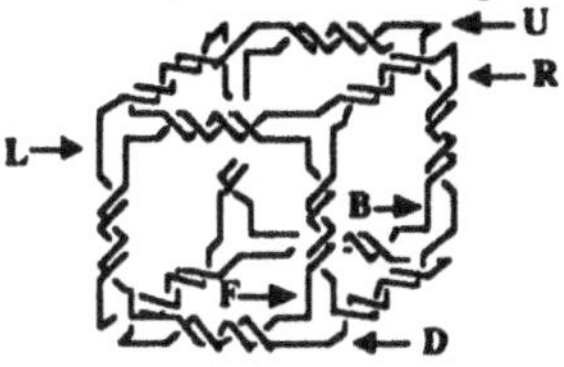

Figure 3. *A DNA Molecule Whose Helix Axes Have the Connectivity of a Cube.* The molecule consists of six cyclic strands catenated together in this particular arrangement. They are labeled by the first letters of their positions, Up, Down, Front, Back, Left and Right.

There is an integral number of turns in each edge, so every face of the cube corresponds to a single cyclic strand. Thus, the object in Figure 3 is a complex catenane of 6 single-stranded cyclic molecules, each doubly linked to its nearest neighbors. The catenated nature of the strands highlights a key point about DNA: The braiding of the strands about the helix axis is a critical property of DNA molecules. Topologically bonded molecules are more easily controlled than hydrogen bonded complexes. The linking of DNA strands means that closed structures will be complex catenanes, and infinite constructs can contain polycatenated mesh structures, like chain mail.

The synthesis of the cube was performed in solution. This led to numerous problems, arising from the combination of (i) the flexibility of the branched junctions whose energetic symmetry led to by-products that had to be purified, (ii) the difficulty of separating products from catalysts, reactants and byproducts, and (iii) the differences between the optimal concentrations for intermolecular additions and for intramolecular cyclizations. Control of DNA ligation in solution derives solely from the ability to phosphorylate sticky ends selectively. This is insufficient for the synthesis of complex objects, so we have developed a new methodology based on the use of a solid support[9]; this permits convenient removal of reagents and catalysts from the growing product. Each ligation cycle adds a component that creates a covalently-closed, topologically-bonded intermediate. A single edge of an object is formed at a time, thereby eliminating by-products that result from isoenergetic flexible conformers. Control derives from the restriction of hairpin loops that form each side of the new edge. The isolation of growing objects on the support removes cross-reactions between growing molecules. Restriction sites are destroyed when an edge forms.

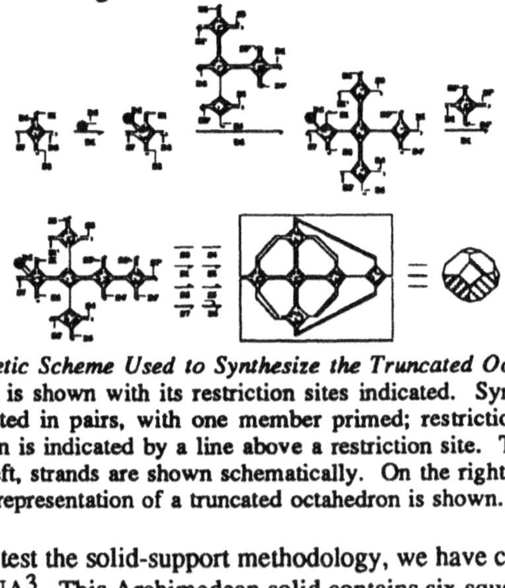

Figure 4. ***The Synthetic Scheme Used to Synthesize the Truncated Octahedron.*** **Each square in the of the diagram is shown with its restriction sites indicated. Symmetric restriction sites are named 'S', indicated in pairs, with one member primed; restriction sites cut distally are named 'D'. A reaction is indicated by a line above a restriction site. The product is shown in two forms. On the left, strands are shown schematically. On the right, a slightly rotated front view of a polyhedral representation of a truncated octahedron is shown.**

In order to test the solid-support methodology, we have constructed a truncated octahedron from DNA[3]. This Archimedean solid contains six squares flanking its 4-fold symmetry axes, and eight hexagons that surround its 3-fold symmetry axes. There are two turns of DNA per edge, so the final object is a complex 14-catenane. The truncated octahedron is a 3-connected figure, but it has been constructed from 4-arm junctions. The entire molecule contains 2550 nucleotides, and has a molecular weight ca. 790 kd.

In this construction, the objects added to the support are squares and square groupings, as indicated in Figure 4. The structure on the lower left of Figure 4 is a heptacatenane hexasquare complex. The square strands are intact in this construct, and the hexagons are all formed from the outer strand: They result from successive intramolecular closures of the sticky ends associated with the restriction sites, S1-S1'...S7-S7'. Whereas the molecules are isolated from each other on the support, it is possible to use symmetrically-cleaving restriction pairs. The synthesis is demonstrated by showing that all six cyclic strands corresponding to the square molecules are in the heptacatenane, followed by digesting the final product to the tetracatenanes that flank the squares.

Four Knotted Topologies from A Single Molecule.

An intimate relationship exists between catenanes and knots[10]: Removal of a node by switching strands, yet maintaining local symmetry and strand polarity, converts a catenane to a knot, and a knot to a catenane. This relationship is important here, because it appears that cloning these DNA structures will be most readily achieved by getting single-stranded DNA molecules to fold up into complex knots, whose restriction will lead to the desired stick figures[11].

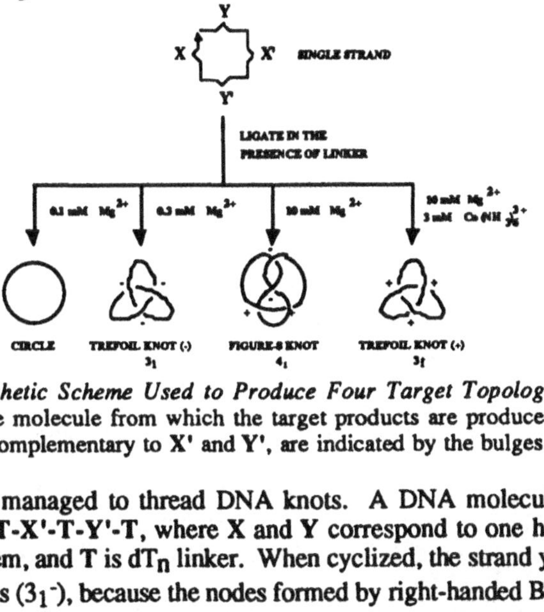

Figure 5. *The Synthetic Scheme Used to Produce Four Target Topologies.* The top of this scheme indicates the molecule from which the target products are produced. The four pairing regions, X and Y, complementary to X' and Y', are indicated by the bulges from the square.

We have managed to thread DNA knots. A DNA molecule can contain the sequence **X-T-Y-T-X'-T-Y'-T**, where **X** and **Y** correspond to one helical turn, **X'** and **Y'** complement them, and **T** is dT_n linker. When cyclized, the strand yields a trefoil knot with negative nodes (3_1^-), because the nodes formed by right-handed B-DNA are negative topological nodes[2]. However, there is a left-handed form of DNA, Z-DNA, that is formed by special sequences, under the control of solution conditions[12]. To make a molecule with 4 topologies, both X and Y must be capable of forming Z-DNA under different conditions. Their sequences cannot be the same, or only the circle will be formed. In Figure 5, the X domain contains methylated cytidines, in an alternating dCpdG motif; the Y domain combines methylated cytidines with two separated A-T pairs, which can be converted to Z-DNA, but only when the driving force is strong. Figure 5 shows that mild Z-promoting conditions produce an amphichiral figure-8 (4_1) knot (in methylated DNA), but strong Z-promoting conditions will produce a trefoil knot with positive nodes (3_1^+). The circle is prepared by ligating the molecule at an ionic strength too low to support the synthesis of a knot.

Symmetry is Antithetical to Control

We have shown that control can be derived by eliminating symmetry. In the case of branched junction assembly, an otherwise disfavored structure is formed by minimizing the sequence of the strands. In DNA object synthesis, the energetic symmetry inherent in different structures of flexible molecules has been overcome by a methodology using different sticky ends and protected 'valences' on a solid support. In knot synthesis, asymmetric sequences, with different responses to conditions, have been used to fashion four different topologies from a single DNA precursor. A key use envisioned for DNA arrays[4,11] is to function as macromolecular zeolites, serving as hosts for macromolecular species, to aid crystallography. The inherent translational symmetry of lattices will require new means to build these symmetric objects.

Acknowledgments

This research has been supported by grants N00014-89-J-3078 from the Office of Naval Research and GM-29554 from the NIH. The support of Biomolecular Imaging on the NYU campus by the W. M. Keck Foundation is gratefully acknowledged.

References

1. Drexler, K.E. (1981), Molecular Engineering, *Proc. Natl. Acad. Sci. (USA)* **78**, 5275-5278.
2. Seeman, N.C., Chen, J., Du, S.M., Mueller, J.E., Zhang, Y., Fu, T.-J. , Wang, H., Wang, Y. & Zhang, S. (1993), Synthetic DNA Knots and Catenanes, *New J.Chem.* **17**, 739-755.
3. Zhang, Y. and Seeman, N.C. (1994), The Construction of a DNA Truncated Octahedron, *J. Am. Chem.Soc.* **116**, 1661-1669.
4. Seeman, N.C. (1982) Nucleic Acid Junctions and Lattices, *J. Theor. Biol.* **99**, 237-247.
5. Petrillo, M.L., Newton, C.J. Cunningham, R.P., R.-I. Ma, Kallenbach,N.R.and Seeman, N.C. (1988)Ligation of 4-Arm Junctions,*Biopolymers* **27**, 1337-1352.
6. Hagerman, P.J.(1988), Flexibility of DNA, *Ann. Rev. Biophys. & Biophys. Chem.* **17**, 265-286.
7. Chen, J., Kallenbach, N.R. and Seeman, N.C. (1989), A Specific Quadrilateral Synthesized from DNA Branched Junctions, *J. Am. Chem. Soc.* **111**,6402-07.
8. Chen, J. and Seeman, N.C. (1991), The Synthesis from DNA of a Molecule with the Connectivity of a Cube, *Nature (London)* **350**, 631-633.
9. Zhang, Y. and Seeman, N.C. (1992),A Solid-Support Methodology for the Construction of Geometrical Objects, *J. Am. Chem.Soc.* **114**, 2656-2663.
10. White, J.H., Millett, K.C., and Cozzarelli, N.R.(1987), Description of the Topology of DNA Catenanes and Knots *J. Mol. Biol.* **197**, 585-603.
11. Seeman, N.C. (1991),The Construction of 3-D Stick Figures From Branched DNA, *DNA and Cell Biology* **10**, 475-486.
12. Behe, M. and Felsenfeld, G.(1981), Effects of Methylation of a Synthetic Polynucleotide, *Proc. Natl. Acad. Sci. (USA)* **78**, 1619-1623.

MOLECULAR RECOGNITION BY CATENATED STRUCTURES

DAVID B. AMABILINO, J. FRASER STODDART
School of Chemistry
University of Birmingham
Edgbaston, Birmingham B15 2TT, UK

ABSTRACT: Catenanes and rotaxanes have provided considerable challenges to the 'artists' that practise chemical synthesis. In certain instances, where the mutual recognition between the components of these interlocked molecules is strong, their self-assembly becomes easy – at least, once the components have been constructed! The challenge that faces the supramolecular chemist is to take advantage of existing knowledge and seek inspiration from the information processing systems of the natural world, so that the molecular recognition present in wholly synthetic compounds can be used and the information contained in them can be relayed in a manner similar to that taking place in biological systems.

1. Introduction

The 'art' of supramolecular chemistry[1] is now being applied to the synthesis of a wide range of interlocked and intertwined chemical structures and superstructures.[2] The production of these molecular 'sculptures' is providing a considerable challenge in the area of chemical synthesis. It is being met by the involvement of template-directed[3] and self-assembly[4] procedures. Yet beyond the achievement associated with the construction process lies a much greater challenge – that is to access the chemical information that has brought about their assembly in the first place and to find functions for the mechanically-interlocked components. The store of genetic information is a chiral intertwined double helical structure, which can be 'read' into proteins. It is becoming clear that part of the cycle that goes into the replication of DNA, which also includes the intervention of catenated and knotted DNA,[5] requires the recognition of the intertwined structure *via* the minor and major grooves in the double helix. Proteins act as exo-receptors to the DNA. Lehn and coworkers[6] have already constructed a synthetic double helix – with convergent binding sites – that appeals to molecular recognition around its core. In this article, we speculate about chemical functions for catenated structures, as well as draw attention to the the possibilities resulting from the improved chemical information content of the components present in these interlocked ring systems.

2. The Two-Step Self-Assembly of Olympiadane

A linear pentacatenane is an extremely appealing synthetic objective to tackle, using the powers of self-assembly. Attention was drawn to the array of five interlocked rings as early as 1960 by van Gulick,[8] who actually proposed the name 'olympiadane' for such an assembly. After a relatively unsuccessful attempt[9] at the synthesis of a [5]catenane, the macrocyclic polyether tris-1,5-naphtho-57-crown-15 (**TNP57C15**) was employed for the self-assembly (*Scheme 1*) of the [3]catenane **1**.$4PF_6$ in 6% yield. Undaunted by the low yield in this first step, this [3]catenane was combined with an excess of the components of cyclobis(paraquat-*p*-phenylene). The reaction mixture afforded[10] the [4]catenane **2**.$8PF_6$ in 31% yield and the [5]catenane **3**.$12PF_6$ in 5%

J. S. Siegel (ed.), Supramolecular Stereochemistry, 33–40.

yield. The tiny yield of the [5]catenane **3**.12PF$_6$ leaves a little to be desired when compared with the yield of – from its self-assembling components – 58% for the [2]catenane **4**.4PF$_6$ incorporating **TNP57C15** and cyclobis(paraquat-*p*-phenylene)[11]. The critical step in the self-assembly of olympiadane is the formation of the [3]catenane **1**.4PF$_6$. Simply increasing the efficiency of this step could result in the formation of a [5]catenane in a single step, *i.e.* a molecular structure incorporating three **TNP57C15** rings and two large tetracationic cyclophanes. Indeed, this reaction also afforded a [2]catenane incorporating one **TNP57C15** ring as well as the large tetracationic cyclophane. Higher catenated materials were also observed during column chromatography, but so far they have escaped isolation and characterization owing to the very small quantities of material present in the reaction mixture.

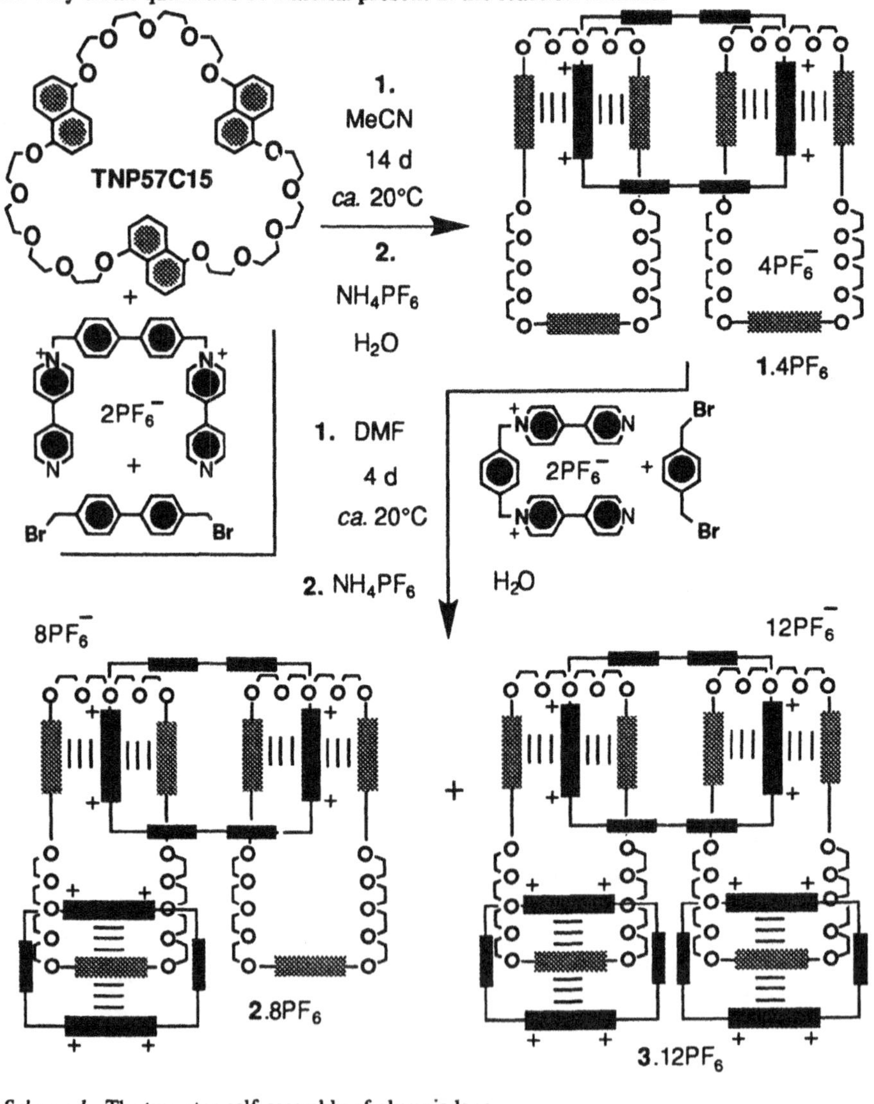

Scheme 1. The two-step self-assembly of olympiadane

Clearly, the template-directed synthesis of the [5]catenane $3.12PF_6$ – and the ultimate goal of self-assembling polycatenanes – requires considerably enhanced template direction by increasing the mutual recognition properties between the component parts. The yield of the [3]catenane analogous to $1.4PF_6$, but incorporating[12] two 1,5-dinaphtho-38-crown-10 macrocycles (**DNP38C10**) around the large tetracationic cyclophane, was 30%. Even this much higher yield is not going to be enough, if we are to stand any chance of constructing higher catenated structures – an issue we shall come back to in the next Section.

3. Stereochemical Aspects of Catenanes

When all the dynamic processes in simple [2]catenanes are 'frozen' out, including the rocking process associated with the angled orientation of the component rings with respect to one another, the result is the existence of two enantiomers. These enantiomers are evident in the solid state structures of these compounds. Invariably, [2]catenanes, incorporating macrocyclic polyethers with two π-electron rich units form donor-acceptor stacks which are directional and chiral, *i.e.* they contain only one enantiomer of the compound. However, the crystals as a whole contain both enantiomers in parallel stacks[13] with respect to each other. No spontaneous resolution upon crystallization has yet been observed, *cf.* the crystallization of a trefoil knot.[14] In the crystals of the self-assembled [2]catenane, incorporating **TNP57C15** and cyclobis(paraquat-*p*-phenylene), the molecules also self-organize themselves into a highly ordered superstructure. In the solid state (*Figure 1*), all the aromatic units of the macrocyclic polyether move into positions where they form π–π stacking interactions with the tetracationic cyclophane. Each of the aromatic units in the molecule are located in different environments. Unlike the [2]catenane, incorporating **DNP38C10**, a quilt-like packing arises from the arrays of π–π stacked aromatic rings adopted in the solid state. A careful inspection of the packing diagram reveals the presence of two enantiomerically-related catenanes in the solid state structure.

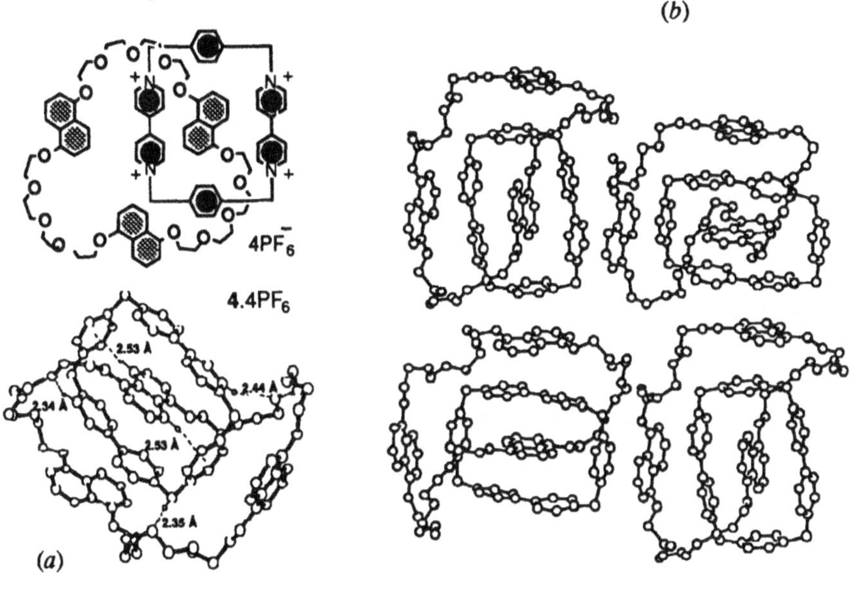

Figure 1. The X-ray crystal structure of the [2]catenane $4.4PF_6$. (*a*) A ball-and-stick representation of one molecule and (*b*) the packing of four molecules in the solid state.

So far, single crystals of the [5]catenane 3.12PF$_6$ have proved to be unsuitable for X-ray crystallography. However, we can dream a little! Inspection of CPK space-filling molecular models of olympiadane shows that its macrocyclic polyethers are quite large and free to move. Thus, it is possible to locate the 'unoccupied' 1,5-dioxynaphthalene unit into a position where it resides between the π-electron deficient bipyridinium residues of the two different and neighboring tetracationic cyclophanes. This extended π-stack, which is symmetrical about its center as it is drawn in *Figure 2*, is four nanometers in length. The polyether chain that links the two aromatic units 'inside' the cavities of neighboring tetracationic cyclophanes appears to be a little too short, however. Homologation of this polyether chain to, say, a heptaethylene glycol unit might allow the coiling depicted in *Figure 2* to occur in solution at low temperatures. As it is, when a CD_3CN solution of olympiadane was cooled down, no well-defined structure was evident in solution, as indicated by the ^1H NMR spectrum.

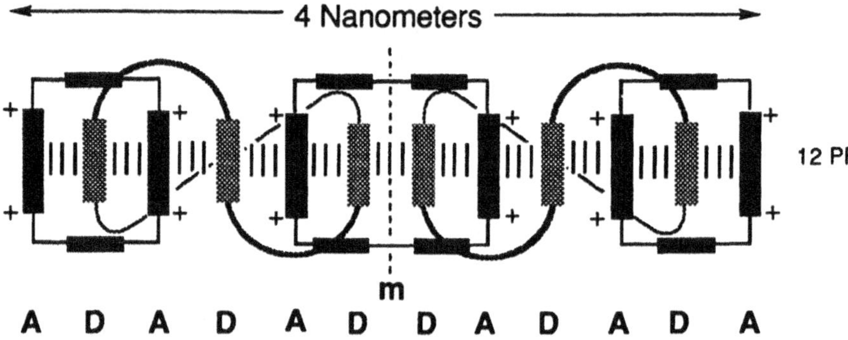

Figure 2. A schematic drawing of the folded conformation of the [5]catenane 3.12PF$_6$

Presumably, the most thermodynamically favorable situation for the π-π stack of aromatic units would be that of a donor-acceptor array. The efficiency of the templating reaction that leads to the [3]catenane 1.4PF$_6$ might be improved by extending the spacers in the large tetracationic cyclophane such that a π-electron deficient template could slip in between the two included π-electron rich aromatic rings. The rate of the reaction associated with the self-assembly of the [3]catenane 1.4PF$_6$ is much slower than the template-directed reactions that produce the catenanes incorporating the smaller tetracationic cyclophane. Perhaps increasing the strengths of the interactions between the spacer unit of the tetracationic cyclophane and the included aromatic rings might lead to an increased rate of ring closure. Such interactions might be introduced by functionalizing the two component macrocycles with mutually compatible groups.

4. Opportunities for Molecular Recognition by Catenanes

The synthesis of olympiadane relies on the availability of recognition sites in the [3]catenane 1.4PF$_6$ for its formation. There are always two π-electron rich residues located outside the cavity at each end of the tetracationic cyclophane, so even allowing for an 'alongside' interaction of one of these rings with the outer π-faces of the bipyridinium units, there is always one π-electron rich site available to act as a template for building on another tetracationic cyclophane. The [2]catenane (*Figure 3*), that was isolated as a by-product in the self-assembly of 1.4PF$_6$, also has the potential to act as a receptor, not only for π-electron deficient guests, but also for π-electron rich guests in the site between the bipyridinium unit of the large tetracationic cyclophane and the included 1,5-dioxynaphthalene ring, *i.e.* it has the potential to act as a ditopic receptor.

Indeed, preliminary results indicate that it can complex π-electron rich aromatic guests in one of its cavities.

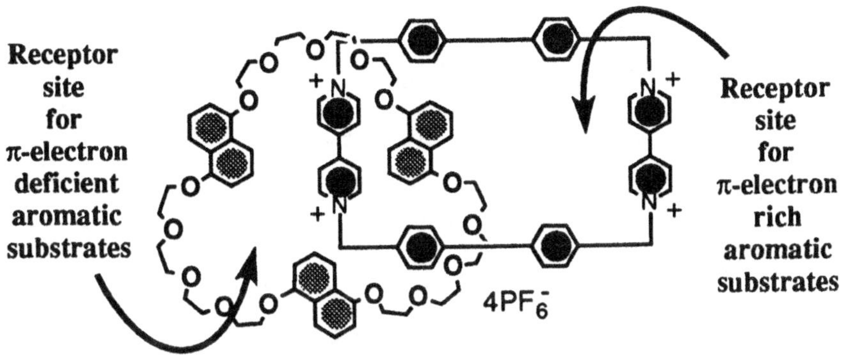

Figure 3. A [2]catenane which could behave as a ditopic receptor

Capitalizing upon the potential illustrated in *Figure 3*, and assuming we could build direction into the two component rings of [2]catenanes of this kind, we would have a ditopic *topologically chiral* receptor.[15] For instance (*Figure 4*), the incorporation of a functional group into the cyclophane's spacer unit would give the tetracation direction. In addition, this site could be endowed with functionality to do work on an included guest. Directionality could be introduced into the neutral component by the incorporation of three different aromatic rings into the macrocycle, for instance 1,5- and 2,6-dioxynaphthalene units along with a hydroquinone ring. In addition, this catenane would possess the property of translational isomerism, in so far as any of the three π-electron rich aromatic residues could be included inside the large cavity of the tetracationic cyclophane.

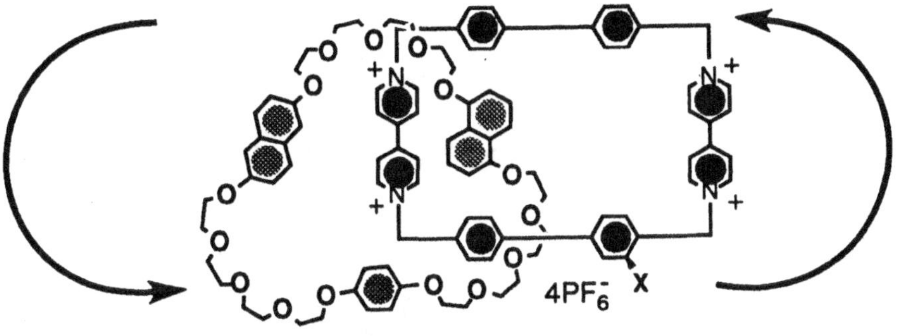

Figure 4. A [2]catenane which could behave as a ditopic topologically-chiral receptor

The incorporation of functional groups on to one or both of the two ring components is an extremely appealing prospect, for it could give rise ultimately to systems which are capable of catalysis. The receptor cyclobis(paraquat-*p*-phenylene) binds π-electron rich residues very well, even perhaps too well! One would expect that the [2]catenanes in which the larger cyclophane is a component would bind π-electron rich aromatic units less well. Then, there is the mechanical movement of the two components relative to each other. The topoisomerase that is responsible for catenane formation in DNA is believed to use a "hinge" mechanism.[16] In the type of [2]catenane shown in *Figure 4*, we would have an example of an axle mechanism. If the [2]catenane were able to promote a polymerization process, assisted by the relative rotations of the rings and a corresponding passage of the reactive threads through the cavities of the interlocked rings, then a molecular extruder would be at work!

5. Branching out in the Self-Assembly of Catenanes

The self-assembly of olympiadane has excited our imaginations since it has opened a doorway to a realm of higher, more exotic interlocked ring systems (*Figure 5*). For example, provided the 'alternative olympiadane' could be synthesized with **TNP57C15** rings present at the termini, we could form a mechanical link with the large tetracationic cyclophane between two pentacatenane molecules to form an undecacatenane. Imagine the same [5]catenane reacting under high dilution conditions in the presence of suitable templates. One can easily conjure up the thought of a cyclic [6]catenane – an intriguing compound indeed!

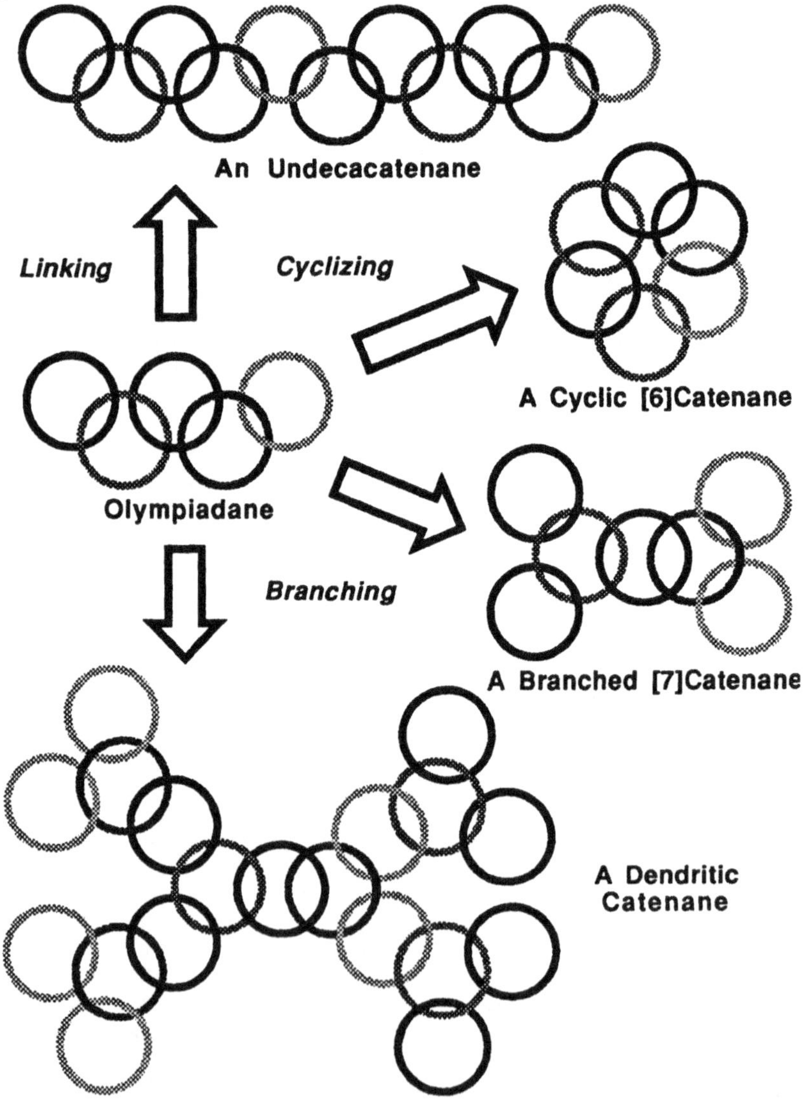

Figure 5. From olympiadane to dendritic catenanes

The preparation of olympiadane at room pressure does not allow the clipping of more than one cyclobis(paraquat-*p*-phenylene) on to each of the **TNP57C15** components of the [3]catenane **1**.4PF$_6$. If we were able to access the remaining π-electron rich sites in **TNP57C15**, it would lead to the realization of a branched [7]catenane. Further, the templating of a larger cyclophane and a macrocyclic polyether simultaneously on to the two **TNP57C15** rings of **1**.4PF$_6$ would provide a route to an prospect – dendritic catenanes! The suggestion of such a synthesis was completely unthinkable not so long ago, yet now it is not too difficult to imagine. However, the creation of such molecular structures will require a major breakthrough in our understanding of the self-assembly mechanisms at work in these systems in order that we can increase greatly the efficiencies of molecular self-assembly processes.

References

1. (a) Lehn, J.-M. (1988) Supramolecular chemistry - scope and perspective, *Angew. Chem. Int. Ed. Engl.* **27**, 89-112. (b) Lehn, J-M. (1990) Perspectives in supramolecular chemistry, *Angew. Chem. Int. Ed. Engl.* **29**, 1304-1319.

2. (a) Sauvage, J.-P. and Dietrich-Buchecker, C. (1991) Interlocked and knotted rings in biology and chemistry, *Bioorg. Chem. Front.* **2**, 195-248. (b) Constable, E.C. (1992) Oligopyridines as helicating agents, *Tetrahedron* **48**, 10013-10059. (c) Hunter, C. (1992) Synthesis and structure elucidation of a new [2]catenane, *J. Am. Chem. Soc.* **114**, 5303-5311. (d) Vögtle, F., Meier, S. and Hoss, R. (1992) One-step synthesis of a fourfold functionalized catenane, *Angew. Chem., Int. Ed. Engl.* **31**, 1619-1622.

3. (a) Busch, D.H. (1992) Structural definition of chemical templates and the prediction of new and unusual materials, *J. Inc. Phenom.* **12**, 389-395. (b) Anderson, S., Anderson, H.L. and Sanders, J.K.M. (1993) Expanding roles for templates in synthesis, *Acc. Chem. Res.* **26**, 469-475. (c) Hoss, R. and Vögtle, F. (1994) Template syntheses, *Angew. Chem. Int. Ed. Engl.* **33**, 375-384.

4. (a) Lindsey, J.S. (1991) Self-assembly in synthetic routes to molecular devices. Biological principles and chemical perspectives : a review, *New J. Chem.* **15**, 153-180. (b) Whitesides, G.M., Mathias, J.P. and Seto, C.T. (1991) Molecular self-assembly and nanochemistry: A chemical strategy for the synthesis of nanostructures, *Science*, **254**, 1312-1319.

5. Wasserman, S.A. and Cozzarelli, N.R. (1985) Biochemical topology : Applications to DNA recombination and replication, *Science* **232**, 951-960.

6. Koert, U., Harding, M.M. and Lehn, J.-M. (1990) DNH deoxyribonucleo-helicates: Self-assembly of oligonucleosidic double-helical metal complexes, *Nature* **346**, 339-342.

7. (a) Walba, D.M. (1985) Topological stereochemistry, *Tetrahedron* **41**, 3161-3212. (b) Chambron, J.-C., Dietrich-Buchecker, C., and Sauvage, J.-P. (1993) From classical chirality to topologically chiral catenands and knots, *Top. Curr. Chem.* **165**, 131-162.

8. van Gulick, N. (1993) Theoretical aspects of the linked ring problem, *New J. Chem.* **17**, 619-625. See also the preface to this paper by Walba, D.M. (1993) Preface to van Gulick paper, *New J. Chem.* **17**, 618.

9. Amabilino, D.B., Ashton, P.R., Reder, A.S., Spencer, N. and Stoddart, J.F. (1994) The two-step self-assembly of [4]- and [5]-catenanes, *Angew. Chem. Int. Ed. Engl.* **33**, 433-437.

10. Amabilino, D.B., Ashton, P.R., Reder, A.S., Spencer, N. and Stoddart, J.F. (1994) Olympiadane, *Angew. Chem. Int. Ed. Engl.* **33**, 1286-1290.

11. Amabilino, D.B., Ashton, P.R., Stoddart, J.F., Menzer, S. and Williams, D.J. (1994) The solid-state self-organisation of a self-assembled [2]catenane, *J. Chem. Soc., Chem. Commun.* 2475-2478.

12. Ashton, P.R., Brown, C.L., Chrystal, E.J.T., Goodnow, T.T., Kaifer, A.E., Parry, K.P., Slawin, A.M.Z., Spencer, N., Stoddart, J.F. and Williams, D.J. (1991) Self-assembling [3]catenanes, *Angew. Chem. Int. Ed. Engl.* **30**, 1039-1042.

13. Amabilino, D.B., Stoddart, J.F. and Williams, D.J. (1994) From solid-state structures and superstructures to self-assembly processes, *Chem. Mater.* **6**, 1159-1167.

14. Dietrich-Buchecker, C.O., Guilhem, J., Pascard, C. and Sauvage, J.-P. (1990) Structure of a synthetic trefoil knot coordinated to two copper(I) centers, *Angew. Chem. Int. Ed. Engl.* **29**, 1154-1156.

15. Chambron, J.-C., Mitchell, D. and Sauvage, J.-P. (1992) Synthesis, characterization, and a proton NMR study of topologically chiral copper(I) [2]catenates and achiral analogues, *J. Am. Chem. Soc.* **114**, 4625-4631.

16. Lima, C.D., Wang, J.C. and Mondragón, A. (1994) Three-dimensional structure of the 67K N-terminal fragment of *E. coli* DNA topoisomerase I, *Nature* **367**, 138-146.

NOVEL PROTEIN TOPOLOGIES

KURT MISLOW & CHENGZHI LIANG
Department of Chemistry, Princeton University
Princeton, New Jersey 08544, USA

Are the polypeptide chains of native proteins ever tied into knots or links (catenated loops)? This question had been repeatedly addressed over the past 20 years [1-10], and it was, until recently, the general consensus that "linked loops and trefoil knots are not found in tertiary structure of proteins known to date" [7]. Typically, it was noted in 1993 that "True topological linking does not occur in any of the proteins currently represented in the PDB [Brookhaven Protein Data Bank] structure database" [8], that "The PDB structural database has been searched for knotted polypeptides ··· No examples of topological knotting were found in this case-by-case search" [8], and that "structures of globular proteins are 'topologically simple' and free of topologically complicated constructs such as knots and links and other types of entanglements" [9]. In none of these investigations, however, had any consideration been given to the possibility that catenation or knotting of the polypeptide chain might be effected through crosslinking by a combination of disulfide bonds and covalently bound cofactors.

In 1994 we provided the first example of "topological complexity" in a protein: a topological link formed by three disulfide bonds and the covalently bound tryptophan-derived quinone cofactor in the light chain of quinoprotein methylamine dehydrogenase (MADH) [11]. We here report our finding [12] that the structures of the metalloproteins ascorbate oxidase (AOase) [13] and human lactoferrin (hLf) [14], as well as closely related conjugated proteins, contain trefoil knots and two-component links. These extraordinary topological features depend on the presence of the covalently bound metal atoms and disulfide bonds, which are integral constituents of the knots and links.

In the multi-copper enzyme AOase, two copper atoms, each of which is bonded to the nitrogen atoms of two histidine residues, and two disulfide bonds form part of a trefoil knot (Figure 1). Additional trefoil knots may be constructed from different combinations of copper atoms, disulfide bonds, and polypeptide chain segments.

Human lactoferrin is structurally characterized by two domains, the N- and the C-lobe. The two lobes exhibit similar folding [14] and contain the same topological tangle (Figure 2). In the C-lobe, one iron atom, which is bonded to the oxygen atoms of aspartate and tyrosine residues, and one disulfide bond are also part of a trefoil knot (Figure 3).

J. S. Siegel (ed.), Supramolecular Stereochemistry, 41–46.

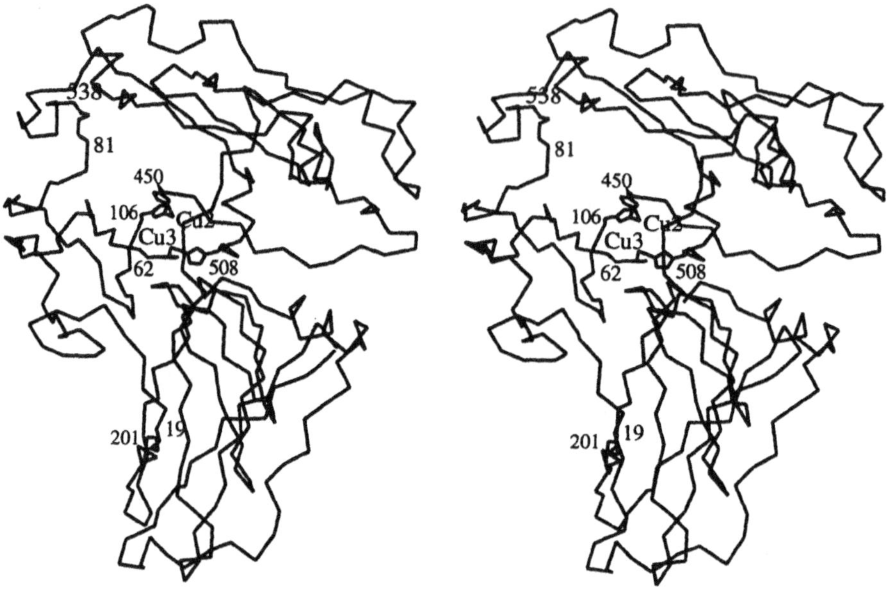

Figure 1. Stereoview (Cα trace plus crosslinks) of a trefoil knot contained within ascorbate oxidase (native oxidized form). The two copper atoms are part of the knot: Cu2 is covalently bonded to the nitrogen atoms of His106 and His450, and Cu3 to the nitrogen atoms of His62 and His508. Cys81-Cys538 and Cys19-Cys201 are disulfide crosslinks.

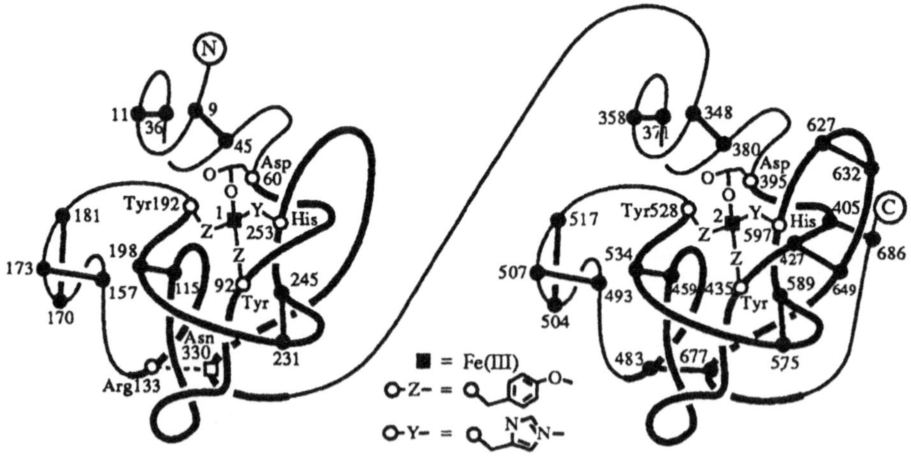

Figure 2. Condensed schematic drawing of hLf [14]. The α-carbons of cysteine (or half-cystine) residues are symbolized by filled circles and the α-carbons of some non-cysteine residues by open circles. Intrachain disulfide linkages are shown as heavy lines. The carbonyl carbon of Asn330 is symbolized by an open square. Unlabeled vertices represent carbon atoms, and hydrogen atoms are suppressed for clarity. The N-lobe is shown on the left and the C-lobe on the right. The topological tangles are indicated by heavy curves.

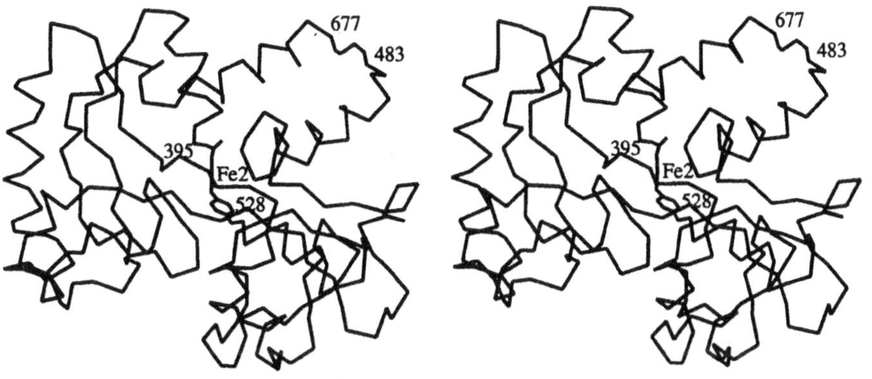

Figure 3. Stereoview (Cα trace plus crosslinks) of a trefoil knot contained within the C-lobe of hLf. The iron atom (Fe2) is covalently bonded to the oxygen atoms of Asp395 and Tyr528, and Cys483-Cys677 is a disulfide crosslink.

According to a general convention developed for topologically chiral knots [15], the knots in AOase and in the C-lobe of hLf are assigned L and D configurations, respectively.

The structures of the two proteins also contain a variety of two-component topological links, depending on the choice of metal atoms, disulfide

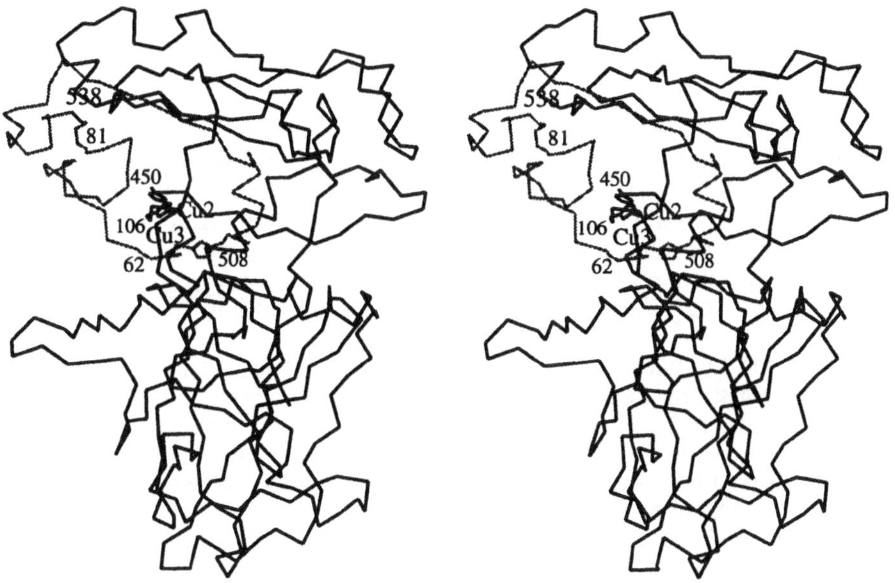

Figure 4. Stereoview (Cα trace plus crosslinks) of a two-component link contained within AOase. Cu2, covalently bonded to the nitrogen atoms of His106 and His450, is part of one component (black trace), while Cu3, covalently bonded to the nitrogen atoms of His62 and His508, together with the Cys81-Cys538 crosslink, is part of the other (grey trace).

bonds, and polypeptide chain segments; examples of such topological links in AOase and hLf are shown in figures 4 and 5. The incorporation of metal atoms and disulfide bonds in all of these links is in accord with our previous observation [11] that topologically complex substructures in proteins are most likely formed from a combination of covalently bound cofactor and disulfide crosslinks. From the limited data available to us so far, it would appear that, in general, where there are knots in proteins there are also links, although, as shown in the case of MADH [11], the converse is evidently not true.

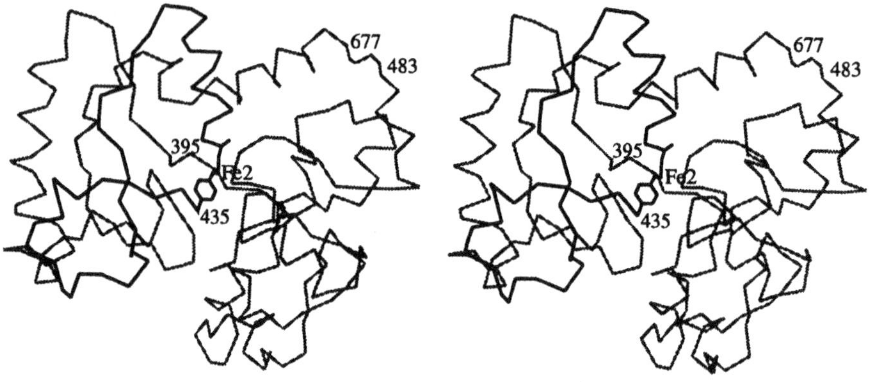

Figure 5. Stereoview (Cα trace plus crosslinks) of a two-component link contained within the C-lobe of hLf. Fe2, which is covalently bonded to the oxygen atoms of Asp395 and Tyr435, is part of one component (black trace), while the polypeptide segment containing the Cys483-Cys677 crosslink is part of the other (grey trace).

In a recent development, we have found that the β-subunit of human chorionic gonadotropin (hCG) [16] contains two catenated substructures, in each of which two closed loops (rings) are linked as shown in Figure 6. This is the first example of a native protein whose structure contains a topological link *without* a cofactor: the two links are made up entirely of amino acid residues, including two cystine disulfide bridges in each of the component rings. Note that one of the rings, (9-23-72-57-9), is common to both links.

In the knotted and linked protein substructures described above, the atoms that constitute these curious topological constructs are all connected by covalent bonds. It needs to be emphasized, however, that the choice of bonds regarded as topologically significant depends on the molecular model under consideration [17], and, given the extensive network of hydrogen bonds in proteins, the inclusion of such bonds as edges in the molecular graph of protein molecules [18] is therefore bound to furnish many additional catenated and knotted substructures. Of particular interest is the observation that the N-lobe of hLf (Figure 2) also contains a trefoil knot (D configuration) if the previously unremarked [14] hydrogen bond between Arg133 and Asn330 (d(N···O) = 2.56 Å) is included in the molecular graph. This knot bears a striking resemblance to the companion knot in the C-lobe: the only significant difference is the replacement of the covalent S-S bond in the C-lobe by the N-H···O bond in the N-lobe.

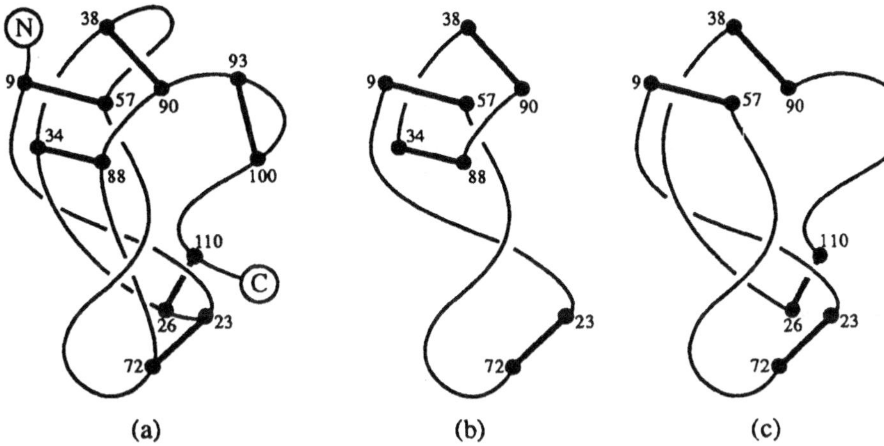

Figure 6. (a) Condensed schematic drawing of the β-subunit of hCG. The α-carbons of cysteine (or half-cystine) residues are symbolized by filled circles. Intrachain disulfide linkages are shown as heavy lines. (b) A topological link derived from (a) in which disulfide linkage (9-57) penetrates the ring (34-38-90-88-34) formed from two polypeptide segments and two disulfide linkages. (c) Another topological link, in which the two component rings in (a) penetrate one another through their polypeptide segments instead of their disulfide linkages.

In light of the present work and of the roughly exponential growth in new high-resolution protein structures that are solved every year [19], it now appears likely that links and knots will in time become familiar topological features of conjugated protein structures.

Acknowledgement

We thank the National Science Foundation for support of this work.

References

1. Crippen, G.M. (1974) Topology of Globular Proteins, *J. Theor. Biol.* **45**, 327-338.
2. Crippen, G.M. (1975) Topology of Globular Proteins II, *J. Theor. Biol.* **51**, 495-500.
3. Connolly, M.L., Kuntz, I.D., and Crippen, G.M. (1980) Linked and Threaded Loops in Proteins, *Biopolymers* **19**, 1167-1182.
4. Klapper, M. H. and Klapper, I. Z. (1980) The 'Knotting' Problem In Proteins -- Loop Penetration, *Biochim. Biophys. Acta* **626**, 97-105.
5. Kikuchi, T., Némethy, G., and Scheraga, H.A. (1986) Spatial Geometrical Arrangements of Disulfide-Crosslinked Loops in Proteins, *J. Comput. Chem.* **7**, 67-88.

6. Kikuchi, T., Némethy, G., and Scheraga, H.A. (1989) Spatial Geometrical Arrangements of Disulfide-Crosslinked Loops in Nonplanar Proteins, *J. Comput. Chem.* **10**, 287-294.
7. Mao, B. (1989) Molecular Topology of Multiple-Disulfide Polypeptide Chains, *J. Am. Chem. Soc.* **111**, 6132-6136.
8. Benham, C.J. and Jafri, M.S. (1993) Disulfide bonding patterns and protein topologies, *Protein Science* **2**, 41-54.
9. Mao, B. (1993) Topological chirality of proteins, *Protein Science* **2**, 1057-1059.
10. Mansfield, M.L. (1994) Are there knots in proteins? *Nature Structural Biology* **1**, 213-214.
11. Liang, C. and Mislow, K. (1994) Topological Chirality of Proteins, *J. Am. Chem. Soc.* **116**, 3588-3592.
12. Liang, C. and Mislow, K. (1994) Knots in Proteins, *J. Am. Chem. Soc.* in press.
13. Messerschmidt, A., Ladenstein, R., Huber, R., Bolognesi, M., Avigliano, L., Petruzzelli, R., Rossi, A., and Finazzi-Agró, A. (1992) Refined Crystal Structure of Ascorbate Oxidase at 1.9 Å Resolution, *J. Mol. Biol.* **224**, 179-205.
14. Anderson, B.F., Baker, H.M., Norris, G.E., Rice, D.W., and Baker, E.N. (1989) Structure of Human Lactoferrin: Crystallographic Structure Analysis and Refinement at 2.8 Å Resolution, *J. Mol. Biol.* **209**, 711-734.
15. Liang, C. and Mislow, K. (1994) A left-right classification of topologically chiral knots, *J. Math. Chem.* **15**, 35-62.
16. (a) Lapthorn, A.J., Harris, D.C., Littlejohn, A., Lustbader, J.W., Canfield, R.E., Machin, K.J., Morgan, F.J., and Isaacs, N.W. (1994) Crystal structure of human chorionic gonadotropin, *Nature* **369**, 455-461. (b) Wu, H., Lustbader, J.W., Liu, Y., Canfield, R.E., and Hendrickson, W.A. (1994) Structure of human chorionic gonadotropin at 2.6 Å resolution from MAD analysis of the selenomethionyl protein, *Structure* **2**, 545-558.
17. Liang, C. and Mislow, K. (1994) Classification of topologically chiral molecules, *J. Math. Chem.* **15**, 245-260.
18. Mao, B., Chou, K.-C., and Maggiora, G.M. (1990) Topological analysis of hydrogen bonding in protein structure, *Eur. J. Biochem.* **188**, 361-365.
19. Lattman, E.E. (1994) Protein Crystallography for All, *PROTEINS: Struct., Funct., and Genet.* **18**, 103-106. Rose, G.D. and Creamer, T.P. (1994) Protein Folding: Predicting Predicting, *Ibid.* **19**, 1-3.

A NEW MOLECULAR ENCODING SYSTEM

Steven A. Benner, Jennifer Horlacher, Johannes Vögel, and Ulrike von Krosigk
Department of Chemistry, E.T.H. Zürich
CH-8092 Zürich, Switzerland

Recognition between the base pairs in complementary oligonucleotide strands is a model for efforts to understand intermolecular complexation and to design intermolecular recognition systems generally. We have asked whether synthetic organic chemistry, enzymology, and molecular biology can be applied together to obtain an expanded molecular recognition system, modeled on nucleic acids, that:

(a) is one-dimensional, and therefore is easy to synthesize.

(b) is capable of duplication via principles of complementarity analogous to those used by DNA and RNA.

(c) can encode information concerning the structure of other biopolymers.

We have then asked whether a full complement of synthetic tools can be developed for this alternative encoding system, including:

(a) polymerases that allow the synthesis of large amounts of product from single template molecules.

(b) translation systems that permit encoded information to guide the synthesis of other biopolymers.

(c) a polymerase chain strategy that allows amplification of the alternative molecular recognition system.

In each case, the answer has proven to be "yes".

Recognition in nucleic acids is mediated by base pairing. The pairing that encodes genetic information is the Watson-Crick base pair, which is governed by two rules of complementarity: size complementarity (a large purine pairs with a small pyrimidine) and hydrogen bonding complementarity (hydrogen bond donors are matched with hydrogen bond acceptors). In addition, both hydrophobicity and planarity in the bases are believed to be important for the stability of the double helical structure. Hydrophobicity allows transfer of the bases from water to the hydrophobic core of the double helix to make an energetic contribution to duplex formation. Planarity allows sequence-independent stacking of the bases to form an aperiodic crystalline structure necessary for duplex stability.

J. S. Siegel (ed.), Supramolecular Stereochemistry, 47–55.

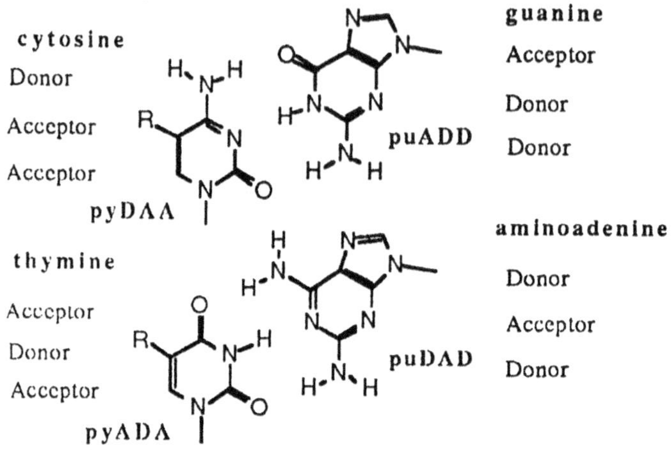

Figure 1: Idealized Watson-Crick Base Pairs

Some time ago, we noted that the Watson-Crick formalism could readily be extended to include 12 independently replicatable bases joined in 6 base pairs by mutually independent hydrogen bonding patterns, provided that some bases were joined to the sugar via a carbon-carbon bond (a "C-glycoside") (Figure 1) [1]. If this were possible, and if the molecular recognition properties of oligonucleotides could be maintained with the expanded set of bases, the information density in an oligonucleotide chain would be greatly increased.

Here, analogs of pyrimidines are designated by the prefix "pu", while analogs of purines are designated by the prefix "pu". Following the prefix is the order, from the major groove to the minor groove, of acceptor (A) and donor (D) groups. In natural nucleosides, the A:T base pair is incomplete.

In principle, a nucleic acid containing any or all of the 12 bases could be accurately copied, as each base pair is joined by mutually exclusive hydrogen bonding patters. To implement the idea requires, however, a combination of synthetic organic chemistry, physical chemistry, enzymology, and molecular biology. We must:

(a) Synthesize the non-standard bases

(b) Develop the synthetic chemistry technology that allows synthesis of oligonucleotides containing the non-standard bases.

(c) Optimize the properties of the non-standard bases, especially those relevant to their participation in a molecular encoding system.

(d) Develop polymerase technology that allows the amplification of oligonucleotides

(e) Develop the ribosome-based technology that allows oligonucleotides containing non-standard bases to encode something.

The pyADA, puDAD, pyDAA, pyADD, puADA, and puDDA bases are all natural products (thymine, diaminopurine, cytosine, guanine, xanthine, and

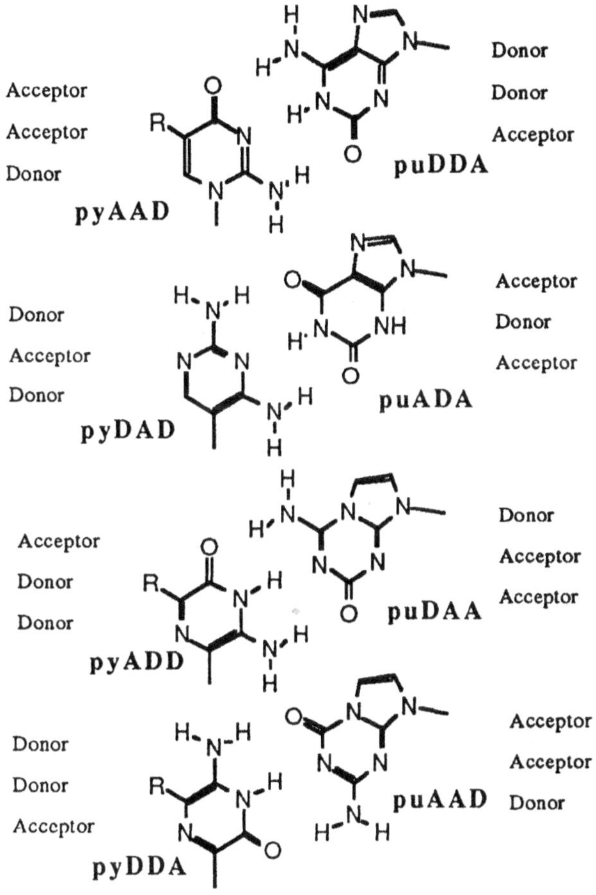

Figure 2: The New Non-standard Base Pairs

crotonoside). The pyDAD, pyAAD, and puAAD bases had been reported in the voluminous literature on nucleoside analogs. This left only three ring systems where syntheses needed to be developed *de novo*.

Two new pyrazine syntheses were developed. The first was developed by Ulrike von Krosigk [2], to yield the pyDDA base. The second pyrazine synthesis was developed by Johannes Vögel [3], yielding the pyADD base (Figure 4), and new chemistry was developed by Johannes Vögel to prepare the puDAA nucleoside (Figure 5) [4].

For each non-standard base, both ribo- and deoxyribo- derivatives were prepared, as were protected forms suitable for the solid phase synthesis of oligonucleotides incorporating the non-standard bases.

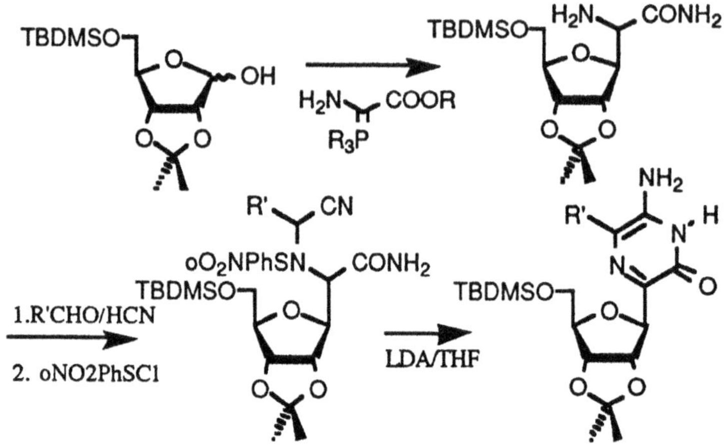

Figure 3: Synthesis A of the Pyrazine-containing Non-standard Base Pairs

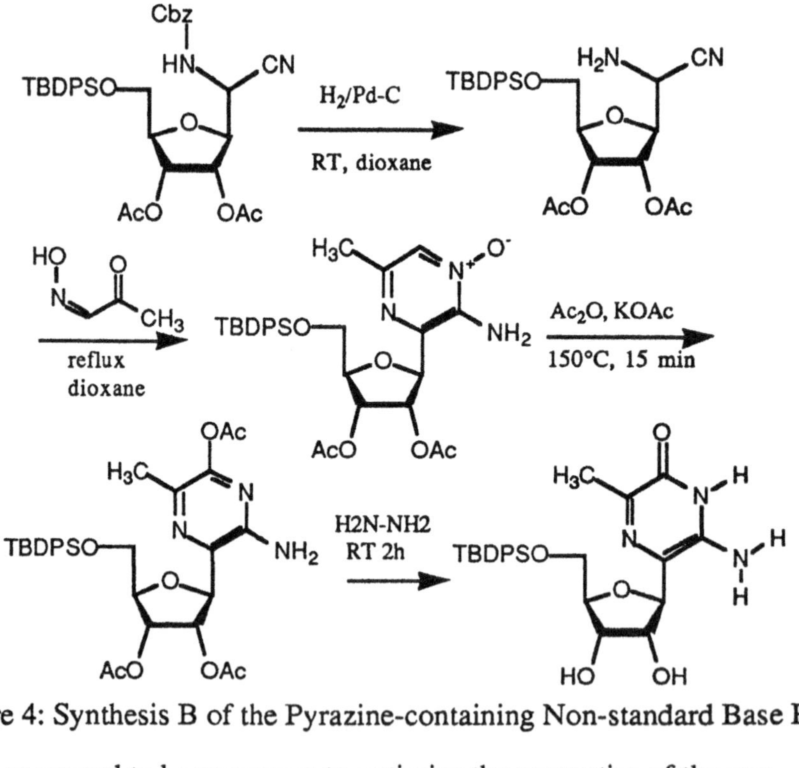

Figure 4: Synthesis B of the Pyrazine-containing Non-standard Base Pairs

It then proved to be necessary to optimize the properties of the non-standard bases. Many of the undesirable properties of the ring systems that present non-standard hydrogen bonding patterns can be removed by tinkering

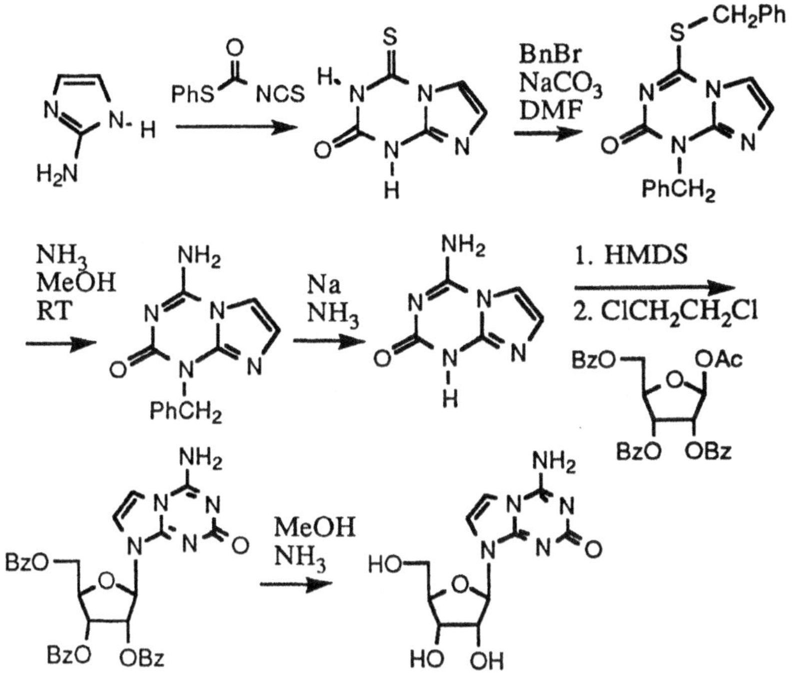

Figure 5: Synthesis of puDAA Nucleosides

with the heterocycle. For example, the pyAAD base deaminates more rapidly than the natural pyDAA base (cytosine). Introducing a methyl group at position 5 increased its stability.

The pyDAD, pyDDA, and pyADD systems based on a pyridine nucleus proved to be easily oxidized and rather basic [5]. We presumed that this arose because the diaminopyridine system was too electron rich. Therefore, a nitrogen was added to the ring system, making the target structures pyrimidines and pyrazines.

Finally, the pyDDA and the pyADD bases undergo a slow, specific acid-catalyzed epimerization by the mechanism in Figure 6 [6]. This is being correcting by preparing carbocyclic analogs of the nucleosides bearing these heterocycles.

Some undesirable properties of the non-standard bases proved to be intrinsic in the hydrogen bonding pattern of the ring system, however. For example, the puDDA ring has a cross-conjugated 5-ring. This creates special properties in the heterocycle that, as we later found, dominate the properties of this non-standard base in a molecular encoding system.

The non-standard base pairs contribute to duplex stability in the expected fashion, both in DNA-DNA duplexes [7] and in DNA-RNA duplexes. The

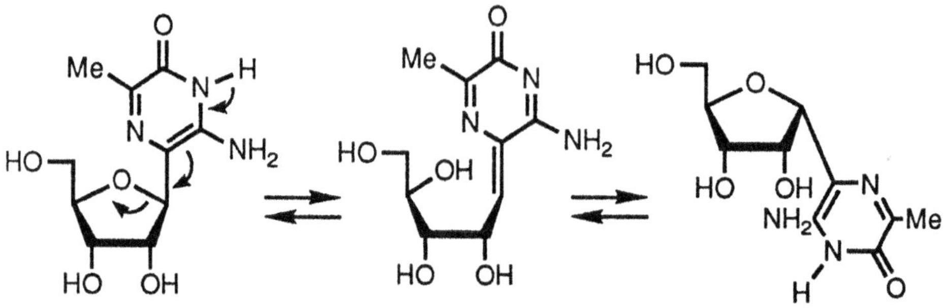

Figure 6: Epimerization of a pyADD Nucleoside

significant mismatches appear to be wobbles.Two sets of data are collected in Table 1 and Table 2 below.

Table 1. Stability of DNA-DNA Duplexes Containing Non-Standard Base Pairs

	N=	pyADA	puDA-	pyDAA	puADD	pyDAD
M=		T	A	C	G	κ
puDA-	A	40				
pyADA	T		37		26	
puADD	G	21		38		
pyDAA	C				42	
puADA	π	29		19		35
pyDAD	κ		22		18	

Melting temperatures of 5'-CAAAMAAAG-3' and 5'-CTTTNTTTG-3'

Table 2. Stability of DNA-RNA Duplexes Containing Non-Standard Base Pairs

	X=	pyDAA	pyADA	pyADD
Y=		C	U	V
puADD	G	36	17	<5
puDA-	A	5	28	17
puDAA	J	<5	13	27

Melting temperatures of 5'-r(GAACXAAA)-3' and 5'-d(TTTYGTTC)-3'

More about the physical properties of the non-standard bases was learned through efforts to identify DNA and RNA polymerases that could synthesize oligonucleotides containing the non-standard bases under the direction of a template. The experiments are relatively simple to execute. A template is prepared containing one non-standard base. The template is then incubated with a polymerase and a primer in the presence of the complementary non-standard triphosphate. Label either the primer or the final base to be

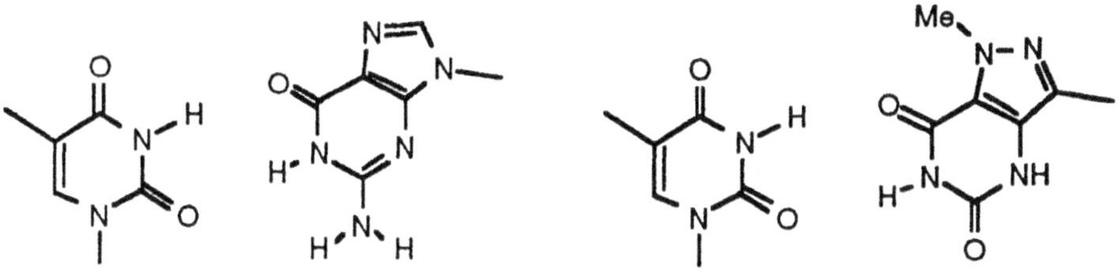

pyADA - puADD wobble pyADA - puADA wobble

Figure 7. Wobble Base Pairs involving Standard and Non-standard Bases

incorporated is used to detect the reaction products, which are isolated and identified by gel electrophoresis. The results for two sets of experiments are summarized in Tables 3 and 4.

Table 3. Polymerase and the pyAAD-puDDA pair

enzyme	*result*
Incorporation of (puDDA)TP opposite pyAAD in template	
Klenow fragment, pol I	successful incorporation
T4 DNA polymerase	no incorporation
T7 RNA polymerase	successful incorporation
AMV reverse transcriptase	successful incorporation
Incorporation of (pyAAD)TP opposite puDDA in template	
Klenow fragment, polymerase I	incorporates pyAAD and pyADA
T7 RNA polymerase	incorporates only pyADA

A minor tautomer of the puDDA base explains the misincorporation observed (Figure 8). This proves to be an intrinsic problem of this purine, and serves as a potential model for tautomerism as a mechanism for spontaneous mutation in natural DNA. A more complete set of polymerases have now been identified, including several that catalyze the incorporation of the pyDAD-puADA base pair in both directions.

To demonstrate that the expanded genetic alphabet can serve an encoding role, we have used it to increase the number of amino acids that can be incorporated into a polypeptide by template-directed polymerization.

In collaboration with Jim Bain and Prof. A. Richard Chamberlin at the University of California at Irvine, the pyAAD-puDDA base pair was chosen for the new codon-anticodon pair [8]. The codon was (pyAAD)AG; the anticodon was CU(puDDA). The stop codon UAG was used as a control. A tRNA molecule was synthesized and charged with iodotyrosine. Two mRNA molecules were synthesized, one containing the UAG stop codon, the other

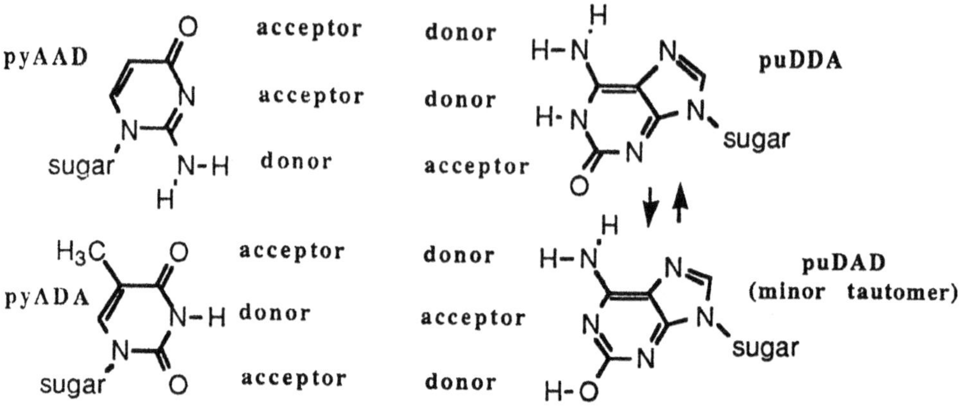

Figure 8. Tautomerism of the puDDA Base

containing the (pyAAD)AG nonsense codon. Efficient read-through and peptide synthesis were observed.

Table 4. Polymerase and the pyAAD-puDDA pair

Incorporation of (puADA)TP opposite pyDAD in template

enzyme	result
Klenow fragment, pol I	good incorporation
Calf polymerase α	no incorporation
Calf polymerase β	no incorporation
Calf polymerase ε	no incorporation
T7 RNA polymerase	good incorporation
HIV reverse transcriptase	good incorporation

Incorporation of (pyDAD)TP opposite puADA in template

enzyme	result
Klenow fragment, pol I	no incorporation
Calf polymerase α	one base added, then stops
Calf polymerase β	no incorporation
Calf polymerase ε	one base added, then stops
HIV reverse transcriptase	some incorporation

The ease with which an expanded coding system can be generated raises some questions about natural history on planet earth. At some point following the emergence of life in the RNA world, a decision was taken to invent a new polymer, encoded by nucleic acids, but built from a new building block, amino acid. This decision was presumably made because the repertoire of encoded functional groups in nucleic acids was rather small, permitting oligonucleotides only poor catalytic powers. The post-transcriptional modification of tRNAs might be viewed as a vestige of the RNA world's efforts to expand the catalytic power of nucleic acids by increasing their functional group diversity.

It is clear from this work here that this decision could have been made differently. Rather than inventing proteins to obtain biopolymers with encoded functionality, life could also have expanded the number of encoded bases, and append functional groups to these. This could have conferred enormous catalytic power on nucleic acids, comparable to that seen with proteins. The invention of translation would not have been necessary.

In many ways, such an approach seems to be simpler than the one used by contemporary life on earth. Further, with the higher density of information per unit length of the encoding strand, it would have been easier to build "nanostructures" with such a biomolecule, analogous to those being constructed by Seeman and his colleagues using natural DNA.

It is extremely difficult to speculate why translation was chosen as a route to increase the catalytic power of biomacromolecules as opposed to an expanded genetic alphabet. However, two things are clear:

(a) Life elsewhere in the universe need not have made the same decision.

(b) In the laboratory, we may use the expanded genetic alphabet, bearing catalytic functional groups, to increase the chances of obtaining self-replicating nucleic acids.

References

1. Benner, S. A., Allemann, R. K., Ellington, A. D., Ge, L., Glasfeld, A., Leanz, G. F., Krauch, T., MacPherson, L. J., Moroney, S. A., Piccirilli, J. A., and Weinhold, E. (1987) *Cold Spring Harbor Sym. Quant. Biol.* **52**, 53-63.
2. Krosigk, U. v. (1993) Entwurf und Synthese eines neuen Basenpaars für das genetischen Alphabet, ETH Diss. Nr. 10164.
3. Vögel, J. J., Krosigk, U. v. and Benner, S. A. (1993) *J. Org. Chem.* **58**, 7542-7547.
4. Vögel, J. J., Altorfer, M. M., and Benner, S. A. (1993) *Helv. Chim Acta* **76**, 2061-2069.
5. Piccirilli, J. A., Krauch, T., MacPherson, L. J. and Benner, S. A. (1991) *Helv. Chim. Acta* **74**, 397-406.
6. Vögel, J. J., and Benner, S. A. (1994) *J. Am. Chem. Soc.* **116**, 6929-6930.
7. Piccirilli, J. A., Krauch, T., Moroney, S. E. and Benner, S. A. (1990) *Nature* **343**, 33-37.
8. Bain, J. D., Chamberlin, A. R., Switzer, C. Y. and Benner, S. A. (1992) *Nature* **356**, 537-539.

RECOGNITION OF B VS. Z-FORM DNA USING NICKEL AND COBALT COMPLEXES

CYNTHIA J. BURROWS, JAMES G. MULLER, HUI-CHEN SHIH and STEVEN E. ROKITA
Department of Chemistry, State University of New York, Stony Brook, NY 11794

INTRODUCTION

Nickel(II) and cobalt(II) coordination compounds are remarkably versatile in their interactions with nucleic acids. The number and nature of the ligands surrounding the metal ion control the kinetics and thermodynamics of DNA binding as well as the metal's redox properties. These metals undergo rapid ligand exchange in the +2 oxidation state, and slower exchange as octahedral +3 complexes. Strong donor ligands such as nitrogen bases stabilize the +3 oxidation state. For nickel(II), 14-membered tetraazamacrocycles allow the complex to be only 4-coordinate, whereas both nickel(III) and cobalt(III) complexes are 6-coordinate.

Recent studies in our laboratories showed that certain nickel(II) complexes such as NiCR (**1**) and Ni(cyclam) (**2**) were valuable tools for the study of guanine stacking in DNA and RNA.[1] In the presence of a peracid oxidant such as $KHSO_5$, a reaction occurred at guanine bases that were present in highly accessible conformations rather than those in regular Watson-Crick G-C pairs.[2] Strand scission was observed at these sites after an alkaline treatment.[3] Examples of reactive guanine sites include bulges, mismatches, and hairpin loops (such as the deoxyoligonucleotide **3**) in DNA oligomers,[2] $tRNA^{Phe}$, and *Tetrahymena* group I intron.[4]

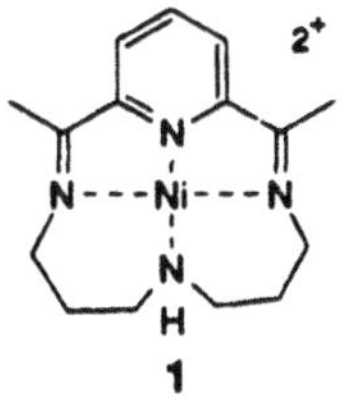

1

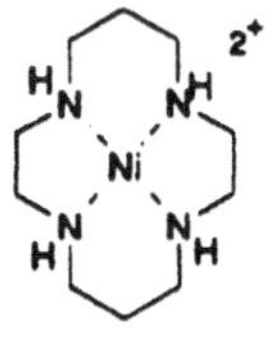

2

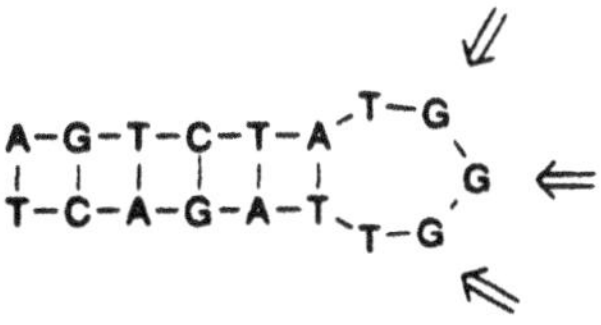

3

J. S. Siegel (ed.), Supramolecular Stereochemistry, 57–62.

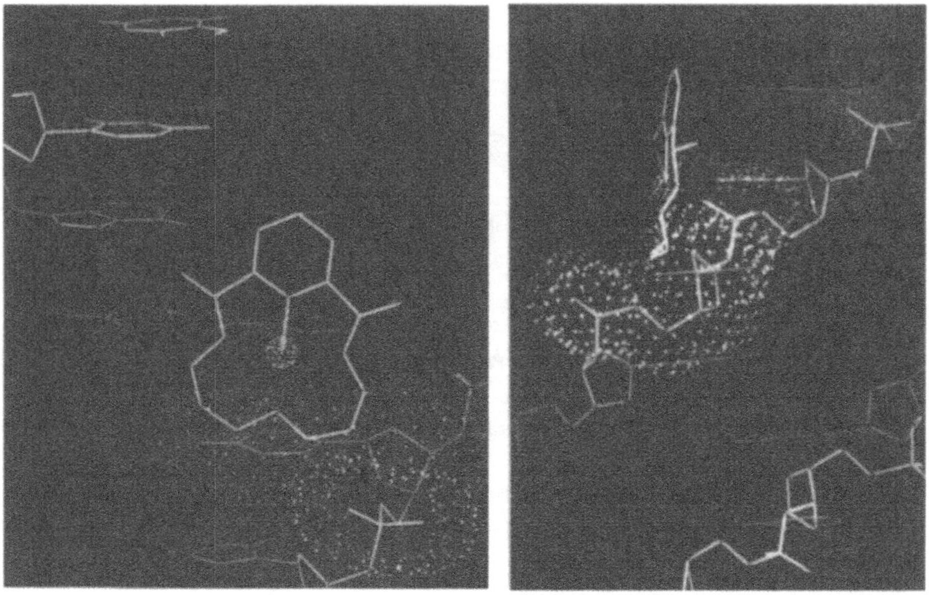

A B

Figure 1. Views of NiCR docked to guanine in major groove of B-DNA with Ni...N7 distance of 2.0Å. (A) Face view. (B) Side view.

ACCESSIBILITY OF GUANINE TO METAL ION BINDING IN DUPLEX DNA

The high selectivity of the NiCR/HSO_5^- reaction for accessible guanine residues begs explanation. The proposed mechanism of NiCR-mediated guanine oxidation calls for initial formation of a nickel(III) complex which then coordinates to guanine N7 via direct metal-nitrogen bonding.[5] Guanine N7 is the best transition metal binding site in DNA and is, for example, the site at which platinum drugs bind. Further support of N7 binding by NiCR is gained from the excellent correlation of nickel reactivity with the surface accessibility of N7's lone pair as seen in the crystal structure of $tRNA^{Phe}$.[6]

Why doesn't NiCR bind to guanine N7 in a regular B helix? The answer lies in the difference in coordination geometry between square planar platinum(II) complexes which do bind to G's in a B helix and octahedral nickel(III) complexes, the proposed reactive intermediates of NiCR-based oxidation. Whereas PtL_4 binds edge-on to G-N7 in the major groove by ligand substitution, NiCR approaches face-on to the helix and attempts to take G-N7 as an axial ligand. Modeling studies suggest that NiCR cannot penetrate deeply enough into the major groove to obtain a normal 2.0Å bond distance. Fig. 1 shows that to do so would cause major steric interactions between the macrocyclic ligand and the sugar-phosphate backbone, especially toward the 5' side of the G.

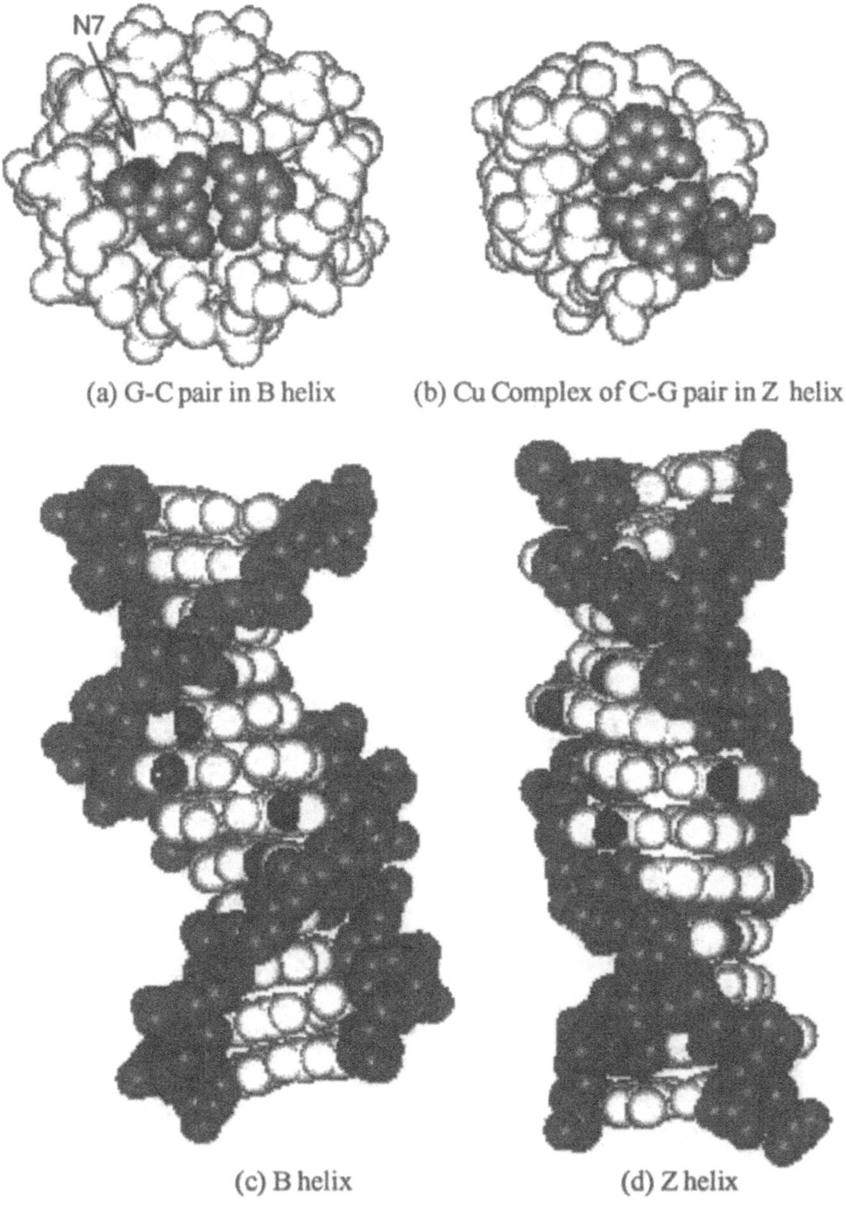

Figure 2. Views parallel (a+b)[7] and perpendicular (c+d)[8,9] to the helix axis of B (a+c) and Z (b+d) DNA showing the location of guanine N7 in black. View (b) shows ligation of $[Cu(H_2O)_5]^{2+}$.[8]

METALS INDUCE B→Z TRANSITIONS

Under what conditions might nickel complexes such as NiCR or Ni(cyclam) be able to bind to G-C pairs? One needs to consider unusual conformations of DNA or RNA in which guanine N7 is highly accessible on the surface of the molecule. This condition is satisfied with the *left*-handed helical form of DNA, or Z-DNA. Inspection of the Z helix in comparison to the B helix shows that G-C pairs are oriented around the periphery of the helix in the Z-form while they lie closer to the helix axis in B-DNA. As a consequence, G-N7 in Z-DNA is located near the outer cylindrical surface. (See Fig. 2.) Indeed, crystal structures confirm that metal ions such as copper(II)[8] and cobalt(II)[12] can bind directly to G-N7 in Z-DNA.

Metal ion binding to G-N7 helps promote the transition from the B to the Z form of DNA, particularly for alternating G, C sequences. This effect has been particularly well studied for $[Ni(H_2O)_6]^{2+}$ and $[Co(H_2O)_6]^{2+}$.[10] In our laboratories, we found that nickel(III)cyclam was unusually effective at promoting the B→Z conversion of poly d(GC). Using circular dichroism, one can observe this transition by addition of a 51μM solution of the nickel(III) complex while the corresponding nickel(II) complex had little effect up to concentrations of 180μM. The strong tendency of the nickel(III) complex to seek additional ligands presumably accounts for this conversion in which the metal complex can better approach and bind to guanine's N7 when DNA is in the Z form.

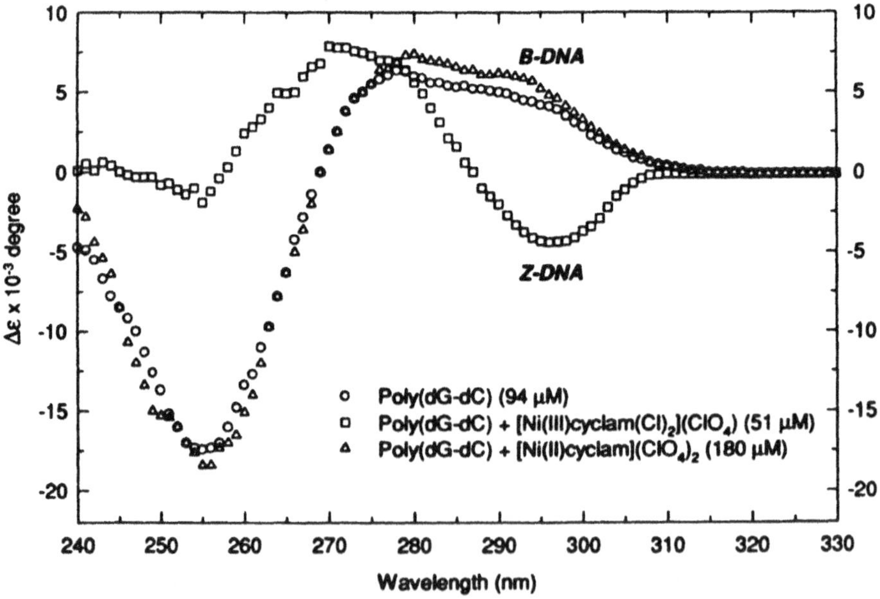

Figure 3. Circular dichroism studies of poly d(GC) with nickel(II) vs. nickel(III) cyclam showing B→Z conversion.

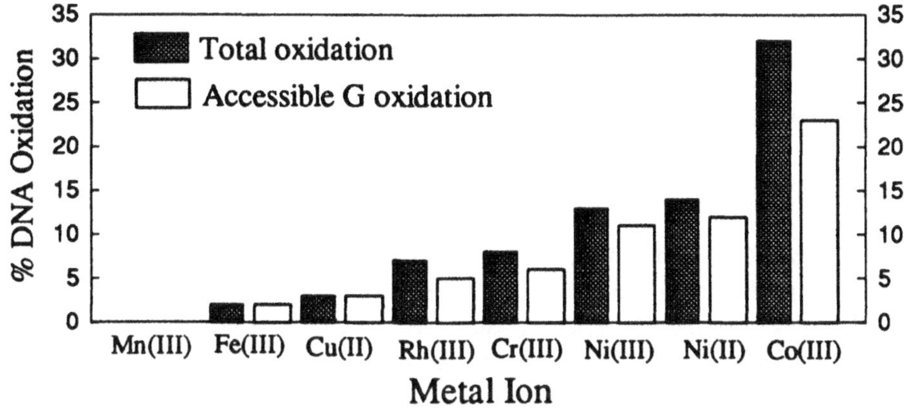

Figure 4. Reactivity of various metal cyclam complexes with hairpin 3 in the presence of HSO_5^- and piperidine work-up.[11] Accessible G oxidation (white bars) refers to reaction at hairpin G's compared to total oxidation at all sites.

COMPARISON OF METAL REACTIVITY WITH DNA

The ability of Co^{2+}, Ni^{2+}, and Cu^{2+} to bind to G-N7 in Z-DNA coupled with the redox activity of these metals suggests that they would be good candidates for structural probes of B vs. Z-DNA via selective recognition and reaction at G's in Z regions. We tested the reactivity of all readily available metal cyclam complexes with the hairpin oligodeoxynucleotide 3 under oxidative conditions.[11] This structure allows evaluation of both the total oxidation at any site in the strand as well as the selectivity of the reaction for accessible G's in the loop region. Surprisingly, cobalt(II) cyclam was even more reactive than the nickel(II) complex, and the copper(II) complex showed negligible reaction. Cobalt complexes appear to be less selective than their corresponding nickel analogues, but their high reactivity is a benefit to their application to Z-DNA.

Z-Form DNA can be obtained under a variety of conditions. Most commonly, alternating G,C sequences are used under high salt conditions, such as 3-4M NaCl. Unfortunately, high concentrations of Cl^- also render nickel(II) complexes unreactive, probably because of strong axial binding of Cl^- ions. Using $NaClO_4$ did not solve the problem for nickel--under conditions that caused 41% of oligo 3 to react with NiCR at 100mM $NaClO_4$, the reactivity dropped to zero when [$NaClO_4$] was increased to 2M. In contrast, cobalt(II) complexes showed only a gradual decrease in reactivity at higher salt conditions: 66% at 100mM $NaClO_4$, 50% at 2M, and 30% at 4M. Thus, we have found reaction conditions in all criteria are met for metal-mediated reaction of G's in Z-DNA. High salt conditions can be used to induce formation of Z-DNA, cobalt complexes are able to bind to G-N7 in DNA,[12] and these complexes are sufficiently reactive and selective under the high salt conditions to effect DNA strand scission.

ACKNOWLEDGEMENTS

We thank Prof. E. Westhof (Strasbourg) for assistance with molecular modeling and Fig. 1. Support of this work by the NSF and the NIH is gratefully acknowledged.

REFERENCES

1. Burrows, C. J. and Rokita, S. E. (1994) Probing guanine structure in nucleic acid folding using nickel complexes, *Acc. Chem. Res.* **27**, 0000.
2. Chen, X., Burrows, C. J., and Rokita, S. E. (1992) Conformation specific oxidation of guanosine in DNA: ends, mismatches, bulges and loops, *J. Am. Chem. Soc.* **114**, 322-325.
3. Chen, X., Rokita, S. E., and Burrows, C. J. (1991) DNA modification: intrinsic selectivity of nickel(II) complexes, *J. Am. Chem. Soc.* **113**, 5885-5886.
4. Chen, X., Woodson, S. A., Burrows, C. J., and Rokita, S. E. (1993) A highly sensitive probe for guanine N7 in Folded Structures of RNA: Application to $tRNA^{Phe}$ and *Tetrahymena* group I intron, *Biochemistry*, **32**, 7610-7616.
5. Rokita, S. E., Zheng, P., Tang, N., Cheng, C.-C., Yeh, R.-H., Muller, J. G., and Burrows, C. J., Nickel complexes in modification of nucleic acids, in B. Sarkar (ed), *Metals and Genetics*, M. Dekkar, New York, in press.
6. Lavery, R. and Pullman, A. (1984) A new theoretical index of biochemical reactivity combining steric and electrostatic factors, *Biophys. Chem.* **19**, 171-181.
7. Westhof, E. (1987) Re-refinement of the B-dodecamer d(CGCGAATTCGCG) with a comparative analysis of the solvent in it and in the Z-hexamer d(5BrCG5BrCG5BrCG), *J. Biomolec. Struct. Dynam.* **5**, 581-600.
8. Kagawa, T. F., Geierstanger, B. H., Wang, A. H.-J., and Ho, P. S. (1991) Covalent modification of guanine bases in double-stranded DNA *J. Biol. Chem.*, **266**, 20175-20184.
9. Wang, A. H.-J., Quigley, G. J., Kolpak, F. J., van der Marel, G., van Boom, J. H., and Rich, A., (1981) Left-handed double helical DNA: Variations in the backbone conformation, *Science* **211**, 171-176.
10. Taboury, J. A., Bourtayre, P., Liquier, J., and Taillandier, E. (1984) Interaction of Z form poly(dG-dC).poly(dG-dC) with divalent metal ions: localization of the binding sites by I.R. spectroscopy, *Nucleic Acids Res.* **12**, 4247-4258.
11. Muller, J. G., Chen, X., Dadiz, A. C., Rokita, S. E., and Burrows, C. J. (1993) Macrocyclic nickel complexes in DNA recognition and oxidation, *Pure Appl. Chem.* **65**, 545-550.
12. Gao, Y.-G., Sriram, M., and Wang, A. H.-J. (1993) Crystallographic studies of metal ion-DNA interactions, *Nucleic Acids Res.* **21**, 4093-4101.

PRINCIPLES OF ANTIBODY CATALYSIS[1]

DONALD HILVERT
Departments of Chemistry and Molecular Biology
The Scripps Research Institute
10666 North Torrey Pines Road
La Jolla, California 92037 USA

Abstract

A wide range of chemical transformations can be catalyzed by antibody molecules elicited with rationally designed transition state analogs. The development of catalytic antibodies consequently represents one of the most versatile and general strategies for creating new enzymes to emerge in the last several years. Recent advances in the production and characterization of these agents are reviewed.

1. Introduction

The mammalian immune system is an unparalleled source of tailored receptor molecules. In response to antigen challenge, antibody molecules are produced over the course of several weeks or months that exhibit extraordinarily high affinity and specificity for their ligands [2-4]. It is estimated that more than 10^8 binding specificities are accessible in the immune system's primary repertoire, and this population can be further expanded and refined through a microevolutionary process involving somatic mutation and selection [5,6]. Harnessing this remarkable protein-generating system for the production of tailored antibody catalysts has emerged in recent years as a versatile and powerful approach for creating novel enzymes [7].

All antibodies share a common scaffold (Figure 1) [8]. Two heavy and two light chains combine to form a homodimeric, Y-shaped molecule of Mr 160,000. The antigen-binding sites are located in the N-terminal variable

J. S. Siegel (ed.), Supramolecular Stereochemistry, 63–81.

domains of the protein and are formed by six peptide loops contributed by the heavy (H1 to H3) and the light (L1 to L3) chains. Antibody diversity derives from the hypervariability in length and sequence of these so-called complementarity determining regions (CDRs). The combinatorial arrangement of six CDR loops allows an extensive range of shapes to be constructed — from deep pockets, clefts, and grooves to flatter, more modulating surfaces — in response to a wide array of structurally diverse antigens [8,9].

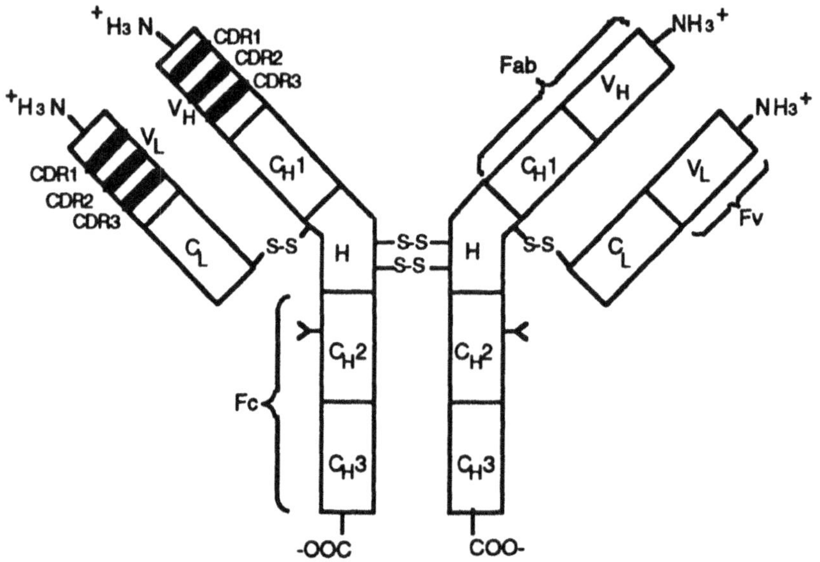

Figure 1. Schematic drawing of an antibody molecule. The active site is formed by the complementarity determining regions (CDRs) of the variable domains of the heavy and light chains.

The notion that antibodies might serve as highly selective catalysts originated with Jencks in 1969 [10]. Enzymes and antibodies exhibit many common features [11]. Not only are the basic sizes and shapes of their binding pockets similar, but ligand recognition is mediated in both cases by a combination of hydrogen bonding, electrostatics, dispersion, and hydrophobic interactions. However, whereas enzymes have evolved for maximum stabilization of the ephemeral transition state of the reaction they catalyze, antibodies are raised in response to a ground state molecule. Jencks' key insight [10] was that antigens mimicking the essential stereoelectronic features of the transition state for a reaction might elicit antibody binding sites with properties suitable for catalysis.

This strategy works remarkably well. Well over 50 chemical transformations have been promoted by antibodies since 1986 [7]. These include acyl transfers, eliminations, pericyclic processes, isomerizations, and redox reactions. Antibody catalysts not only accelerate these reactions 10^2 to 10^6-fold over background, they also exhibit a high degree of substrate specificity, regio- and stereoselectivity. Because these properties correlate well with the structural features of the transition state analog, the development of catalytic antibodies provides the means to create tailored agents for a wide range of applications in chemistry and biology. Furthermore, detailed mechanistic and structural studies of individual catalytic antibodies have the potential to reveal a great deal about the fundamental principles of protein catalysis.

Much effort is currently directed toward defining both the scope and limitations of catalytic antibody technology. Expansion of the repertoire of reactions amenable to antibody catalysis is one important objective. Chemical transformations that cannot be achieved efficiently or selectively with available chemical methods are particularly attractive targets. Elucidation of the structural basis of antibody catalysis is another major goal. Improved design of transition state analogs and optimization of the catalytic efficiency of first-generation catalytic antibodies via protein engineering will require a detailed understanding of the structure-function relationships that govern catalysis. Structural studies are also important for determining what if any limitation antibody structure places on the nature and efficiency of the reactions that can be carried out within the combining site. As outlined below, considerable progress has recently been made on each of these fronts.

2. Structural basis of antibody catalysis

The rearrangement of chorismate **1** into prephenate **3** is a useful transformation for illustrating the basic strategies for creating catalytic antibodies and the lessons that can be learned through detailed study of their structure and properties. This simple unimolecular reaction is formally a Claisen rearrangement and does not require general acid-base or nucleophilic catalysis. It is, moreover, a rare example of a biologically relevant pericyclic reaction: in plants and lower organisms, it is the branchpoint in the shikimate pathway leading to the essential aromatic amino acids tyrosine and phenylalanine. *In vivo* this step is catalyzed by the enzyme chorismate mutase with a rate acceleration $> 10^6$ over background [12].

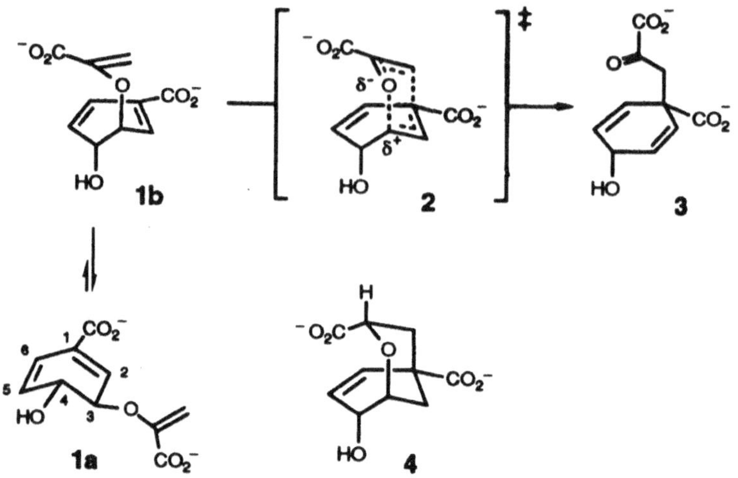

How chorismate mutase speeds the rearrangement has been a matter of debate for some time. Labeling studies and inhibitor experiments indicate that both the catalyzed and spontaneous reactions proceed via a conformationally restricted chair-like transition state **2** [13,14]. Solvent and kinetic isotope effects show that the reaction is asynchronous with carbon-oxygen bond cleavage preceding carbon-carbon bond formation [15,16]. Some of the observed rate acceleration may therefore result from stabilization of charge separation in the transition state through specific hydrogen bonding or electrostatic interactions at the active site [17]. However, entropic contributions must also be important. The flexible chorismate molecule adopts an extended pseudo-diequatorial conformation in solution (**1a**) and must undergo an energetically unfavorable conformational change to the diaxial conformer **1b**, in which the enol pyruvate side chain is oriented properly for the ensuing reaction [16]. Given this, the enzyme might act as an 'entropy trap' [18] to stabilize **1b** and freeze out the rotational degrees of freedom of the side chain in the transition state [12,19]. The favorable activation entropy of the enzyme-catalyzed process compared with the spontaneous thermal rearrangement ($\Delta\Delta S^{\ddagger}$ = 13 cal K^{-1} mol^{-1}) is consistent with this notion [19].

Strong inhibition of chorismate mutase by the conformationally restricted oxabicyclic transition state analog **4** [20] also suggests that the enzyme has an active site congruent to the compact transition state structure **2**. Compound **4** binds ≈200-times more tightly to the enzyme than chorismate [20], capturing a significant fraction of the energy available for stabilizing the metastable species. This compound has also proved effective as a template for producing antibodies with substantial chorismate mutase activity [21,22]. Because small molecules are generally not immunogenic, it was necessary to couple **4** to a carrier protein to generate an effective immune response. The coupling strategy is dictated by the need for recognition of both the enolpyruvate side chain and the cyclohexadiene ring of chorismate. Attachment of **4** to keyhole limpet hemocyanin via its secondary alcohol ensures this by making the two carboxylates the primary determinants for immune recognition.

This approach has yielded antibodies that catalyze the chorismate to prephenate reaction in a highly enantioselective fashion with rate accelerations in the range 10^2-10^4 fold over background [21-24]. The best catalyst (11F1-2E9) is only two orders of magnitude less efficient than natural chorismate mutase, and like the natural enzyme, has a very favorable entropy of activation compared to the thermal rearrangement [24]. Surprisingly, the less active antibody catalyst (1F7) achieves its 200-fold rate acceleration entirely by lowering the enthalpy of activation [21]. The entropy of activation in this case is actually 10 cal K^{-1} mol^{-1} less favorable than the uncatalyzed reaction. While it is difficult to interpret activation parameters even under the best of circumstances, the differences exhibited by 1F7 and 11F1-2E9 illustrate the capacity of the immune system to solve a specific chemical problem in several ways.

The unfavorable entropy of activation observed for 1F7 raises interesting questions regarding the extent to which the antibody populates and stabilizes the diaxial substrate conformer **1b**. This issue has been investigated by NMR spectroscopy [25]. Diagnostic transferred nuclear Overhauser effects (TRNOEs) between protons on the enol pyruvate side chain and protons on the cyclohexadiene ring demonstrate that a significant fraction of the chorismate bound at the active site of 1F7 does adopt the reactive geometry 'programmed' by the conformationally restricted transition state analog. Although additional, smaller TRNOE signals suggest that the antibody may not completely reduce the rotational degrees of freedom of the bound substrate, the NMR results rule out simple preorganization of the substrate as the major contributor to the high entropy barrier for the antibody-catalyzed reaction.

Crystallographic studies of 1F7 complexed with the transition state analog **4** [26] complement and extend the spectroscopic experiments. As shown in Figure 2, hapten recognition in 1F7 is achieved through a combination of hydrophobic interactions, hydrogen bonds and electrostatics. Indeed, the combining site of the antibody exhibits excellent overall shape and charge complementarity to **4**. It is this complementarity that presumably dictates preferential binding of the correct substrate enantiomer in a conformation appropriate for reaction. Moreover, the orientation of the transition state analog within the active site reflects the coupling strategy

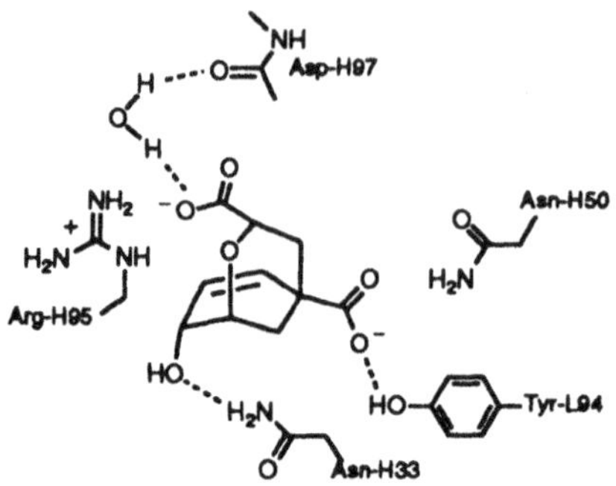

Figure 2. Schematic drawing illustrating the hydrogen bonding and electrostatic interactions between the chorismate mutase antibody 1F7 and the transition state analog **4**.

adopted for the purposes of immunization: the two hapten carboxylates are the primary recognition elements, while the C-4 alcohol group sticks out of the cavity. The antibody's structural properties thus accurately mirror those of the transition state analog.

Why is 1F7 10^4-times less active than chorismate mutase? The antibody molecule's distinctive architecture does not appear to impose intrinsic structural limitations on catalysis. Indeed, comparison of the active sites of the antibody and the monofunctional enzyme from *Bacillus subtilis* [27] suggests that the differences between them are more a matter of degree than of kind. Upon complex formation, the hapten is buried to a similar extent in both proteins, and similar types of interactions are available for orienting the flexible substrate correctly for reaction. Furthermore, the enzyme and the antibody seem to promote the rearrangement of chorismate via the same concerted transition state as the uncatalyzed reaction. Other formal mechanistic possibilities, such as a two-step heterolytic process assisted by an enzymic nucleophile, can be ruled out by the lack of appropriate functional groups in the respective active sites [26,27].

Unlike 1F7, chorismate mutase appears to have more extensive interactions with the hapten's alcohol moiety than with the tertiary carboxylate [27], but this is unlikely to account for the observed differences in rate. In conjunction with the overall shape complementarity of the protein pocket, recognition of either the C-1 carboxylate or the C-4 alcohol of chorismate should be sufficient for correctly positioning the cyclohexadiene ring for reaction. Consistent with this idea, antibody 11F1-2E9, which was elicited in the same way as 1F7, has remarkably high catalytic activity [22]. Detailed structural data for 11F1-2E9 are eagerly awaited.

A more plausible reason for the relatively low efficiency of 1F7 is that the antibody is a poorer 'entropy trap' than chorismate mutase. The antibody exploits far fewer specific hydrogen bonding and electrostatic interactions for ligand recognition than the enzyme [26]. This point is illustrated by the interactions available for fixing the enol pyruvate side chain in place. In the enzyme, the carboxylate moiety of the side chain is hydrogen bonded to a tyrosine and two arginine residues [27]. In the antibody, this group binds loosely in a relatively hydrophobic pocket. ArgH95 is probably important for charge neutralization, but it is too far away to donate a hydrogen bond to either of the carboxylate oxygens. A water molecule takes on this role, serving as a bridge between the carboxylate and the protein backbone. However, the water molecule is likely to provide only poor control over the placement of the side chain in the flexible substrate during the reaction. In addition to being an ineffective means of freezing out rotational degrees of freedom in the bound substrate, fixation of a water molecule from bulk solvent will be entropically quite costly [28].

In contrast to the enzyme, the antibody also appears relatively poorly equipped to accommodate charge separation in the transition state [26]. The enzyme possesses a dipolar binding site near the substrate carbon-oxygen bond that breaks during rearrangement [27]: Arg95 is poised to stabilize the partial negative charge that accumulates on the ether oxygen, and the side chains of Glu78, Cys75, and possibly even Phe57 may help to neutralize any positive charge that develops on the cyclohexadiene ring. The side chain of ArgH95 occupies a similar position in the antibody as Arg95 in the enzyme, but given its role in neutralizing a full negative charge on the substrate, it will be much less effective in accommodating additional negative charge on the ether oxygen at the transition state. The antibody has no groups analogous to Glu78 or Cys75. Of course, the lack of a dipolar binding site in the antibody is not surprising. Although hapten **4** mimics the conformationally restricted geometry of **2** very well, it models the polarized character of the transition state poorly.

If 1F7's relatively low catalytic efficiency is due to its short evolutionary history and imperfect transition state analog design, rather than to an intrinsic architectural bias against catalysis, rational modifications of the strategy that yielded this antibody may increase the probability of finding better catalysts. Expanded screening of anti-**4** antibodies is one possibility. A relatively small number of antibodies that bind **4** ($\approx$ 50) have been tested for chorismate mutase activity [21,22], representing but a tiny fraction of the immune response. By assaying more antibodies, even better catalysts than 11F1-2E9 might be found. Because the hapten was racemic, it may also be possible to identify catalysts that preferentially rearrange the unnatural isomer of chorismate. Recently developed methods for displaying antibody fragments on phage [29], in conjunction with powerful biological screening and selection techniques, may facilitate such a search by providing improved access to the diversity within the immunological repertoire.

While more extensive screening can increase the chances of finding rare but highly active catalysts, improvements in hapten design could substantially increase the presence of such catalysts in the population of candidate clones. Knowledge of the critical interactions between chorismate mutase and **4** [27] will be invaluable in this regard. For example, derivatives of **4** in which the C-4 alcohol is replaced with an amine or in which the secondary carboxylate is replaced with a phosphonate might be much better enzyme inhibitors because of improved electrostatic complementarity with the active site. For the same reason, they might also be more effective at producing antibodies that resemble natural chorismate mutase.

The availability of structural information on the antibody-hapten complex also makes possible the redesign of the first generation catalyst by protein engineering. Thus, engineering a dipolar binding site into 1F7 via site-directed mutagenesis is conceivable. Alternatively, the water molecule that makes a crucial interaction with the hapten's secondary carboxylate might be replaced with an amino acid side chain through redesign of the CDR3 loop of the heavy chain. These tasks will be greatly aided by the availability of a biological selection assay that has been developed for 1F7. This catalyst has been shown to function *in vivo,* conferring a substantial growth advantage to a chorismate mutase-deficient yeast strain under auxotrophic conditions [30]. The ability to select for this phenotype provides the means to identify 1F7 derivatives with enhanced potency. Large numbers of variants, generated by random mutagenesis, can be assayed directly for their ability to complement the chorismate mutase deficiency of the host organism. Analogous experiments in engineered strains of *Escherichia coli* lacking chorismate mutase appear feasible as well (P. Kast, unpublished results).

How general are the lessons learned through study of 1F7? Although investigation of the mechanistic and structural basis of antibody catalysis is only beginning, striking parallels between a number of antibodies and analogous enzymes are apparent [31,32]. Moreover, to the extent they have been investigated, the mechanisms of action of most antibody catalysts reflect the structural properties of the transition state analog quite well. In some cases, the mechanisms are actually more complex than envisaged. For example, a remarkably active hydrolytic antibody (43C9) appears to operate via a two-step mechanism involving transient acylation of the catalyst reminiscent of the serine proteases [33], even though it was generated with a simple phosphonamidate transition state analog [34]. As more crystal structures of catalytic antibodies become available, rules that make one hapten better than another may become apparent. Elucidation of such rules will be essential if highly efficient first-generation antibody catalysts are to be created as a routine exercise.

3. Controlling chemical reactivity and selectivity with antibodies

The power of the catalytic antibody approach for creating new enzymes is its generality. Virtually any reaction should be susceptible to antibody catalysis, as long as it can be carried out in a protein microenvironment in aqueous buffer and its energy barrier for converting substrates to products does not exceed the ca. 10 to 20 kcal mol^{-1} of binding energy available from an antibody. As in enzymes, substrate destabilization, proximity effects, nucleophilic and general acid-base catalysis, and catalytic cofactors can all be exploited to help reduce the activation barrier for many different reaction types.

The importance of substrate destabilization is illustrated by the antibody-catalyzed decarboxylation of 3-carboxybenzisoxazoles (**5**) to give salicylonitriles (**7**) [35]. This reaction is extraordinarily sensitive to its solvent microenvironment, with rate enhancements up to 10^8-fold observed upon transfer of the reactant from aqueous buffer to aprotic dipolar solvents [36]. Desolvation of the negatively charged carboxylate group greatly destabilizes the substrate, while the charge delocalized transition state (**6**) may be stabilized by dispersion interactions with solvent. Similar factors are believed

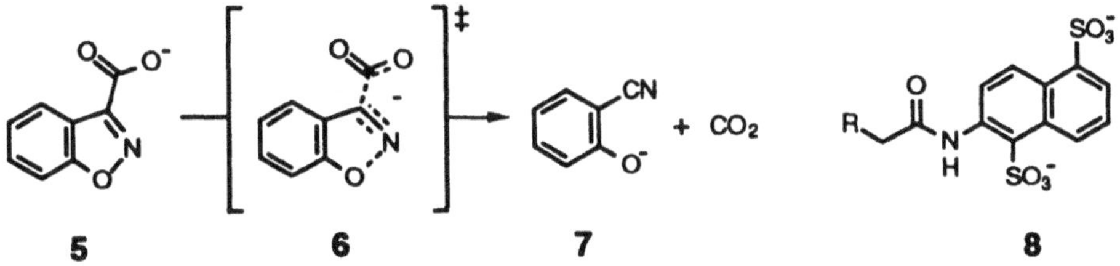

5 **6** **7** **8**

to contribute to the efficiency of a number of decarboxylase enzymes. The 1,5-naphthalene disulfonate hapten **8** was used to generate antibodies capable of catalyzing the decarboxylation of 5-nitro-3-carboxybenzisoxazole [35]. Rate accelerations up to 2 x 10^4 over background were observed, and because the model reaction is completely insensitive to general acid-base chemistry, the catalytic advantage provided by these proteins can be ascribed entirely to medium effects. In essence, the antibody provides a 'rigid' and chemically well-defined solvent microenvironment for investigating how dipolarity, electrostatics and hydrogen bonding influence chemical reactivity. Kinetic and spectroscopic studies [35, 37] suggest that the hapten's large aromatic ring and negatively-charged substituents yielded an appropriately-sized apolar pocket containing positively-charged residues for sequestering the anionic species that occur along the reaction coordinate.

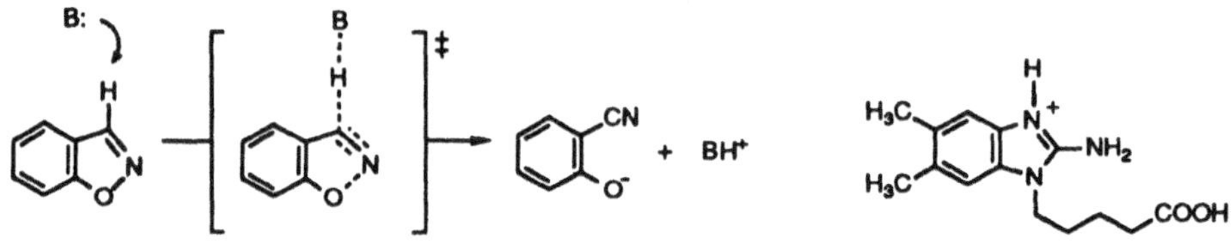

9 10 7 11

Many important chemical transformations, including aldol condensations, S_N2 substitutions, E2 eliminations, and even some hydrolytic reactions, are solvent sensitive and should be highly susceptible to antibody catalysis. The base-promoted deprotonation of benzisoxazoles **9** to give salicylonitriles **7** [38] has recently been used as a model system to probe how microenvironmental factors influence general base catalysis in an antibody active site [39]. The benzimidazolium hapten **11** ($pK_a = 7.8$) was designed to capitalize on charge complementarity, which has been exploited in a number of instances to induce general acids or bases in an antibody binding pocket. Because it is planar, **11** also correctly mimics all reacting bond geometries of the transition state **10**. Hapten **11** was an unusually effective hapten, eliciting highly active antibody catalysts. The activity of antibody 34E4 depends on an active site base, which chemical modification studies have identified as a carboxylate with a perturbed pK_a. The rate acceleration observed for this catalyst is in excess of 10^8 $[(k_{cat}/K_m)/k_{AcO^-}]$, with an effective molarity for the active site carboxylate of ca. 40,000 M. By way of contrast, effective molarities rarely exceed 10 M for intramolecular general base catalysis in model systems or for general catalysis in other antibodies. Moreover, 34E4 achieves more than 10^3 turnovers per active site as a consequence of the dissimilarity in shape and charge of product **7** and hapten **11**. The extraordinary activity of 34E4 shows that very large effects can be achieved by strategic use of haptenic charge.

Bringing two molecules together in the proper orientation for reaction — catalysis by approximation — is one of the easiest things an antibody can do. As the benzisoxazole elimination shows, large rates can be achieved by juxtaposing a substrate and a catalytic group. Bimolecular Diels-Alder

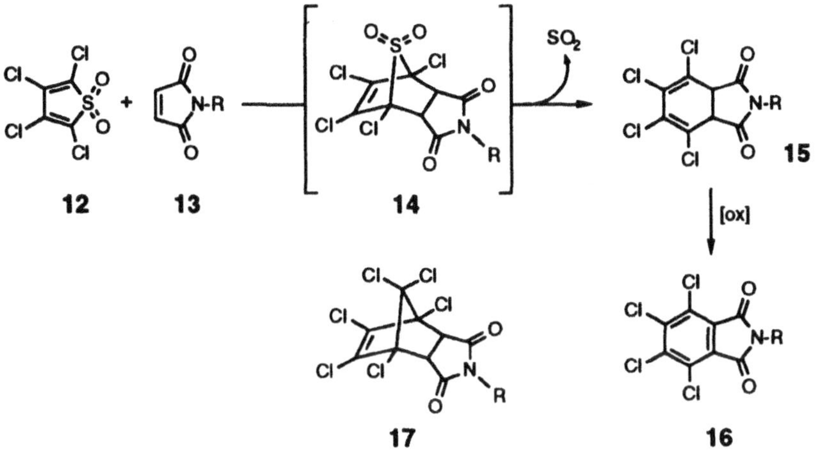

reactions are also susceptible to proximity effects. Antibody catalysis of cycloadditions [40-42] represents a particularly significant accomplishment, given that analogous enzymes are as yet unknown. The boatlike transition state geometry for these processes has been successfully mimicked with bicyclic compounds. For example, the hexachloronorbornene derivative **17** resembles the transition state for the cycloaddition between **12** and **13** as well as the transition state for the subsequent chelotropic elimination of SO_2. An anti-**17** antibody (1E9) catalyzes this two-step transformation with high efficiency (effective molarities > 10^2 M) and multiple turnovers [40]. The large geometrical change accompanying the reaction minimizes problems associated with product inhibition. This feature was 'programmed' into the catalyst through the design of the transition state analog, which binds to 1E9 five to six orders of magnitude more tightly than substrate (**13**) or products (**15** and **16**), respectively (K. Hill, unpublished results). While structure-function relationships must await detailed crystallographic studies, sequence data show that the antibody is closely related to a family of steroid-binding antibodies [43]. Models based on the structure of an anti-progesterone antibody (whose structure is available [44]) show a hydrophobic active site with multiple polar interactions that could be used to orient the diene and dienophile and facilitate breakdown of the unstable intermediate **14**. The development [44] of a powerful immunoassay (catELISA) for this reaction in which catalytic activity can be detected directly in tissue culture supernatant has led to the discovery of even more potent catalysts than 1E9 and provides the means to systematically screen large numbers of alternative scaffolds.

The Diels-Alder reaction between tetrachlorothiophene dioxide and maleimide yields a symmetrical product. If the diene and dienophile are asymmetrically substituted, several diastereomeric products can be formed. In principle, antibody catalysts could be prepared for each of the possible stereochemical pathways by synthesizing the corresponding transition state analogs. The preparation of antibodies that steer a Diels-Alder cycloaddition down a normally disfavored *exo* pathway demonstrates the feasibility of this proposition [42]. The substituted bicyclo[2.2.2]octane derivative **20**, corresponding to the transition state for *exo* addition of acrylamide to diene **18**, served as a template for generating antibodies that catalyze formation of the *exo* adduct **19** [42]. Catalysts for the *endo* pathway were prepared with the epimer of hapten **20**. In both cases, a single enantiomer of the cyclohexene product was produced.

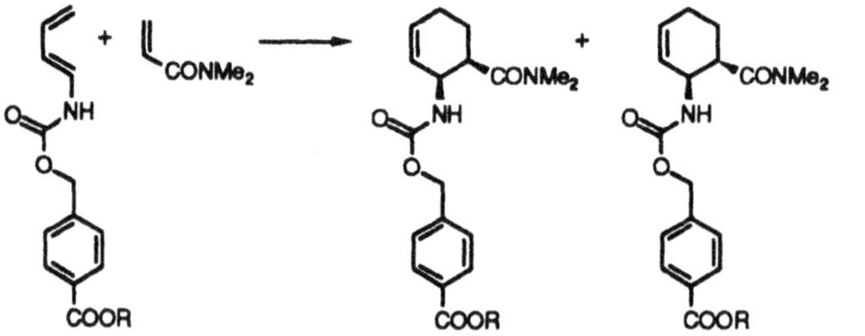

18 **19** *endo* **19** *exo*

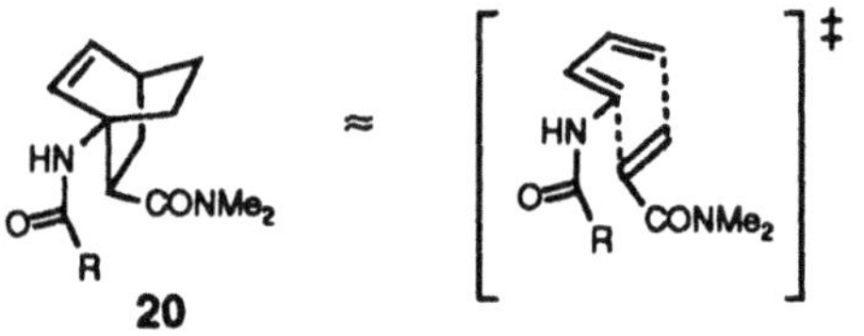

20

The control of chemical selectivity is a classical problem in organic chemistry and an antibody's ability to route reactions via disfavored pathways, as in the asymmetric Diels-Alder reaction, has important practical ramifications. Not surprisingly, extension of this approach to the catalysis of other unselective or energetically demanding reactions has become an important focus of research. Syn eliminations [46], cationic olefin cyclizations [47], and enol ether hydrolyses [48] are among the processes that have been investigated. For the purposes of the current discussion, however, the regio- and stereoselective reduction of the diketone **21** [49] illustrates what

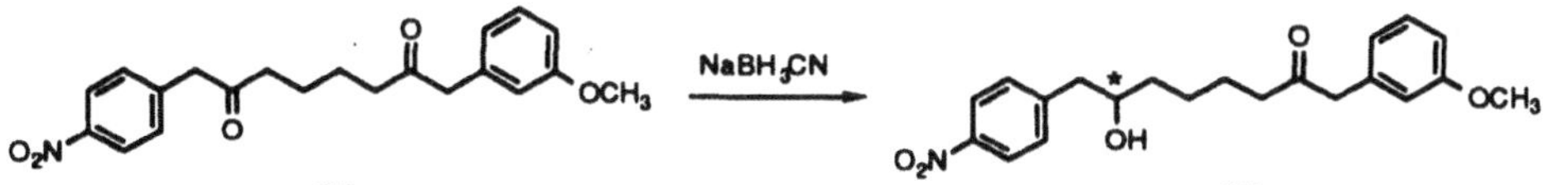

21 **22**

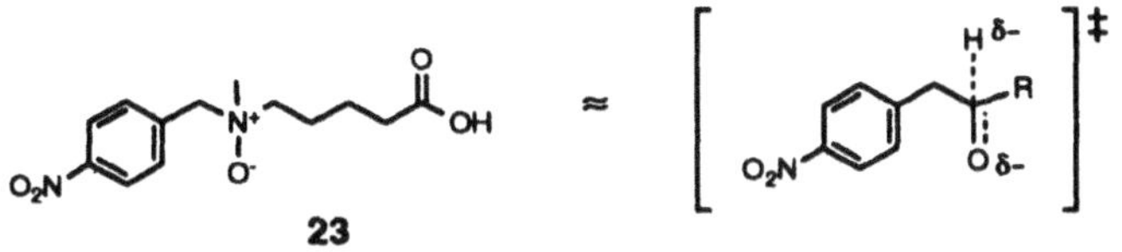

23

can be accomplished. In the presence of an antibody raised against *N*-oxide **23**, **21** is reduced by $NaCNBH_3$ in high yield and greater than 175:1 regioselectivity at the *p*-nitrobenzyl-substituted ketone. Comparable selectivity would be difficult or impossible to achieve with more standard chemical methods. The cyclization of oxirane **24** provides another impressive example of how an antibody can alter the balance between competing reaction pathways [50]. In the absence of catalyst, tetrahydrofuran **26** is the exclusive cyclization product, but an antibody raised against the *N*-oxide hapten **27** overwhelms the normal preference for formation of five-membered rings to give enantiomerically pure **25**. Calculations [51] suggest that the catalyst must stabilize the disfavored 6-*endo* transition state by 3.6 kcal/mol more than the 5-*exo* transition state.

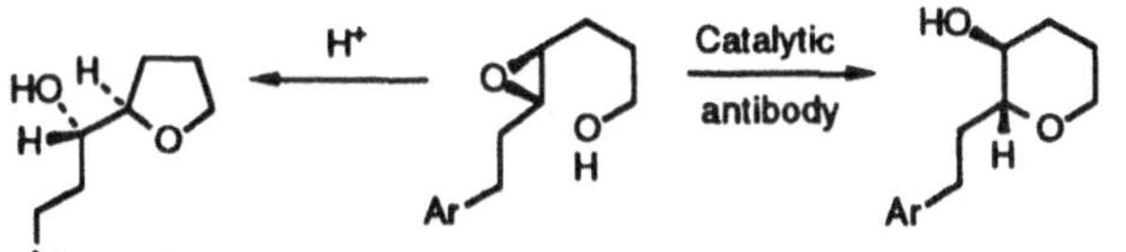

26 **24** **25**

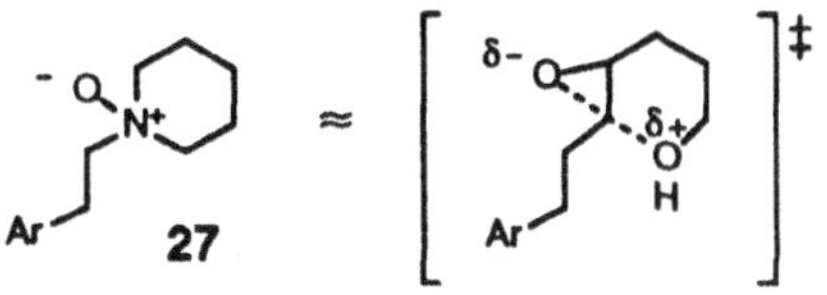

27

While antibodies are clearly capable of controlling chemical reactivity and selectivity, the low specific activities (compared to enzymes) that are typically observed is of concern and could limit practical applications. However, an antibody-catalyzed process need not be particularly fast to be useful to the synthetic chemist, as long as it is highly selective and the competing background processes are relatively slow. The gram-scale conversion of cyclic enol ether **28** to ketone **29** in 60-65% overall yield and 86% ee with a modestly active catalytic antibody (raised against hapten **30**) establishes the synthetic utility of these agents [52]. Alternative catalysts for the stereoselective hydrolysis of **28** are unknown.

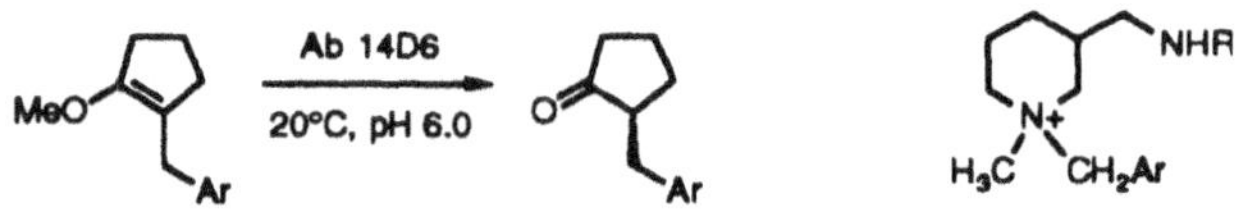

28 **29** **30**

The scope of antibody catalysis has clearly expanded dramatically since the discovery of immunoglobulins with modest esterolytic activity in 1986 [53,54]. Nevertheless, many challenges remain. Aldol condensations, S_N2 displacements, and phosphoryl and glycosyl group transfers are among the important reaction types that have not yet been catalyzed by antibodies. Moreover, the strategies employed in the past to generate catalysts for some reactions, like amide hydrolysis, may not be easily generalized. For example, despite the availability of excellent transition state analogs, only two monoclonal antibodies with amidase activity have been prepared to date [34,55]. The difficulties encountered may reflect the intrinsic stability of the amide bond and the low probability of eliciting an effective constellation of catalytic groups within the antibody combining site. Whatever the reason, the production of antibodies that can serve as site-specific proteases remains an as yet unrealized goal. Overcoming these difficulties will require the development of more sophisticated transition state analogs and the implementation of new methods for generating and screening large immunoglobulin libraries [see e.g. 29,45,56]. Mutagenesis and genetic selection, on the other hand, are potentially powerful tools for optimizing the activity of first-generation antibody catalysts [30,57].

4. Outlook

1994 was the centennial anniversary of Emil Fischer's famous 'lock and key' description of enzyme action [58-61]. This beautiful metaphor can also be applied to the field of catalytic antibodies in which a rationally designed transition state analog is used as a key to unlock the catalytic potential of the mammalian immune system. This simple yet elegant approach to tailored catalysts complements other strategies for creating new enzymes, including the *de novo* synthesis of proteins, the reengineering of existing enzymes through site-directed mutagenesis and chemical modification, ribozyme selection, and the chemical synthesis of low molecular weight macrocyclic or cleft-like molecules. The generation of catalytic antibodies is an inherently multidisciplinary undertaking, involving chemical synthesis, immunology, molecular biology and genetics. By successfully integrating these various disciplines, increasingly active catalysts for increasingly difficult chemical transformations may become available on demand.

5. Acknowledgments

Research in the author's laboratory has been generously supported by grants from the National Institutes of Health, the American Cancer Society, the National Science Foundation, the Army Research Office, and the Office of Naval Research.

6. References

1. This contribution originally appeared in the NATO ASI series volume entitled "Chemical Synthesis: Gnosis to Prognosis" (C. Chatgilialoglu and V. Snieckus, eds.), 1995.
2. Pressman, D.; Grossberg, A. (1968) *The Structrual Basis of Antibody Specificity*, Benjamin, New York.
3. Nisonoff, A., Hopper, J., and Spring, S. (1975) *The Antibody Molecule*, Academic Press, New York.
4. Kabat, E.A. (1976) *Structural Concepts in Immunology and Immunochemistry*, Holt, Reinhart and Winston, New York.
5. Alt, F.W., Blackwell, T.K., and Yancopoulos, G.D. (1987) Development of the primary antibody repertoire, *Science* **238**, 1079.
6. Rajewsky, K., Förster, I., and Cumang, A. (1987) Evolutionary and somatic selection of the antibody repertoire in the mouse, *Science* **238**, 1088.
7. Lerner, R.A., Benkovic, S.J., and Schultz, P.G. (1991) At the crossroads of chemistry and immunology: Catalytic antibodies, *Science* **252**, 659.
8. Davies, D.R., Padlan, E.A., and Sheriff, S. (1990) Antibody-antigen complexes, *Ann. Rev. Biochem.* **59**, 439.

9. Wilson, I.A. and Stanfield, R.L. (1993) Antibody-antigen interactions, *Current Opinion in Struct. Biol.* **3**, 113.
10. Jencks, W.P. (1969) *Catalysis in Chemistry and Enzymology*; McGraw-Hill: New York, p. 288.
11. Pauling, L. (1948) Chemical achievement and hope for the future, *Am. Sci.* **36**, 51.
12. Andrews, P.R., Smith, G.D., and Young, I.G. (1973) Transition-state stabilization and enzymic catalysis. Kinetic and molecular orbital studies of the rearrangement of chorismate to prephenate, *Biochemistry* **12**, 3492.
13. Sogo, S.G., Widlanski, T.S., Hoare, J.H., Grimshaw, C.E., Berchtold, G.A., and Knowles, J.R. (1984) Stereochemistry of the rearrangement of chorismate to prephenate: Chorismate mutase involves a chair transition state, *J. Am. Chem. Soc.* **106**, 2701.
14. Andrews, P.R., Cain, E.N., Rizardo, E., and Smith, G.D. (1977) Rearrangement of chorismate to prephenate. Use of chorismate mutase inhibitors to define the transition state structure, *Biochemistry* **16**, 4848.
15. Addadi, L., Jaffe, E.K., and Knowles, J.R. (1984) Secondary tritium isotope effects as probes of the enzymic and nonenzymic conversion of chorismate to prephenate, *Biochemistry* **22**, 4494.
16. Copley, S.D. and Knowles, J.R. (1987) The conformational equilibrium of chorismate in solution: Implications for the mechanism of the non-enzymic and the enzyme-catalyzed rearrangement of chorismate to prephenate, *J. Am. Chem. Soc.* **109**, 5008.
17. Severence, D.L. and Jorgensen, W.L. (1992) Effects of hydration on the Claisen rearrangement of allyl vinyl ether from computer simulations, *J. Am. Chem. Soc.* **114**, 10966.
18. Westheimer, F.H. (1962) Mechanisms related to enzyme catalysis, *Adv. Enzymol.* **24**, 441.
19. Görisch, J. (1978) On the mechanism of the chorismate mutase reaction, *Biochemistry* **17**, 3700.
20. Bartlett, P.A., Nakagawa, Y., Johnson, C.R., Reich, S.H., and Luis, A. (1988) Chorismate mutase inhibitors: Synthesis and evaluation of some potential transition-state analogues, *J. Org. Chem.* **53**, 3195.
21. Hilvert, D., Carpenter, S.H., Nared, K.D., and Auditor, M.-T.M. (1988) Catalysis of concerted reactions by antibodies: The Claisen rearrangement, *Proc. Natl. Acad. Sci. USA* **85**, 4953.

22. Jackson, D.Y., Jacobs, J.W., Sugasawara, R., Reich, S.H., Bartlett, P.A., and Schultz, P.G. (1988) An antibody-catalyzed Claisen rearrangement, *J. Am. Chem. Soc.* **110**, 4841.
23. Hilvert, D. and Nared, K.D. (1988) Stereospecific Claisen rearrangement catalyzed by an antibody, *J. Am. Chem. Soc.* **110**, 5593.
24. Jackson, D.Y., Liang, M.N., Bartlett, P.A. and Schultz, P.G. (1992) Activation parameters and stereochemistry of an antibody-catalyzed Claisen rearrangement, *Angew. Chem. Int. Ed. Engl.* **31**, 182.
25. Campbell, A.P., Tarasow, T.M., Massefski, W., Wright, P.E., and Hilvert, D. (1993) Binding of a high-energy substrate conformer in antibody catalysis, *Proc. Natl. Acad. Sci. USA* **90**, 8663.
26. Haynes, M.R., Stura, E.A., Hilvert, D., and Wilson, I.A. (1994) Routes to catalysis: Structure of a catalytic antibody and comparison with its natural counterpart, *Science* **263**, 646.
27. Chook, Y.M., Ke, H., and Lipscomb, W.N. (1993) Crystal structures of the monofunctional chorismate mutase from *Bacillus subtilis* and its complex with a transition state analog, *Proc. Natl. Acad. Sci. USA* **90**, 8600.
28. Dunitz, J. (1994) The entropic cost of bound water in crystals and biomolecules, *Science* **264**, 670.
29. Garrard, L.J. and Zhukovsky, E.A. (1992) Antibody expression in bacteriophage systems: The future of monoclonal antibodies? *Current Opinion in Biotechnology* **3**, 474.
30. Tang, Y., Hicks, J.B., and Hilvert, D. (1991) *In vivo* catalysis of a metabolically essential reaction by an antibody, *Proc. Natl. Acad. Sci. USA* **88**, 8784.
31. Stewart, J.D., Liotta, L.J., and Benkovic, S.J. (1993) Reaction mechanisms displayed by catalytic antibodies, *Acc. Chem. Res.* **26**, 396.
32. Golinelli-Pimpaneau, B., Gigant, B., Bizebard, T., Navaza, J., Saludjian, P., Zemel, R., Tawfik, D.S., Eshhar, Z., Green, B.S., and Knossow, M. (1994) Crystal structure of a catalytic antibody Fab with esterase-like activity, *Structure* **2**, 175.
33. Stewart, J.D., Roberts, V.A., Thomas, N.R., Getzoff, E.D., and Benkovic, S.J. (1994) Site-directed mutagenesis of a catalytic antibody: An arginine and a histidine residue play key roles, *Biochemistry* **33**, 1994, and references therein.
34. Janda, K.D., Schloeder, D., Benkovic, S.J., and Lerner, R.A. (1988) Induction of an antibody that catalyzes the hydrolysis of an amide bond, *Science* **241**, 1188.

35. Lewis, C., Krämer, T., Robinson, S., and Hilvert, D. (1991) Medium effects in antibody-catalyzed reactions, *Science* **253**, 1019.

36. Kemp, D.S. and Paul, K.G. (1975) The physical organic chemistry of benzisoxazoles. III. The mechanism and the effects of solents on rates of decarboxylation of benzisoxazole-3-carboxylic acids, *J. Am. Chem. Soc.* **97**, 7305.

37. Tarasow, T.M., Lewis, C., and Hilvert, D. (1994). Investigation of medium effects in a family of decarboxylase antibodies. *J. Am. Chem. Soc.* **116**, 7959.

38. Casey, M.L., Kemp, D.S., Paul, K.G., and Cox, D.D. (1973) The physical organic chemistry of benzisoxazoles. I. The mechanism of the base-catalyzed decomposition of benzisoxazoles, *J. Org. Chem.* **38**, 2294.

39. Thorn, S. N., Daniels, R. G., Auditor, M.-T. M. and Hilvert, D. (1995). Large rate accelerations in antibody catalysis by strategic use of haptenic charge. *Nature* **373**, 228.

40. Hilvert, D., Hill, K.W., Nared, K.D., and Auditor, M.-T.M. (1989) Antibody catalysis of a Diels-Alder reaction, *J. Am. Chem. Soc.* **111**, 9261.

41. Braisted, A.C. and Schultz, P.G. (1990) An antibody-catalyzed bimolecular Diels-Alder reaction, *J. Am. Chem. Soc.* **112**, 7430.

42. Gouverneur, V.E., Houk, K.N., Pascual-Teresa, B., Beno, B., Janda, K.D., and Lerner, R.A. (1993) Control of the exo and endo pathways of the Diels-Alder reaction by antibody catalysis. *Science* **262**, 204.

43. Hilvert, D. (1993). Antibody catalysis of carbon-carbon bond formation and cleavage. *Acc. Chem. Res.* **26**, 552.

44. Arevalo, J.H., Stura, E.A., Taussig, M.J., and Wilson, I.A. (1993) Three-dimensional structure of an anti-steroid Fab' and progesterone-Fab' complex, *J. Mol. Biol.* **231**, 103.

45. MacBeath, G. and Hilvert, D. (1994) Monitoring catalytic activity by immunoassay: Implications for screening, *J. Am. Chem. Soc.* **116**, 6101.

46. Cravatt, B.F., Ashley, J.A., Janda, K.D., Boger, D.L., and Lerner, R.A. (1994) Crossing extreme mechanistic barriers by antibody catalysis: Syn elimination to a cis olefin, *J. Am. Chem. Soc.* **116**, 6013.

47. Li, T., Janda, K.D., Ashley, J.A., and Lerner, R.A. (1994) Antibody catalyzed cationic cyclization, *Science* **264**, 1289.

48. Sinha, S.C., Keinan, E., and Reymond, J.-L. (1993) Antibody-catalyzed reversal of chemoselectivity, *Proc. Natl. Acad. Sci. USA* **90**, 11910.

49. Hsieh, L.C., Yonkovich, S., Kochersperger, L., and Schultz P.G. (1993) Controlling chemical reactivity with antibodies, *Science* **260**, 337.

50. Janda, K.D., Shevlin, C.G., and Lerner, R.A. (1993) Antibody catalysis of a disfavored chemical transformation, *Science* **259**, 490.

51. Na, J., Houk, K.N., Shevlin, C.G., Janda, K.D., and Lerner, R.A. (1993) The energetic advantage of 5-exo versus 6-endo epoxide openings: A preference overwhelmed by antibody catalysis, *J. Am. Chem. Soc.* **115**, 8453.

52. Reymond, J.-L., Reber, J.-L., and Lerner, R.A. (1994) Enantioselective, multigram-scale synthesis with a catalytic antibody, *Angew. Chem. Int. Ed. Engl.* **33**, 475.

53. Tramontano, A., Janda, K.D., and Lerner, R.A. (1986) Catalytic antibodies, *Science* **234**, 1566.

54. Pollack, S.J., Jacobs, J.W., and Schultz, P.G. (1986) Selective chemical catalysis by an antibody, *Science* **234**, 1570.

55. Iverson, B.L. and Lerner, R.A. (1989) Sequence-specific peptide cleavage catalyzed by an antibody, *Science* **243**, 1184.

56. Tawfik, D.S., Green, B.S., Chap, R., Sela, M., and Eshhar, Z. (1993) catELISA: A facile general route to catalytic antibodies, *Proc. Natl. Acad. Sci. USA* **90**, 373.

57. Lesley, S.A., Patten, P.A., and Schultz, P.G. (1993) A genetic approach to the generation of antibodies with enhanced catalytic activities. *Proc. Natl. Acad. Sci. USA* **90**, 1160.

58. Fischer, E. (1894) Einfluss der Configuration auf die Wirkung der Enzyme, *Ber. Dt. Chem. Ges.* **27**, 2985.

59. Eschenmoser, A. (1994) One hundred years lock-and-key principle. *Angew. Chem. Int. Ed. Engl.* **33**, 2363.

60. Lichtenthaler, F.W. (1994) 100 Years "Schlüssel-Schloss-Prinzip": What made Emil Fischer use this analogy? *Angew. Chem. Int. Ed. Engl.* **33**, 2364.

61. Koshland, D.E., Jr. (1994) The Key-Lock Theory and the Induced Fit Theory. *Angew. Chem. Int. Ed. Engl.* **33**, 2375.

MECHANISTIC STUDIES ON THE FORMATION OF BZIP•DNA INTERFACES

A Simple Example of Supramolecular Stereochemistry

DAVID N. PAOLELLA,§† C. RODGERS PALMER,#
STEVEN J. METALLO§ & ALANNA SCHEPARTZ§†
§*Department of Chemistry*
#*Department of Molecular Biophysics and Biochemistry*
†*Training Program in Biophysics*
Yale University
225 Prospect Street
New Haven, Connecticut 06511-8118

1. Introduction

Transcription factors that bind DNA as dimers often recognize "half sites" arranged in an inverted repeat such as the sequence ATGACGTCAT. The half sites are the underlined sequences and the CG in the middle is a spacer between the sites. Transcription factor dimers that bind to these sites can recognize not only the sequence of bases within each half site ("base-pair specificity"), but also the spacing between the half sites ("half-site spacing specificity"). Dimeric transcription factor families whose members display half-site spacing specificity include nuclear hormone receptors [1], class-3 zinc-binding domains [2,3], homeodomains [4,5], and basic region-leucine zipper (bZIP) proteins [6].

bZIP proteins employ an atypically simple motif for DNA recognition, and therefore represent an excellent system in which to dissect the structural and energetic basis for half-site spacing specificity. The DNA-binding region of a bZIP protein contains a short, helical, basic segment whose residues participate in DNA contacts, a zipper segment responsible for protein dimerization, and a six residue spacer of variable sequence connecting the two (Figure 1) [7]. X-ray crystallography data from two different bZIP•DNA complexes shows the dimeric protein to contain a pair of uninterrupted α–helices that interact with each other along the length of the zipper segment to form a parallel coiled coil. These α–helices diverge in the vicinity of the nucleic acid and interact with the major groove of the target

J. S. Siegel (ed.), Supramolecular Stereochemistry, 83–89.

DNA [8,9]. bZIP proteins are distinct among known DNA-binding motifs in that the "recognition helix" is isolated and appears not to be stabilized through tertiary interactions with any other region of the protein [8,9].

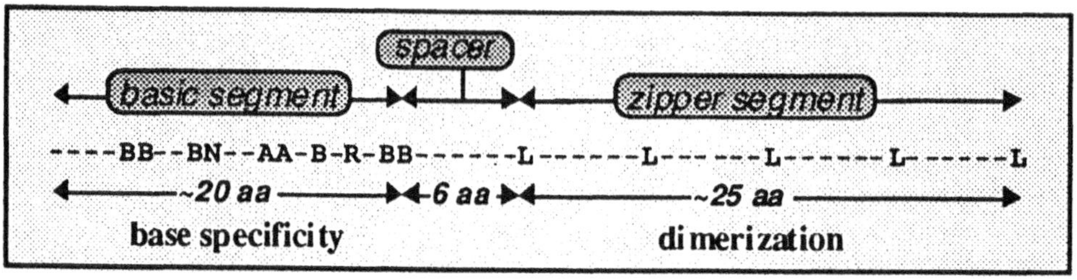

Figure 1. Schematic illustration of a minimal "bZIP element", the 60 amino acid segment that is both necessary and sufficient for DNA recognition by bZIP proteins.

Despite the simplicity of their DNA-binding domains, bZIP proteins discriminate differentially between CRE (ATGACGTCAT) and AP-1 (ATGACTCAT) target sites. These sequences are the nuclear end-points of two different signal transduction pathways, yet they differ by only a single dG•dC base pair that separates the two half sites. bZIP proteins in the AP-1 family prefer the 9 base pair AP-1 target site, whereas members of the CREB/ATF family prefer the 10 base pair CRE target site [10,11]. Within the context of B-form DNA, the extra dG•dC base pair in the center of the CRE target site displaces one of the two half sites by an axial translation of 3.25 Å and a twist angle of 34°, a geometric operation that moves the bases by 4 Å and the phosphates by 7 Å. Despite these presumed structural differences, the bZIP protein GCN4, unlike the AP-1 or CREB/ATF proteins, binds both sites with comparable affinity [12]. Some workers have posited that GCN4 contains a flexible α–helical segment that allows it to readjust to fit both sites [8,13,14]; others hypothesize that GCN4 induces a structural readjustment of one or both of the DNA sites [9]. A third possibility is that GCN4 has an equal affinity for the two DNA target sites because there is an intrinsic deformation of the B-form structure of one site that compensates for the difference in half-site spacing.

2. Results and Discussion

2.1 THE CRE TARGET SITE BENDS INTRINSICALLY

Recently, David Paolella and Rodgers Palmer used helical phasing analysis [15] to provide support for the intrinsic deformation idea [16]. Their results demonstrate that the CRE target site bends towards the major groove by approximately 11° in the absence of any protein. The AP-1 target site, even though it differs from the CRE target site by a single base pair, shows no intrinsic bend. Neither the CRE nor the AP-1 target site undergoes a detectable conformational change upon binding to GCN4. However, upon binding to CRE-BP1 (a CREB/ATF family member), both DNAs bend toward the minor groove. In the case of the CRE target site, this minor groove bend negates the intrinsic major groove bend and in effect "straightens" the DNA (Figure 2). Additional experiments performed with GCN4-CRE-BP1 chimeric peptides demonstrate that this straightening effect is uniformly absent in all bZIP•CRE complexes containing the GCN4 basic segment and present in complexes containing the CRE-BP1 basic segment [16]. Furthermore, induced bending correlates with specificity: Steve Metallo has shown that all bZIP peptides studied which straighten the CRE target site show high CRE target site selectivity [17,18].

These results suggest a model to explain the differential half-site spacing preferences of bZIP proteins. For GCN4, the absence of half-site spacing specificity results from an intrinsic bend in the CRE target site that equates its recognition surface with that of the AP-1 target site. Neither DNA undergoes a detectable change in conformation upon binding to GCN4, and x-ray data indicate that most contacts are conserved between the two structures [8,9]. Therefore it is not surprising that GCN4 binds the CRE and AP-1 sites equally; the protein has evolved to recognize a DNA conformation assumed by both sequences. For CRE-BP1, the presence of a strong half-site spacing preference ($\Delta\Delta G^o$>2.3 kcal•mol^{-1}) is correlated with an ability to overcome the intrinsic curvature in the CRE target site through a mechanism dependent on basic segment residues (Figure 2).

2.2 A CLUSTER OF BASIC RESIDUES AT THE N-TERMINUS OF THE BZIP ELEMENT ARE NECESSARY TO "STRAIGHTEN" THE CRE TARGET SITE.

Several other bZIP family members bend the AP-1 target site when they bind [19]. We compared the sequences of these proteins and noticed a correlation between certain residues in the basic segment and the direction of protein-induced bending [16]. Proteins that induce a minor groove bend contain three basic residues at the amino terminus of the basic segment, whereas proteins that induce a major groove bend contain a conserved pattern of charged and hydrophobic residues in these positions. GCN4, which does not induce a DNA bend, contains the sequence Pro-Ala-Ala, residues which do not contact DNA [8,9].

On the basis of these comparisons, we proposed that the differential interactions of these basic segment residues with DNA regulate the differential bending and binding of bZIP proteins [16]. To test this hypothesis, David

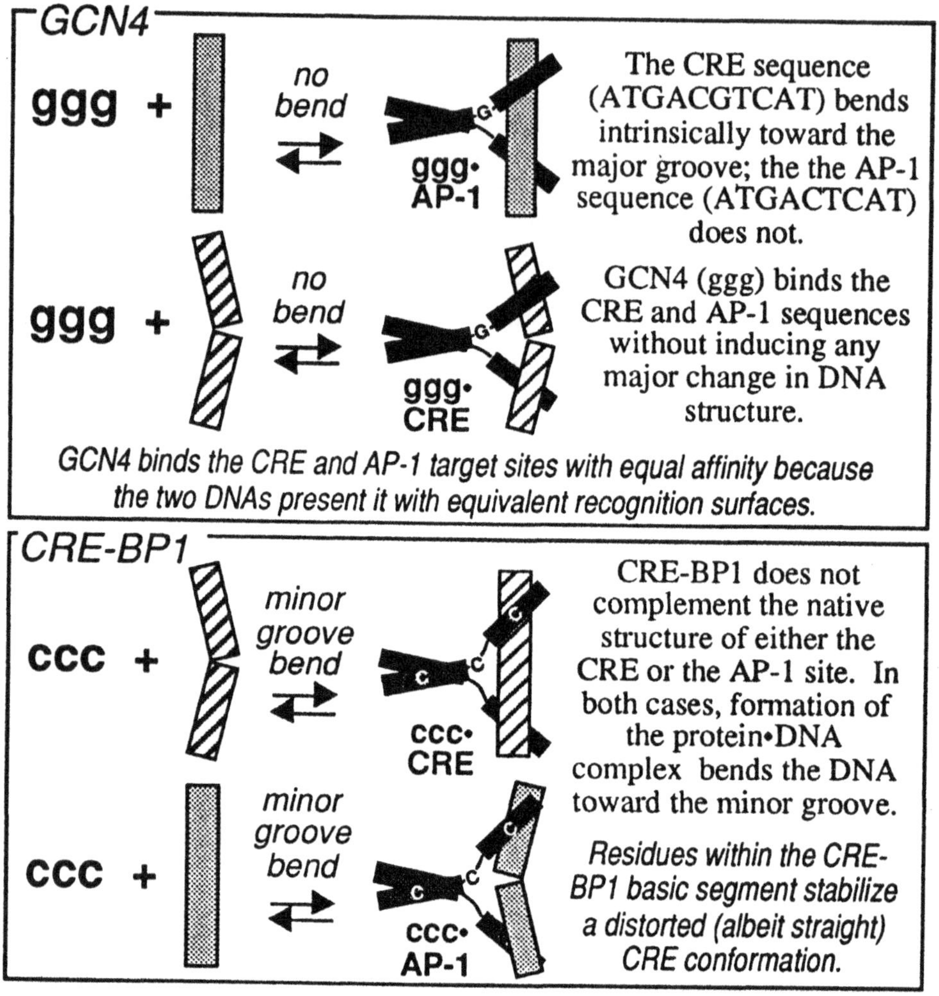

Figure 2. Model proposed to account for the half-site spacing preferences of GCN4 and CRE-BP1.

Paolella constructed two bZIP chimeras: one in which the five amino terminal residues of the CRE-BP1 bZIP element are replaced with the corresponding residues from GCN4 (g5c), and another in which the five amino terminal residues of the GCN4 bZIP element are replaced with the corresponding residues from CRE-BP1 (c5g). The g5c•CRE complex exhibits a major groove bend comparable to that observed in the free CRE target site, illustrating that the Lys-Arg-Arg-Lys-Phe sequence found in CRE-BP1 is necessary to straighten the CRE target site. However, this pentapeptide sequence is not sufficient to straighten the CRE target site, as the c_5g•CRE complexes also retain the major groove bend.

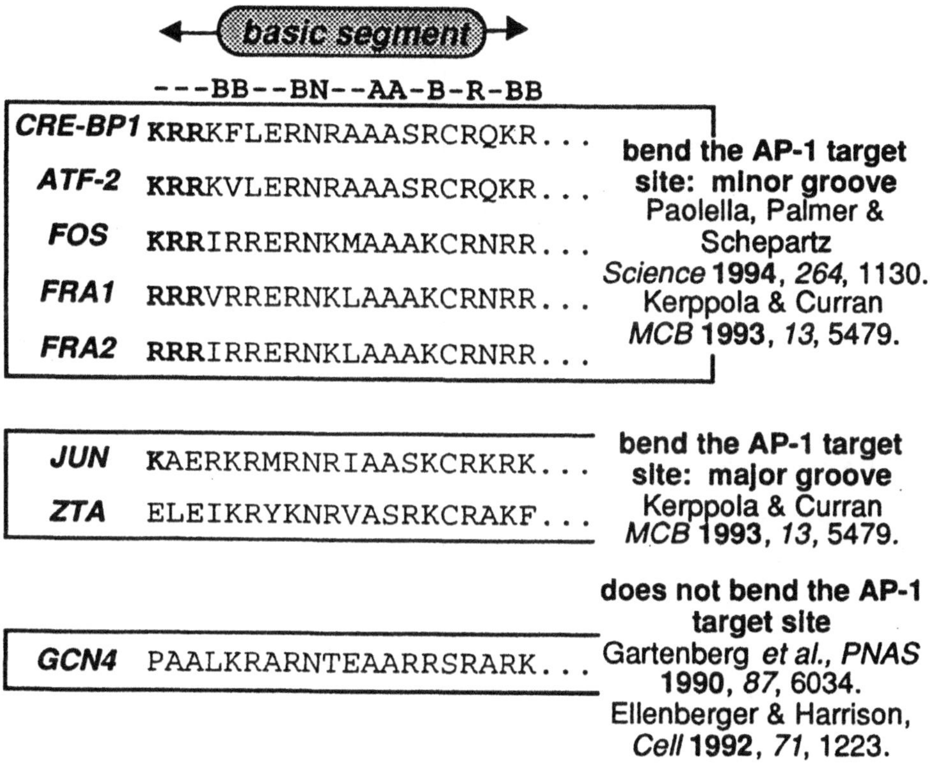

Figure 3. Basic segment sequences of bZIP proteins studied by helical phasing analysis.

2.3 SUMMARY

The discovery of bent A-tract DNA over a decade ago by Crothers and Englund [20] confirmed that certain natural DNA sequences can populate conformations distinct from those defined by the Drew-Dickerson dodecamer. Although several non-A-tract sequences display intrinsic bends, the magnitudes of the bends in these cases are small relative to the bend in a single A-tract. We discovered that the CRE target site bends intrinsically towards the major groove during a study of the half-site spacing selectivities of bZIP proteins [16]. The bend in the CRE target site is similar in magnitude, but opposite in direction, to the bend in a single A-tract. The observation that CRE target site selectivity correlates with the ability to remove this bend suggest that intrinsic DNA distortion, as well as intrinsic

DNA flexibility, can play an important role in determining the stabilities of protein•DNA complexes. Moreover, these results raise interesting questions about the inherent relative flexibility of these two macromolecules; despite the simplicity of the bZIP motif, many family members are unable to present a helical recognition surface suitable for their target site and the DNA must distort to ensure a proper "fit".

3. Acknowledgements

This work was supported by a grant to A.S. from the American Cancer Society, with additional support provided by the National Foundation for Cancer Research, the Lucile Packard Foundation, the National Science Foundation, the Camille and Henry Dreyfus Foundation, the Arthur P. Sloan Foundation and by a 1995 Arthur C. Cope Scholar Award. D.N.P. was supported by an N.I.H. Training Grant in Biophysics. C.R.P. is an NSF Predoctoral Fellow.

4. References

1. Umesono, K. and Evans, R.M. (1989) Determinants of Target Gene Specificity for Steroid/Thyroid Hormone Receptors, *Cell* **57**, 1139-1146.
2. Reece, R.J. and Ptashne, M. (1993) Determinants of Binding-Site Specificity Among Yeast C_6 Zinc Cluster Proteins, *Science* **261**, 909-911.
3. Corton, J.C. and Johnston, S.A. (1989) Altering DNA-Binding Specificity of GAL4 Requires Sequences Adjacent to their Zinc Finger, *Nature* **340**, 724-727.
4. Mendel, D.B., Khavari, P.A., Conley, P.B., Graves, M.K., Hansen, L.P., Admon, A., and Crabtree, G.R. (1991) Characterization of a Cofactor that Regulates Dimerization, *Science* **254**, 1762-1767.
5. Smith, D.L. and Johnson, A.D. (1992) A Molecular Mechanism for Combinatorial Control in Yeast: MCM1 Protein Sets the Spacing and Orientation of the Homeodomains of an α2 Dimer, *Cell* **68**, 133-142.
6. Cuenoud, B and Schepartz, A. (1993) Altered Specificity of DNA-Binding Proteins with Transition Metal Dimerization Domains, *Science* **259**, 510-513.
7. McKnight, S.L. (1991) Molecular Zippers in Gene Regulation, *Sci. Am.* 54-64.
8. Ellenberger, T.E., Brandl, C.J., Struhl, K., and Harrison, S.C. (1992) The GCN4 Basic Region Leucine Zipper Binds DNA as a

Dimer of Uninterrupted α Helices: Crystal Structure of the Protein-DNA Complex, *Cell* **71**, 1223-1237.

9. König, P. and Richmond, T. (1993) The X-ray Structure of the GCN4-bZIP Bound to ATF/CREB Site DNA Shows the Complex Depends on DNA Flexibility, *J. Mol. Biol.* **230**, 139-154.
10. Hai, T., Liu, F., Allegretto, E.A., Karin, M., and Green, M.R. (1988) A family of immunologically related transcription factors that includes multiple forms of ATF and AP-1, *Genes Dev.* **2**, 1216-1226.
11. Hai, T., Liu, F., Coukos, J., and Green, M.R. (1989) Transcription factor ATF clones: an extensive family of leucine zipper proteins able to selectively form DNA-binding heterodimers, *Genes Dev.* **3**, 2083-2090.
12. Sellers, J.W., Vincent, A.C., and Struhl, K. (1990) Mutations That Define the Optimal Half-Site for Binding Yeast GCN4 Activator Protein and Identify an ATF/CREB-Like Repressor That Recognizes Similar DNA Sites, *Mol. Cell. Biol.* **10**, 5077-5086.
13. Kim, J., Tzamarias, D., Ellenberger, T., Struhl, K., and Harrison, S.C. (1993) Adaptability at the Protein-DNA Interface Is an Important Aspect of Sequence Recognition by bZIP Proteins, *Proc. Natl. Acad. Sci. USA* **90**, 4513-4517.
14. Ellenberger, T. (1994) Getting a grip on DNA recognition: structures of the basic region leucine zipper, and the basic region helix-loop-helix DNA binding domains, *Current Biology* **4**, 12-21.
15. Zinkel, S.S. and Crothers, D.M. (1987) DNA Bend Direction by Phase Sensitive Detection, *Nature* **328**, 178-181.
16. Paolella, D.N., Palmer, C.R., and Schepartz, A. (1994) DNA Targets for Certain bZIP Proteins Distinguished by an Intrinsic bend, *Science* **264**, 1130-1133.
17. Metallo, S.J. and Schepartz, A. (1994) Distribution of Labor Among CRE-BP1 bZIP Segments in the Control of DNA Affinity and Specificity, submitted.
18. Metallo, S.J. and Schepartz, A. (1994) Peptide Models of bZIP Proteins. Quantitative Analysis of Affinity and Specificity, *Tech. Protein Chem.*, in press.
19. Kerppola, T.K. and Curran, T. (1993) Selective DNA Bending by a Variety of bZIP Proteins, *Mol. Cell. Biol.* **13**, 5479-5489.
20. Marini, J.C., Levene, S.D., Crothers, D.M., and Englund, P.T. (1982) Bent helical structure in kinetoplast DNA, *Proc. Natl. Acad. Sci. USA* **79**, 7664-7668.

CHIRAL DISCRIMINATION IN CRYSTALLINE ENANTIOMER SYSTEMS : FACTS, INTERPRETATIONS, AND SPECULATIONS

A. COLLET, L. ZIMINSKI, C. GARCIA, (a) AND F. VIGNÉ-MAEDER (b)
École normale supérieure de Lyon, (a) Stéréochimie et Interactions Moléculaires, (b) Chimie Théorique
69364 Lyon cedex 07, France

1. Introduction

Because it implies a perfect organization of symmetry related species, crystallization is one of the simplest and most efficient molecular recognition process. In general, a given molecule will not allow the presence of foreign molecules in its crystal lattice, and this property is widely used in the separation and purification methods based on crystallization.

Chiral molecules do not behave differently in this respect. However, the existence of a mirror image relationship between enantiomers has an important consequence : homochiral, enantiomorphous crystals, made of molecules of the same handedness, as well as heterochiral, racemic crystals, consisting of an ordered array of symmetry related right and left handed molecules, represent equally viable alternatives, from a purely crystallographic point of view.

Experimentally, this behaviour holds for the vast majority of enantiomer systems. The formation of solid solutions (*pseudoracemates*) between enantiomers, resulting from a lack of mutual chiral recognition, is rare and generally predictable (camphor). In most instances a racemic melt, solution, or vapor, will yield, on crystallization, *either* a mixture of crystals of the pure enantiomers (*conglomerate*), *or* a single racemic crystal species (*racemic compound*). This simple structural alternative between the two major racemate types in turn translates into the well known phase diagrams describing the solid-liquid (or solid-vapor) phase equilibria that may take place in enantiomer systems (*Figure 1*).[1]

This homochiral *vs.* heterochiral packing alternative is by no means governed by chance; in a given set of conditions (T, P, solvent), a single type of racemate - conglomerate *or* racemic compound - is thermodynamically allowed for a chiral molecule. Moreover, in most of the cases, the racemate type of a molecule does not change within the usual range of physical conditions to

J. S. Siegel (ed.), Supramolecular Stereochemistry, 91–110.

which it may be submitted. In somes cases, however, transformations of a racemic compound into a conglomerate (or *vice versa*) may occur, as for example in the conglomerate sodium ammonium tartrate (Pasteur salt) which becomes a racemic compound (Scacchi salt) over 27 °C,[2] or in the more recently studied case of 1,1'-binaphthyl.[3]

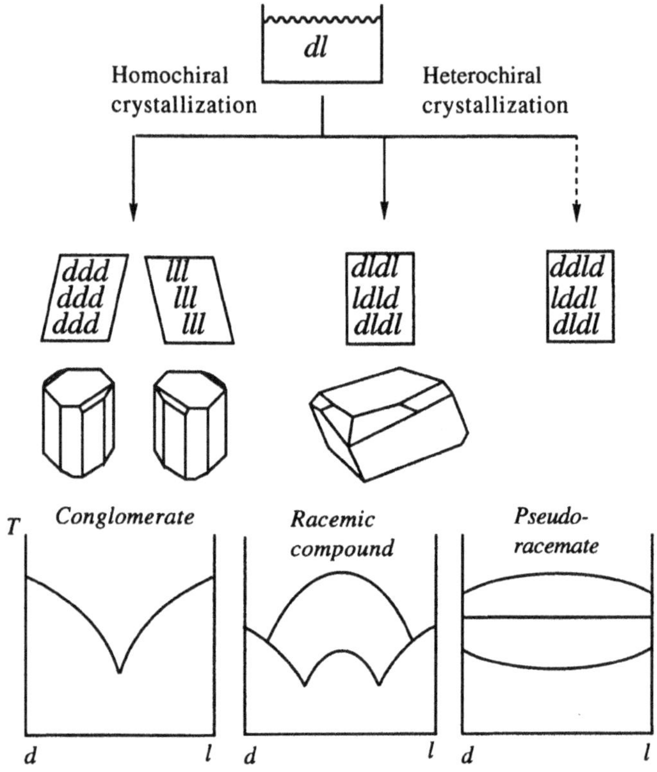

Figure 1. The homochiral *vs*. heterochiral packing alternative and its thermodynamical implications (phase diagrams)

If the existence of conglomerates and racemic compounds is a well established fact, it is also well accepted among the chemist community that the occurrence of these two forms is very unequal. For a long time, the fact that only a small number of conglomerates had been identified as such, has been taken as evidence that racemic compounds must be more stable than crystals made of a single enantiomer, a conclusion which was easy to accept on the basis of the old Wallach's rule (1895),[4] stating (from measurements of 8 pairs of densities) that racemic compounds are denser than their crystalline enantiomers.

Unfortunately for this theory, a number of examples have since been discovered where a racemic compound is less dense than its enantiomers!

Are racemic compounds "generally more stable" than conglomerates and, should this statement be true, would it also imply that heterochiral crystal packings are generally more stable than their homochiral counterparts? Can we quantify the relative occurrence of the two major racemate types, and explain the prevalence of racemic compounds? Can we identify structural features, on a molecular level, that would be responsible for the formation of conglomerates, or would increase their relative frequency within series of closely related molecules? And, finaly, can we expect to alter significantly the balance between the two racemate forms by a drastic change of the physical conditions, such as the application of high pressures? These questions, some of which have recently been the object of controversial views,[5,6] are the main object of this paper.

We wish to begin, however, with another aspect of chiral discrimination in the solid state. What is the degree of tolerance of a homochiral crystal network towards substitution of chiral foreign species for its own molecules?

2. Experiments on the co-crystallization of structurally related chiral molecules

Homochiral crystals, made of molecules of the same handedness, cannot, in general, accept the incorporation of the opposite enantiomer in their lattice, and this is one of the reasons why optical purification of partially resolved mixtures can often be effected conveniently by crystallization. On the other hand, molecules having similar structures and sizes may co-crystallize to form solid solutions (sometimes called mixed crystals). Figure 2 shows the phase diagram corresponding to the formation of a terminal solid solution in which a small quantity of ○ molecules are incorporated in the lattice of the ● molecules.

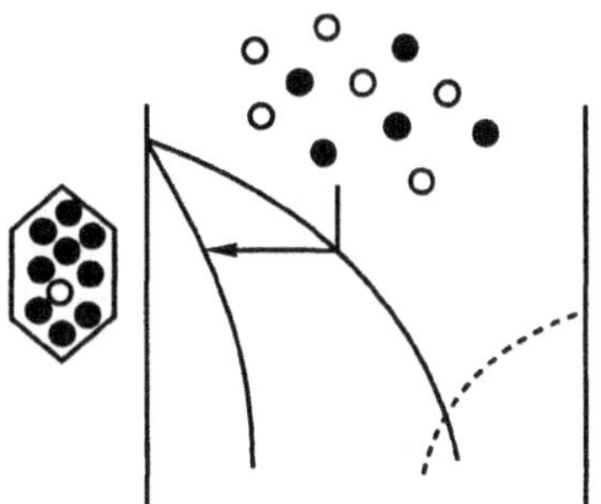

Figure 2. Terminal solid solution

The extension of the solid solution domain in the phase diagram depends on the degree of similarity between the ● and O molecules and on the crystal structures of both partners. According to Kitaigorodskii, the degree of similarity of two molecules can been quantified by means of a "isosterism coefficient" $\varepsilon=1-V_{no}/V_o$ where V_{no} and V_o represent the non-overlap and overlap volume of the two molecules, respectively, when they are superposed. The value of ε is 1 for identical molecules and the formation of a solid solution between two molecules would require that ε be in the range 0.8 to 1. When the crystal structures of the two species are different, only terminal solid solutions, on each side of the phase diagram, are allowed (as in *Figure 2*), whereas *isomorphism* of the two crystal species would allow the existence of a continuous series of solid solution between them if ε is sufficiently close to 1.[7]

The isosterism coefficient can also be used as a *chirality index*, a way to quantify the chirality of a molecule.[8] An example of the calculation of ε for simple tetrahedral enantiomeric molecules is given in *Figure 3*. Experimentally, it is found that *enantiomorphism* generally precludes *isomorphism*; apart for the case of quasi-spherical molecules, enantiomers rarely cocrystallize, and the range of terminal solid solutions which may exist in their phase diagrams is in most cases very narrow, almost insignificant.

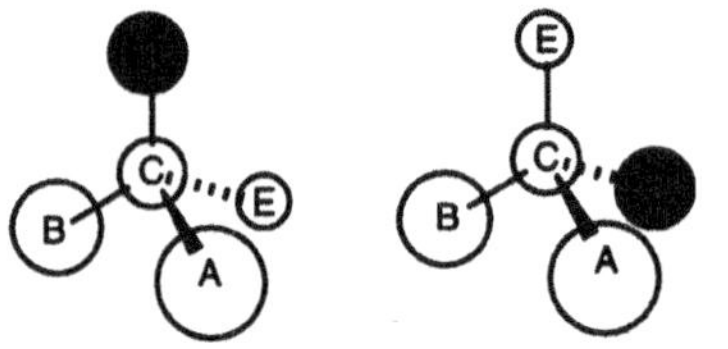

A > B > D > E

$V_o = A + B + C + 2E$

$V_{no} = 2(D\text{-}E)$

$\varepsilon = 1 - V_{no}/V_o$

Figure 3. Isosterism of enantiomers ; the largest groups A and B and the central atom are kept superposed, a choice which is not necessarily the most pertinent

For closely related chiral molecules, such as for example *o*-chloro and *o*-bromomandelic acid, the isosterism coefficient ε should be closer to 1 for the pair of the same configuration than for the pair of quasi-enantiomers (*Figure 4*). Accordingly, co-crystallization is expected to occur more easily between species having the same handedness, than between pairs of quasi-enantiomers (*Figure 5*).[9] Crystallization of R-(+)-*o*-chloromandelic acid (as the host) in the presence of some (±)-*o*-bromo acid (which is 3% larger in size) actually leads to a solid solution of the R-(+) bromo acid (*ee*>98%) into the R-(+)-chloro acid lattice.[10] The crystallization of an enantiopure host in the presence of structurally related racemic guests, to form a solid solution, may thus represent a new optical resolution method.[11] Relevant (preliminary) experiments of this type are assembled in TABLE 1, from which the following conclusions may be drawn: (i) only very small amounts of the foreign species are found in the host crystal in most cases; (ii) the host lattice recognizes the size and configuration of the guest but is also sensitive to its chemical functionalities; (iii) excellent chiral recognition can be achieved, particularly when the guest is slightly too big.

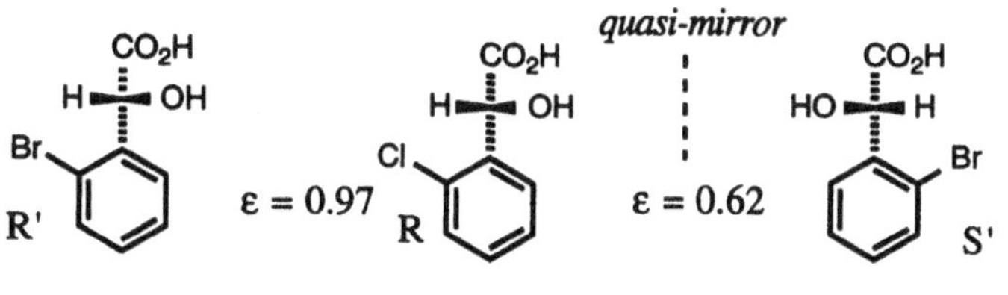

Figure 4. Isosterism between R-*o*-chloromandelic acid and its R or S bromo analogues

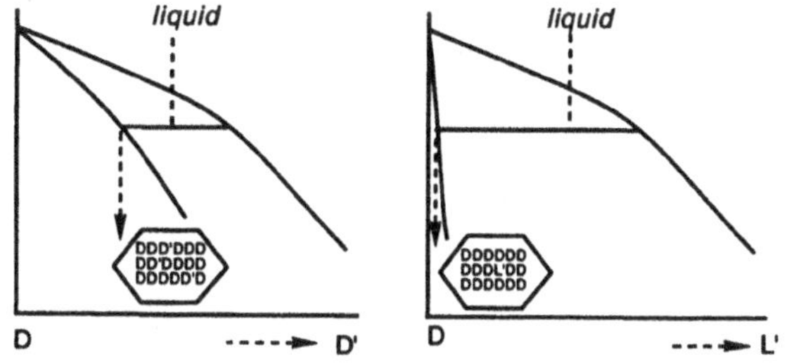

Figure 5. Co-crystallization of configurationally related (left) or opposed (right) analogues.

TABLE 1. Co-crystallization of enantiopure hosts with racemic guests

host crystal	(±) guest	relative size	Selected guest enantiomer	
	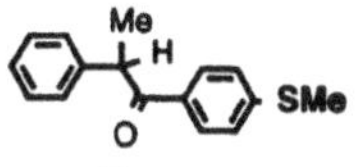	104%	8%	S-(+) >98% ee
	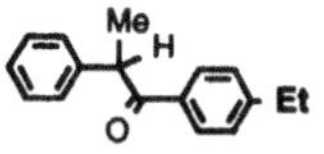	104%	2%	(+) 88% ee
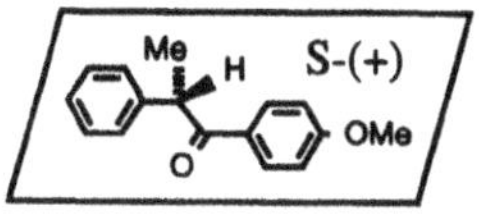	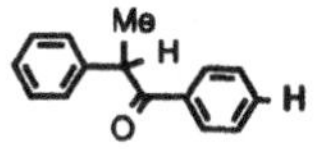	90%	1-2%	(+) 56% ee
	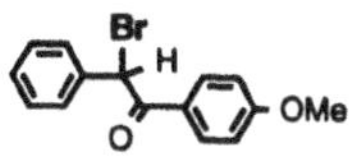	102%	12-15%	(+) >75% ee
	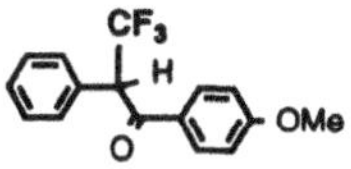	105%	2%	n.d.

These experiments have a general bearing, in the sense that terminal solid solutions may always exist in the phase diagrams of structurally related chiral substances; in spite of their preliminary nature, they clearly suggest that molecular recognition (and chiral discrimination) performed by a rigid box (a molecular site in a crystal lattice) is a highly specific process.

It should be emphasized that *host-guest solid solutions* and conventional *crystalline host-guest complexes*, which are also used for optical resolution purposes,[12] represent basically different systems. In host-guest solid solutions, a generally small and non-stoichiometric amount of guest molecules replace host molecules in the host lattice (site substitution). The formation of such solid solutions requires that the host and the guest be sterically and configurationally related. In conventional host-guest complexes, the guest and host molecules need not be structurally related; the guest molecules fill empty spaces of the host lattice (clathrates) or occupy specific sites of the lattice (in which case stoichiometric addition complexes are obtained).

3. The Homochiral vs. Heterochiral Packing Dilemma

The knowledge and understanding of factors that may govern the crystallization of racemates are not only a matter of theoretical interest: one of the simplest and most efficient methods for preparing optically active materials rests on the direct crystallization of racemates existing as conglomerates.[13,14]

Yet, we must admit that we are still totally unable to predict the racemate type from the molecular structure, just as we are largely unable to predict the crystal structure of a molecule except perhaps in very simple cases. In fact, the homochiral or heterochiral outcome of the crystallization of a racemate is not only governed by the crystal properties, but also depends on the nature of the solid-liquid phase equilibria that occur in enantiomer mixtures. For this reason, both thermodynamic and structural aspects of the question must be taken in consideration here.

3.1. ARE RACEMIC COMPOUNDS GENERALLY MORE STABLE THAN CONGLOMERATES?

This question is not so trivial as it would seem, and deserves some clarification. Let us first recall (Figure 6) that pure enantiomers (melting separately at T_a) can form a conglomerate (melting at T_c) or a racemic compound (melting at T_r). Although T_c is thermodynamically dependent on T_a and ΔH_a, the heat of fusion of a pure enantiomer (usually, T_a-$T_c \approx$ 25-35 °C), this is not the case for T_r which may be lower or higher than T_a. If, however, T_r is lower than T_c, then the racemic compound is thermodynamically less stable than the corresponding conglomerate. Although such racemic compound may occasionally exist as metastable forms, in most known conglomerates they do not exist at all, and this circumstance makes impossible a direct comparison of the homochiral and (hypothetical) heterochiral packings in these instances.

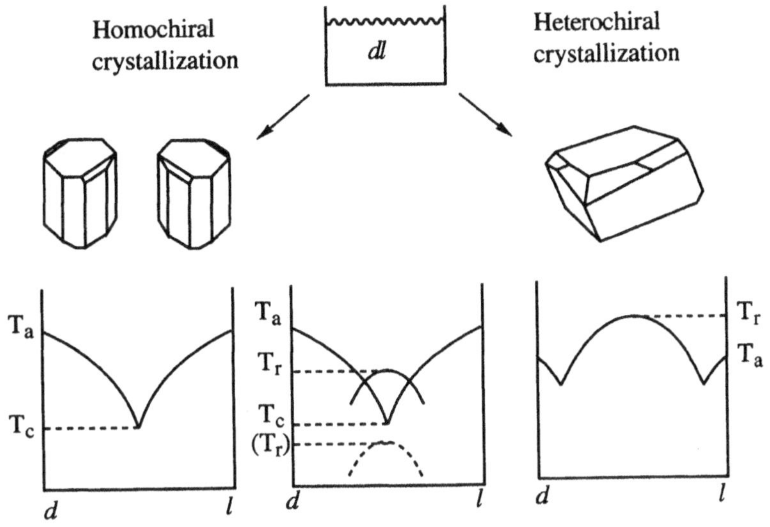

Figure 6. Solid liquid equilibria in conglomerates and racemic compounds

The statement that *racemic compounds are generally more stable than conglomerates* simply means that there exist much more examples of racemic compounds than of conglomerates in the presently known population of chiral molecules. Is that true, and, if yes, can we give an estimate of the relative occurrence of both forms?

These questions do not have obvious answers. If thousands of racemates have been described (who knows how many?), it is only rarely that their type has been identified and recorded in such a way that it can be retrieved. Fortunately, the fact that conglomerates melts ≈30 °C lower than their enantiomers may be a useful guide to identify them from literature data. This requirement naturally restricts the field of investigation to instances where the racemate and at least one enantiomer are described (natural products are virtually eliminated). Along these lines, a careful scrutiny of 1308 racemate and enantiomer pairs listed in Beilstein led to the identification of 83 conglomerates, 6% of the considered sample.[15] A different approach consists of examining the racemates in series of compounds synthesized on purpose. Although this approach may be biased by the choice of the target compounds, it gave a result which is not very different from the preceding one: 194 new compounds, members of 8 different chemical families, yielded 22 conglomerates (11%).[16]

As far as one can rely on such statistical data, they tell us that racemic compounds probably represent 90-95% of crystalline racemates.

3.2. SOME THERMODYNAMICS

What is the magnitude of the stability difference between a racemic compound and the corresponding conglomerate of enantiomers? A simple and certainly exact way to give an answer to this question is to define this stability difference as the free energy change ($\Delta G°$) of the solid state transformation of a conglomerate into its racemic compound at a temperature T.[17] For a stable racemic compound, $\Delta G°$ must be negative, and, as we shall see, this statement can be experimentally verified from the melting points, enthalpies of fusion, and specific heats of the pure enantiomers and of their racemic compound, as indicated in the thermodynamic cycle of *Figure 7* (where for convenience $\Delta G° = \Delta H° - T_r\Delta S°$ is calculated at T_r). Conversely, for a stable conglomerate, $\Delta G°$ as defined in *Figure 7* must be positive; however, it is generally not measurable in this case because the data for the racemic compound are missing (with the exception of metastable racemic compounds).

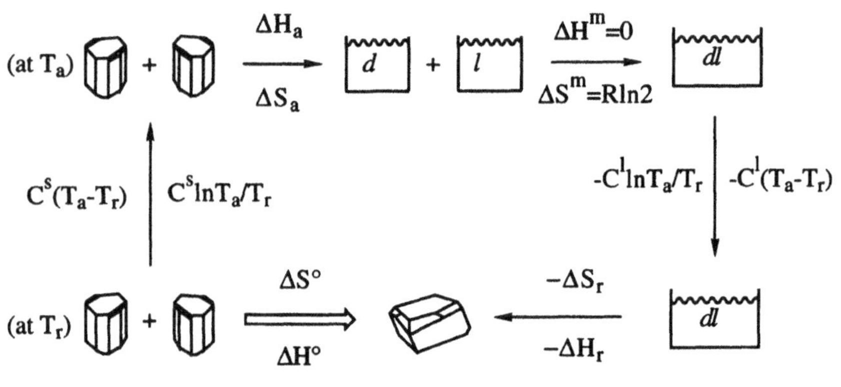

$$\Delta H° = \Delta H_a - \Delta H_r - (C^l - C^s)(T_a - T_r)$$
$$\Delta S° = \Delta S_a - \Delta S_r - (C^l - C^s)\ln(T_a/T_r) + R\ln 2$$

Figure 7. Cycle allowing the calculation of the free energy of formation of racemic compounds from the conglomerate at T_r ($\Delta G° = \Delta H° - T_r\Delta S°$). C^l and C^s represent the specific heats (C_p) of the considered liquid or solid phases.

A plot of $\Delta G°(T_r)$ as a function of the melting point difference T_a-T_r, for a sample of 56 racemic compounds for which the necessary data are available, is shown in *Figure 8*. As expected, $\Delta G°$ is negative and ranges from 0 (for T_a-$T_r \approx$

30 K) to *ca.* -1.8 kcal/mol for the stablest racemic compounds (for which T_a-T_r approaches -80 K). A striking feature of this plot is the observation of a linear correlation between $\Delta G°(T_r)$ and T_a-T_r; using the expressions of $\Delta H°$ and $\Delta S°$ given in Figure 7, $\Delta G°(T_r)$ can be written as

$$\Delta G°(T_r) = \Delta H_a - \Delta H_r - T_r(\Delta S_a - \Delta S_r) - T_r R\ln 2 - (C^l - C^s)((T_a - T_r) - T_r \ln(T_a/T_r));$$

rearranging this expression leads to

$$\Delta G°(T_r) = (T_a - T_r)\Delta S_a - T_r R\ln 2 - (C^l - C^s)((T_a - T_r) - T_r \ln(T_a/T_r));$$

it turns out that the enthalpic and entropic contribution of the specific heats, which cannot be neglected in $\Delta H°$ and $\Delta S°$ when $T_a \neq T_r$, almost cancel each other in the above expression of $\Delta G°(T_r)$ which therefore can be simply written as

$$\Delta G°(T_r) = (T_a - T_r)\Delta S_a - T_r R\ln 2;$$

the slope of the regression line (15.35 cal mol^{-1} K^{-1}) can thus be identified with the mean entropy of fusion ΔS_a of the enantiomers ($\langle \Delta S_a \rangle$ = 15.6, esd 3.4), and T_r similarly represents the average melting point of the racemates in the considered sample ($\langle T_r \rangle$ 411 K, esd 41 K). In effect, the *y* intercept of the correlation (561 cal mol^{-1}) almost exactly equals $\langle T_r \rangle R\ln 2$ (566 cal mol^{-1}).

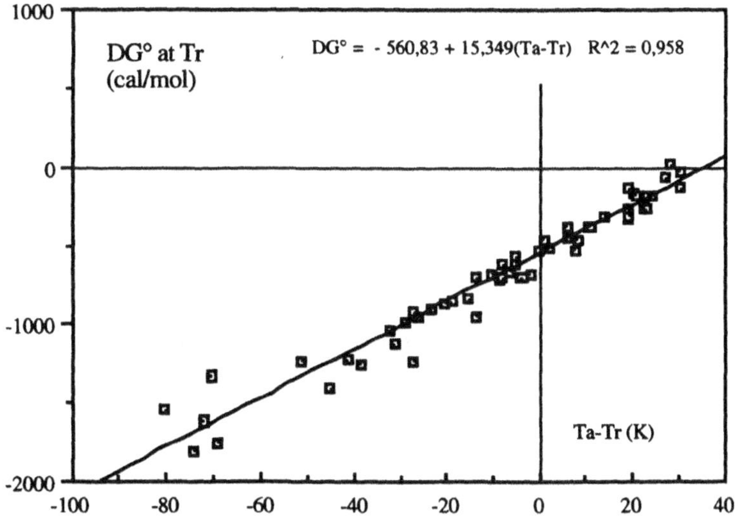

Figure 8. Plot of $\Delta G°(T_r)$ as a function of T_a-T_r for a sample of 56 racemic compounds.

When the pure enantiomers and the racemic compound have the same melting point (then would seem, at a first approximation, equally stable), the magnitude of the free energy of formation of the racemic compound from the conglomerate still amounts to ca. -0.56 kcal mol^{-1}. Here is probably the origin of the overwhelming occurrence of racemic compounds.

To give a physical interpretation of the above equations and particularly of the data distribution in Figure 8 is not an obvious matter. In the above considered sample of 56 pairs, the mean value $<T_r>$ 411(41) K is higher by only 8 K than $<T_a>$ 403(42) K. Assuming that the sample is biased by the fact that 5-10% of conglomerates are missing (for which T_r-T_a > 30 K), the actual difference between $<T_r>$ and $<T_a>$ should even be smaller than 8 K. Based on such melting point criteria, heterochiral crystals do not seem to be stabler than their homochiral counterparts to an extent that would account for their 90-95% prevalence. In our earlier papers,[1,5,18] we proposed that this systematic free energy difference (-TRln2) favoring the formation of the racemic compound was related to the cost of the phase separation which must occur during the crystallization of a conglomerate. This interpretation, in which Rln2, the entropy of mixing of the liquid enantiomers to give the liquid racemate, played a central role (see *Figure 7*), was recently challenged by Pratt Brock, Schweizer and Dunitz,[6] who pointed out that the three considered solids being ordered (enantiomer, conglomerate, and racemic compound) all have essentially the same entropy, and that for this reason the above interpretation would contradict the Third Law of thermodynamics.

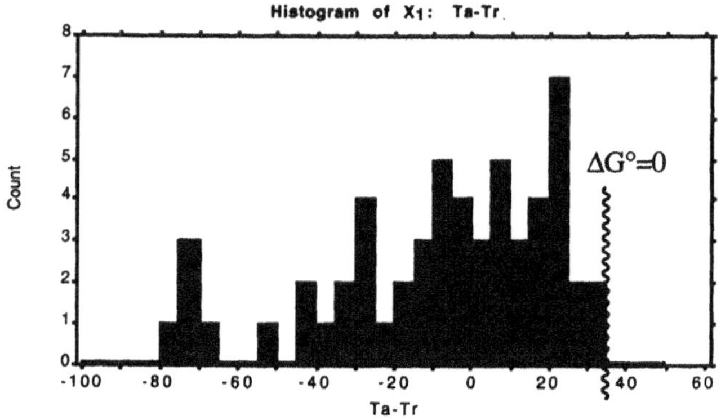

Figure 9. Frequency distribution of T_a-T_r in 56 racemic compounds.

A (possibly not) less controversial interpretation is the following. The racemic compound is in equilibrium with a liquid of the same composition, whereas in a conglomerate each pure enantiomer crystal is in equilibrium with a liquid that does not match its composition (x_d or x_l = 0.5 in the liquid instead of 1 in the solid). For this reason, the equilibrium temperature for a conglomerate in its racemic liquid is lowered by ≈ 30 K with respect to that of the enantiomers in their respective pure liquids. As we have seen above, the mean value of T_r and that of T_a are not very different. On a statistical basis, one may expect that the frequency of the conglomerates will be related to the histogram of Figure 9, where the frequency distribution of T_a-T_r in the 56 considered pairs is shown. The fraction of these racemic compounds having T_a-T_r in the region of 25 to 35 K of the histogram (i.e., those which deviate from the mean value by 23 to 33 K) is indeed very small. Furthermore, the fact that their number decreases rapidly as the negative value of T_a-T_r increases means that the probability of finding racemic compound with positive $\Delta G°$ is low, which is consistent with the observed low frequency of the conglomerates.

3.3. STABILITY, CRYSTAL DENSITY AND WALLACH'S RULE

The fact that racemic compounds are more stable than their conglomerates in 90-95% of the cases does not mean that homochiral packing are generally less stable than their heterochiral variant. This statement can be easily verified if other criteria than the relative occurrence of the two racemate forms are considered. In the three specific examples of hydroxyacids **1-3** assembled in TABLE 2, the racemic compounds are all *less dense* than the enantiomer crystals, and, in **2**, the enthalpy of formation of the racemic compound ($\Delta H°$, as defined in Figure 8) is even *positive*. In this case the homochiral packing is both more compact and enthalpically more stable than the racemic compound, although in equilibrium with its racemic liquid the latter is thermodynamically favored.

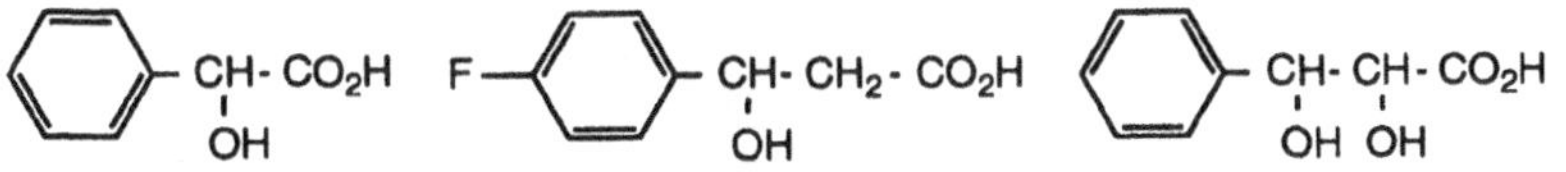

1 **2** **3**

TABLE 2. Thermodynamic data and crystal densities for hydroxyacids **1-3**

	T_a (K)	T_r (K)	$\Delta H°$	$T_r\Delta S°$ (cal/mol)	$\Delta G°(T_r)$	d_a	d_r
1	406	392	-80	+250	-330	1.350	1.320
2	381	362	+325	+470	-145	1.417	1.390
3	371	395	-1240	-320	-920	1.400	1.365

Several compilations of racemic and enantiomer pairs for which crystallographic data are available have been reported; they allow a much more accurate comparison of crystal densities than those obtained by Wallach a century ago. In the first of these lists reported by Jacques, Collet and Wilen,[1] and containing 14 pairs of structures, the average density variation[19] on going from the enantiomer to the racemic crystal $\Delta\%$ is -1.2(9)%, with 9 of the 14 racemic compounds being less dense than their enantiomers. In a similar compilation by Mason,[20] listing 14 pairs which partially overlap the previous list, the average of Δ% is +2.2(13), with 5 of the 14 racemic compound being less dense than their chiral counterparts.

A more recent compilation, based on the Cambridge Structural Database,[21] was reported by Pratt Brock, Schweizer and Dunitz.[6] For 129 pairs, they found 53 racemates less dense than their enantiomers, the average of Δ% being +0.56(22)%, a deviation from zero which was considered to be 'hardly large enough to be convincing, but (...) too large to ignore'. If these studies demonstrate that, *stricto sensu*, Wallach's rule does not hold, they also raise the question of the existence of a relation between the magnitude of Δ% and that of $\Delta G°(T_r)$ for pairs of racemic and enantiomer crystals. That such a relation indeed exists is suggested by reconsidering the data of Pratt Brock *et al.* in the light of the above discussed thermodynamic analysis.

Their sample of 129 pair actually consists of two groups of molecules, Group I (64 pairs) containing rapidly interconverting enantiomers,[22] and Group II (65 pairs) consisting of resolvable enantiomers (*Figure 10*).

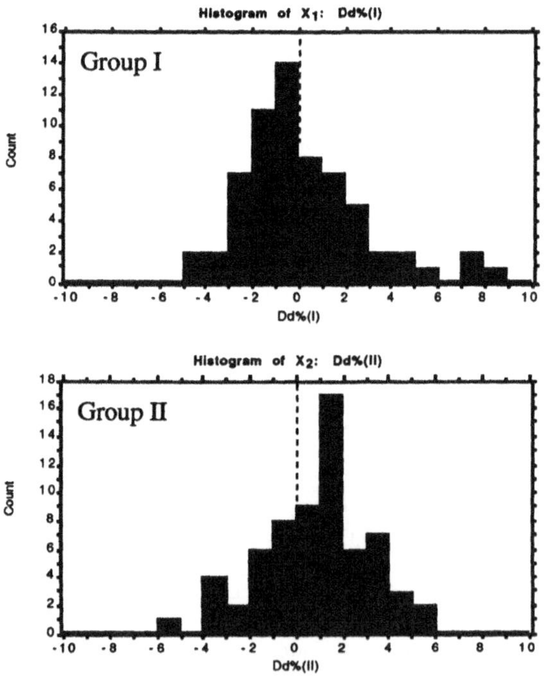

Figure 10. Distribution of Δ% in Group I and Group II subpopulation of 129 racemic and enantiomer crystal pairs (after Pratt Brock *et al.*, ref 6)

In Group I, 32 of the 64 pairs have $d_r<d_a$, and the average of $\Delta\%$ is reduced to +0.20(34)%, i.e., almost zero. In Group II, in which only 21 of the 65 pairs have $d_r<d_a$, the average of $\Delta\%$ is increased to +0.92(29)%. Pratt Brock *et al.* attributed this difference between the two groups to a statistical bias in Group II population, 'since this can contain only pairs in which the racemic crystal is markedly more stable than the chiral one (...) but no pairs in which the racemic crystals are markedly more stable'. We disagree with this interpretation. In our opinion, Group II is not more biased than any other sample of this kind, in the sense that it must be representative for 90-95% of the racemates.

On the other hand, Group I, which Pratt Brock *et al.* consider as non-biased is, in our opinion, totally biased: this group in which enantiomers are rapidly interconverting in the liquid phase can only contain pairs for which enantiomer and racemic crystals have the same stability when they are in equilibrium with the racemic liquid, otherwise one of the forms would not be observable. The equilibrium temperature that must be considered here is therefore that of the eutectic (conglomerate) of the two enantiomers, and the melting points of the racemic compounds of this sample should be equal (or at least close) to those of the corresponding conglomerates. Accordingly, for all racemates of Group I, $\Delta G°(T_r)$ should be close to zero (*Figure 8)*. By contrast, the average of $\Delta G°(T_r)$ for the racemates of Group II should be negative; for comparison, the average $<\Delta G°(T_r)>$ in the 56 pairs of *Figure 8* is -0.69 kcal/mol, a figure which slightly exceeds the boundary of -0.56 kcal/mol below which we consider that the homochiral packing is more stable than the heterochiral one. The fact that the average of $\Delta\%$ increases on going from Group I to Group II subpopulations is entirely consistent with the expected variation of $<\Delta G°(T_r)>$ between them.

3.4. MOLECULAR STRUCTURE AND RACEMATE STABILITY

The existence of series of structurally related compounds in which the occurrence of conglomerates seems to be greater than the average, and, on the contrary, of structural families where no examples of conglomerates have ever been reported, has nothing really astonishing. On a macroscopic level, the bissection of a cube (or a sphere) into halves of C2 symmetry gives two fragments having the same handedness, while a similar cut into halves of C3 symmetry yields two fragments of opposite handedness. Conversely, two homochiral object of C2 symmetry can form a close packed dimer in which their twofold axes are aligned, while a similar close packing of C3 object is heterochiral. This observation,[23] which takes its origin from a long practice of Horeau's Coupe du Roi, has for some time led Jacques and his coworkers to the (hitherto unconfirmed) belief that molecules having a twofold axis would form conglomerates more often that molecules devoid of this symmetry element.[24] Unfortunately, on a molecular level things seems a bit more complicated, perhaps because the magnitude of the energy difference between homochiral and heterochiral packings is small compared with the total lattice energy of a crystal.

Nevertheless, there are perhaps a few structural families where a relatively clear picture seems to emerge: simple carboxylic acids, such as 2-arylpropionic acids or 2-aryloxypropionic acids, which are devoid of H-bonding groups other than the carboxylic function, invariably form very stable racemic compounds. In contrast, their salts show an opposite behaviour, i.e., form racemic compounds of weak stability, and for this reason the formation of salts has been suggested as a way to increase the chance of finding new conglomerates.[15]

The above discussed sample of 56 racemates contains a subset of seven *o*, *m*, *p*-chloro and nitro 2-phenoxypropionic acids and 18 of their salts with achiral amines. A plot of $\Delta G°$ *vs.* (T_a-T_r) for this subset illustrate this trend; the average $<\Delta G°(T_r)>$ is -0.97(27) kcal/mol for the acids, a value reduced to -0.53(30) kcal/mol for their salts (*Figure 11*). Aromatic hydroxy acids (such as **1-3** above and their congeners) exhibit the same behaviour as the salts of these 2-phenoxypropionic acids; a comparatively high frequency of conglomerates has been found among their racemates.[1,25]

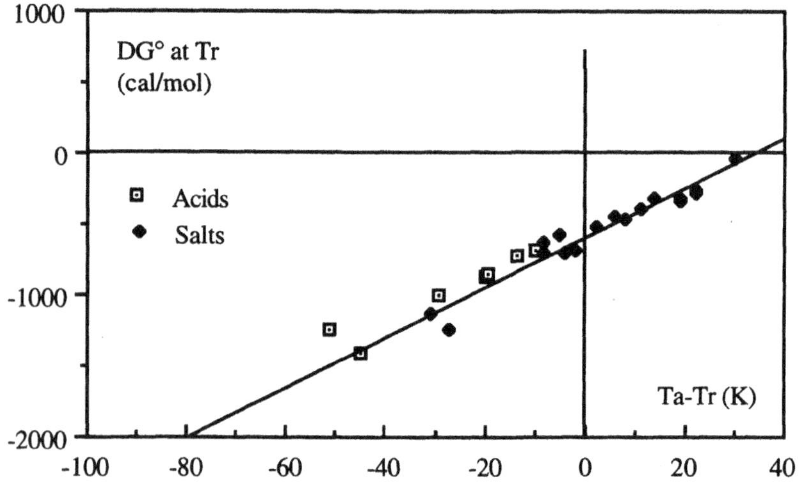

Figure 11. $\Delta G°(T_r)$ *vs.* (T_a-T_r) in subpopulations of the 56 racemates of Figure 8.

A general interpretation of these observations has been advanced.[5,15,18] Enantiomers which can form centrosymmetric pairs on crystallization will always form heterochiral lattices that are more efficiently packed than those of their homochiral counterparts. Simple carboxylic acids typically show this behaviour (for these compounds, Wallach's rule holds). When, on the other hand, the formation of centrosymmetric dimers is made impossible for structural reasons, then homochiral and heterochiral packings will not differ much in energy, and this will result in a shift of the average magnitude of $\Delta G°$ towards lower values, approaching the thermodynamic domain of the conglomerates. In support to this view, it has been observed that racemic aromatic hydroxyacids as well as carboxylic acid salts form crystals dominated by complex H-bonding patterns in which centrosymmetric dimer are generally absent.[25]

In theory, suitable lattice energy calculations may provide new insights into these questions. The first general approach of this kind, addressing the crystallization of chiral molecules, was suggestd by Zorkii in the late sixties,[26] in the line of the ideas of Kitaigorodskii and his school.[27] Given the shape and symmetry of the considered molecule, only a few number of space groups would be permitted, among them variants corresponding to racemic and enantiomer structures; for each variant, the free energy is calculated as a function of structural parameters (lattice dimensions, location and orientation of the molecules in the cell). Calculation of this type have been described in 1990 by Saigo and coworkers,[28] in the cases of alanine, valine, and *trans* 1,2-cyclohexanecarboxylic acid (all forming racemic compounds), and of 1-phenylethylammonium and 1-(4-isopropylphenyl)ammonium cinnamates (both salts forming conglomerates). Even though the calculations correctly predicted the racemate type in all cases, the general validity of such approach rests not only on the elaboration of a suitable methodology for the builing up of hypothetical crystal structures, but also on the availability of reliable procedures for the calculation of intermolecular potential energies in molecular crystals. In this domain much progress have still to be done[29] before crystal structures become predictable from molecular structures alone.[30,31]

4. A Prediction: Crystallization Under High Pressure Should Increase the Occurrence of Conglomerates

The observation that ≈40% of the racemic compounds do not follow Wallach's rule has an interesting physical consequence. Since in these cases the transformation of a racemic compound into its conglomerate would be attended by a reduction of the cell volume, a simple application of Le Chatelier's principle suggests that this process could take place under high pressure. What would be the order of magnitude of the pressure required to achieve this transformation?

Clapeyron equation (Figure 12) tells us that the slope of the solid liquid boundary in a (P, T) phase diagram depends on ΔH^f, the molar enthalpy of fusion (>0), and ΔV^f, the molar volume change on melting. In general, the slope of this boundary is positive because $\Delta V^f = V^l - V^s$ is positive (water is an exception), and this is why the melting point of most compounds actually increases with pressure.

For chiral molecules, the molar volumes of enantiomer and racemic liquids are virtually identical (ideal systems), and hence, for the cases which do not follow Wallach's rule, $(V^l - V^s)_a$, the molar volume increase of an enantiomer on melting, will be greater than $(V^l - V^s)_r$, the corresponding quantity for the racemic compound. The slopes of their respective solid liquid boundaries in the (P, T) diagram will therefore be different from one another (Figure 13). For the same pressure increase, the melting point of the enantiomers, and that of their conglomerate, increase more rapidly than the melting point of the racemic compound, and there exist a transition pressure P_{trans} beyond which the melting point of the latter becomes lower than that of the conglomerate, *which then becomes the stable crystal form* of the considered racemate.

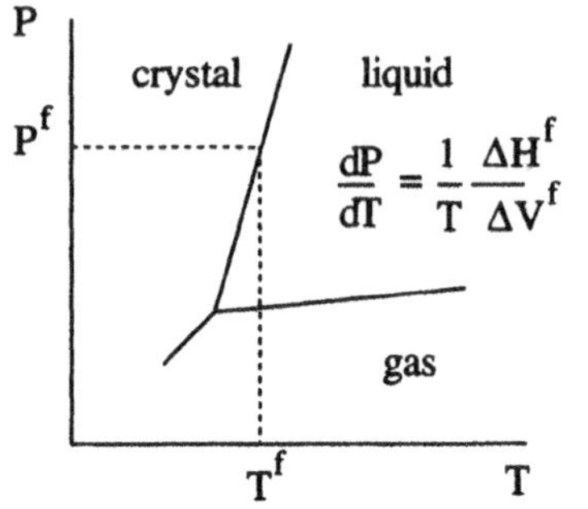

Figure 12. Clapeyron equation

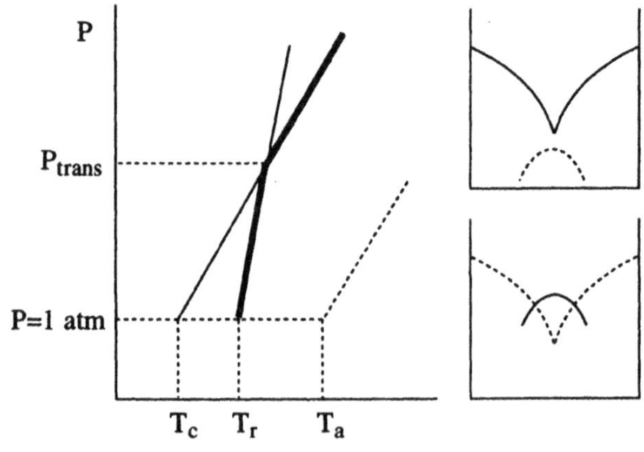

Figure 13. Pressure induced transition of a racemic compound to a conglomerate

In order to evaluate the magnitude of P_{trans}, it is necessary to incorporate into the Schröder-van Laar and Prigogine-Defay equation, which describe the liquidus curves in the binary phase diagrams of enantiomer mixtures,[1] the effect of the pressure; this can be done by means of the Clapeyron equation. Assuming that the enthalpies of fusion (ΔH_a and ΔH_r) as well as the molar volume changes (ΔV_a and ΔV_r) for the pure enantiomer and the racemic compound, respectively, are independent from T and P,[32] we thus arrive at the following modified equations, from which the binary melting points of enantiomer mixtures can be calculated as a function of the applied pressure:

$$\ln(x) = \frac{\Delta H_a}{R}\left(\frac{1}{T_{a(P_0)}} - \frac{1}{T}\right) - \frac{\Delta V_a}{RT}(P - P_0)$$

which describes the liquidus curve of the enantiomers, and

$$\ln(4x(1-x)) = \frac{2\Delta H_r}{R}\left(\frac{1}{T_{r(P_0)}} - \frac{1}{T}\right) - \frac{2\Delta V_r}{RT}(P - P_0)$$

which describes the liquidus of the racemic compound.

The phase diagrams of mandelic acid and *erythro*-phenylglyceric acid (compounds **1** and **3** above) were calculated as a function of pressure using these equations and the data of TABLE 2 (*Figure 14*). The unknown molar volume of the liquids was estimated on the assumption that the average molar increase on melting is 16%.

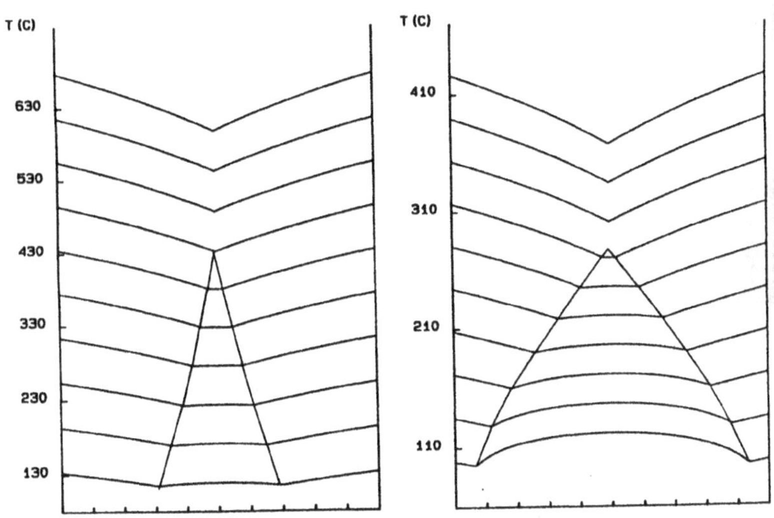

Figure 14. Melting point phase diagrams of mandelic acid **1** (left) in the range of 0-18 kbar, and of phenylglyceric acid **3** (right) in the range of 0-9 kbar (See Table 2).

Mandelic acid, for which the enantiomer crystal is 2.3% denser than the racemic compound at normal pressure, is thus predicted to become a conglomerate at ≈12 kbar, with a conglomerate melting point T_C≈500 °C. Phenylglyceric acid, with a density difference of 2.6%, is similarly predicted to be a conglomerate above 7 kbar, with T_C at the transition pressure around 300 °C.

Although these pressures are easily accessible, e.g., by means of diamond cells, the above predictions do not seem to have given rise to experimental verifications at the time (1994) this paper is written. Should these views appear to be correct, then the question of the behaviour of chiral systems under high temperature and pressure conditions such as those existing in hydrothermal systems would certainly merit study. In such conditions, the occurrence of conglomerates might become as high as 30-40%! Combination of temperature promoted racemization (like in the case of 1,1'-binaphtyl) with pressure induced spontaneous resolution could then represent a new mechanism for the generation of optical activity, that does not seem to have been considered so far.

5. References

1 Jacques J., Collet A. and Wilen S. H., (1981) *Enantiomers, Racemates, and Resolutions*, Wiley, New York.

2 Kuroda R. and Mason S. F. (1981) Crystal structures of dextrorotatory and racemic sodium ammonium tartrate, *J. Chem. Soc., Dalton Trans.*, 1268-1273.

3 Kress R. B., Duesler E. N., Etter M. C., Paul I. C., and Curtin D. A. (1980) Solid-state resolution of binaphthyl: crystal and molecular structure of the chiral (A) form and racemic (B) form and the study of the rearrangement of single crystals. Requirements for the development of hemihedral faces for enantiomer identification, *J. Am. Chem. Soc.*, **102**, 7709-7714 and refs therein.

4 Wallach, O (1895) Zur Kenntniss der Terpene und der ätherischen Oele, *Liebigs Ann. Chem.*, **286**, 90-143 (see p 140).

5 Collet, A. (1990) The homochiral versus heterochiral packing dilemma, in M. Simonyi (ed), *Problems and wonders of chiral molecules*, Akademiai Kiado, Budapest, pp 91-109.

6 Pratt Brock C., Schweizer W. B. and Dunitz J. D. (1991) On the validity of Wallach's rule: on the density and stability of racemic crystals compared with their chiral counterparts, *J. Am. Chem. Soc.*, **113**, 9811-9820.

7 A. I. Kitaigorodskii, *Organic Chemical Crystallography*, Consultant Bureau, New York, 1961, p. 230.

8 Chion B., Lajzerowicz J., Bordeaux D., Collet A. and Jacques J. (1978) Structural aspects of solid solutions of enantiomers, *J. Phys. Chem.*, **82**, 2682-2688. See also: Buda A. B., Auf der Heyde T., and Mislow K. (1992) On quantifying chirality, *Angew. Chem. Intern. Ed. Engl.*, **31**, 989-1007.

9 Stucturally related molecules of opposite configurations may also form quasi-racemates; see ref. 1, p 100, and for example, Fredga A., (1960) *Tetrahedron* , **8**, 126.

10 Garcia C. and Collet A. (1992) unpublished results; presented at the 3rd International Symposium on Chiral Discrimination, Tubingen, Germany.

11 Garcia, C. and Collet A. (1992) New optical resolution method based on the formation of a solid solution, *Tetrahedron:Asymmetry* **3**, 361.

12 Toda F. (1987) Isolation and optical resolution of materials utilizing inclusion crystallization,*Topics Curr. Chem.*, **140**, 43-69.

13 Collet A. (1995) Optical resolution in *Comprehensive supramolecular chemistry,* Vol. 10, Reinhoudt D. N. ed., Pergamon, Oxford (in press).

14 Collet A., Brienne M. J. and Jacques J. (1981) Optical resolution by direct crystallization of enantiomer mixtures, *Chem. Rev.* **80**, 215.

15 Jacques J., Leclercq M. and Brienne M. J. (1981) La formation de sels augmente-t-elle la fréquence des dédoublements spontanés? *Tetrahedron* **37**, 1727-1733.

16 Details will be given elsewhere.

17 Leclercq M., Collet A., and Jacques J. (1976) Étude des mélanges d'antipodes optique-XII. Mesure de la stabilité des racémiques vrais, *Tetrahedron* **32**, 821-828.

18 Gabard J. and Collet A. (1986) Purification d'acides partiellement dédoublés par l'intermédiaire de sels d'amines achirales. Théorie et application aux acides 2-phénoxypropioniques, *Nouv. J. Chim.* **10**, 685-690.

19 Defined as $\Delta\% = 100(d_r - d_a)/[0.5(d_r + d_a)]$ (ref 6).

20 Mason S. F. (1982) *Molecular optical activity and the chiral discriminations*, Cambridge University Press, p 171.

21 Allen F. H., Kennard O., and Taylor R. (1983) *Acc. Chem. Res.* **16**, 146-153.

22 Group I also contain achiral molecules that may form both chiral and achiral crystals. In our opinion these cases should be considered separately, because in these compounds the liquid which is in equilibrium with the solid at the melting point is a pure component, whereas in the other cases of this group the liquid always has a racemic composition.

23 Brienne M. J. and Jacques J. (1975) Remarques sur la chiralité de certains clathrates du type cage, *Tetrahedron Lett.* 2349-2352.

24 Collet A., Brienne M. J. and Jacques J. (1972) Dédoublements spontanés et conglomérats d'énantiomères, *Bull. Soc. Chim. Fr.*, 127-142; ibid., 336-342.

25 Cesario M., Guilhem J., Pascard C., Collet A. and Jacques J. (1978) Structures cristallines et dédoublements spontanés: étude d'une famille d'hydroxyacides aromatiques, *Nouv. J. Chim.* **2**, 343-349.

26 Zorkii P. M. (1968), *Soviet. Phys. Crystallogr.* **13**, 19. See ref 1 p 30.

27 Perstin A. J. and Kitaigorodskii A. I. (1987), *The atom-atom potential model*, Springer-Verlag, Berlin.

28 Kimoto H., Saigo K. and Hasegawa M. (1990) The potential energy calculation for conglomerate crystals, *Chemistry Letters*, 711-714.

29 For example, see: Gavezzoti A. (1990) Packing analysis of organic crystals containing C=O and C≡N groups, *J. Phys. Chem.* **94**, 4319-4325.

30 Gavezzoti A. (1991) Generation of possible crystal structures from the molecular structure for low polarity organic compounds, *J. Am. Chem. Soc.* **113**, 4622-4629.

31 Perlstein J. (1994) Molecular self assemblies. 2. A computational method for the prediction of the structure of one-dimensional screw, glide, and inversion molecular aggregates and implications for the packing of molecules in monolayers and crystals, *J. Am. Chem. Soc.* **116**, 455-470.

32 Strictly speaking these assumptions, and particularly that concerning the enthalpies, are not valid; however, in the present case, we may assume that the variation with pressure of the *differences* ΔH_a-ΔH_a and ΔV_a-ΔV_r are small, and hence that the intersection of the two liquidus curves as calculated by the modified equations are roughly correct.

A CARBOHYDRATE-CARBOHYDRATE INTERACTION IN WATER USING A SYNTHETIC MODEL SYSTEM

S. PENADES
Grupo de Carbohidratos
Instituto de Química Orgánica, CSIC
Juan de la Cierva 3, 28006 Madrid
Spain.

In recent years, designed model receptors have proved to be very useful in the study of non covalent interactions in water. Among the many model receptors, cyclodextrins (CDs) [1] and cyclophanes (CPs) [2] have been shown to be useful tools for the study of apolar binding of neutral organic molecules in water. We are also using model receptors with sugar incorporated to understand the molecular forces that govern the carbohydrate recognition processes.

Much is already known about the structure, interactions and functions of DNA and proteins. However, it is less clear what complex carbohydrates do. The effect of structure on properties and behavior is very keenly felt in carbohydrates: subtil changes in carbohydrate structure may dictate its properties drastically. On the other hand, a larger number of structures can be obtained from simple carbohydrates than from amino acids or nucleosides, so that the information these complex oligosaccharides carry is tremendously larger in comparison with nucleic acids and proteins.

Carbohydrates (CARB) in biological systems appear conjugated to lipids and proteins and it is as glycoconjugates that they develop their biological functions [3]. All these functions imply carbohydrate-protein interactions [4], or, as has been proposed recently, carbohydrate-carbohydrate interactions [5].

The molecular basis for these interactions is very controversial and still not well understood. From studies involving lectins and antibodies Lemieux proposed, a long time ago, lipophilic interactions and hydration effects [6] in addition to the expected hydrogen bonds (the key polar interactions) [7] as the responsible forces for the especificity of lectins for oligosaccharides. The importance of aromatic sidechain/saccharide stacking [8] as well as the role of water in stabilizing the protein-carbohydrate interactions has also been recognized.

In the case of the carbohydrate-carbohydrate interaction mediated by glycosphingolipids, Hakomori has also proposed hydrophobic interaction between complementary lipophilic surfaces [5]. This concept may be quite unexpected for

J. S. Siegel (ed.), Supramolecular Stereochemistry, 111–116.

structures so polar as carbohydrates. However, sugars are not only made up of hydroxyl groups. Even the simplest saccharide is an amphiphile with hydrophobic and hydrophilic surfaces, with its stereochemistry playing an important role in the arrangement of these surfaces. Thus, carbohydrates should be considered as amphiphilic molecules with hydrophilic and hydrophobic recognition elements to approach the understanding of the recognition of these important biological molecules.

In this paper the results obtained with a simple system constituted by a glycophane (**1**), a new type of synthetic receptor, and a series of axial and equatorial 4-nitrophenyl glycosides (Fig. 1) will be discussed and evidence of a lipophilic carbohydrate-carbohydrate interaction in water will be shown.

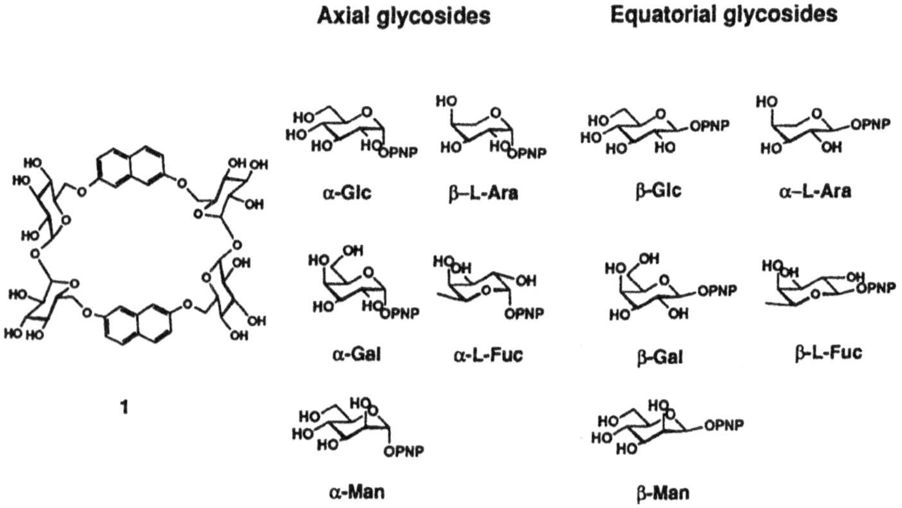

Figure 1. Glycophane **1** and the 4-nitrophenyl (PNP) glycosides.

The amphiphilic nature of glycophan **1** with a hydrophobic domain surrounded by a hydrophilic area is a good model to reproduce CARB-CARB interactions. The substrates were chosen because of their different stereochemistry and hydrophobic character. Simultanously, a comparative study has been carried out with another simple system constituted by the well known CDs and the nitrophenyl glycosides.

The questions we address with these models are essentially:

(a) whether it is possible to show CARB-CARB interaction in water between the sugar of these receptors and simple monosaccharides, even though calorimetric evidence suggests that this interaction is an unfavorable process in water [9].
(b) how the stereochemistry of the interacting sugar influences the interaction.
(c) in the case that the interaction actually takes place, whether it is possible to evaluate the individual contributions of the different intermolecular forces (H-bonds, lipophilic, solvation) to the free energy of the interaction.

The new receptors [10] may be considered as CD-CP hybrids. They are constituted by sugars and aromatic molecules to endow them with the properties of both CDs, which are chiral, neutral, water soluble and have a lipophilic cavity, and those of CPs as far as aromatic interactions are concerned.

The glycophane **1** has been prepared from α,α-trehalose (a disaccharide with C_2 symmetry) and 2,7-dihydroxynaphthalene using conventional but not trivial carbohydrate chemistry [10]. The glycophane **1** is highly symmetric and to some extent rigid around the glycosidic bonds, as would be expected for a disaccharide. In contrast, the naphthalene moieties can have three different orientations, parallel, antiparallel and a T-shape form. These conformational properties were confirmed by molecular mechanics and dynamics simulations. Both rectangular and rhomboid cavities are local minima (Fig. 2). Some of these conformations do not have an adequate cavity for a productive binding, but the rhomboid conformation (Fig 2a) with naphthalene walls c.a. 8 Å apart and the trehalose c.a. 13 Å apart is a well preorganized cavity for the sandwiching an aromatic guest in a face-to-face offset geometry. We have used the ability of the glycophane to bind electron deficient aromatic guests [10] to bring both host and guest carbohydrate moieties in close proximity.

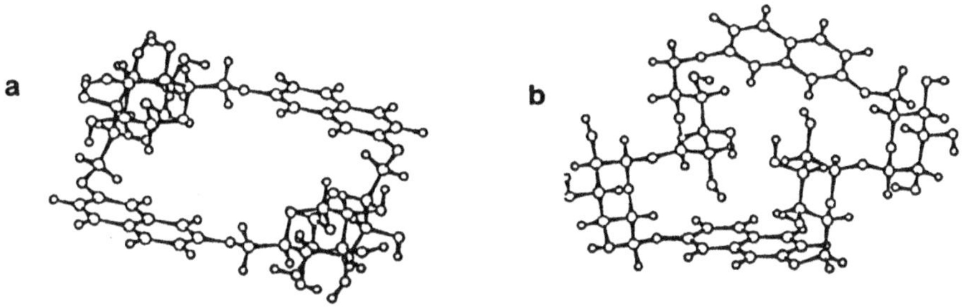

Figure 2. Conformational minima of glycophane **1**.

The stability constants (K), measured by NMR titration, indicate that the glycophane forms complexes in D_2O with all the p-nitrophenyl glycosides (Table 1). The association constants in the axial series are all higher than in the equatorial, with the highest K value corresponding to the most hydrophobic sugar, the αLFuc derivative. In all the complexes a contribution of the CARB-moiety to the binding was observed. This contribution could be estimated from the comparison of the K-values with those of the complexes formed by **1** and the p-nitrophenol (PNP) as well as 1-(p-nitrophenyl)-glycerol (PNPG). All glycosides have K-values higher than those of PNP and PNPG (Table 1), which accounts for a contribution of the CARB-moiety in the equatorial glycosides of ca. 1 $kcal.mol^{-1}$. On the axial glycosides the magnitude of the contribution is higher and changes with the stereochemistry of the sugar. The case of the αLFuc derivative being the highest with a contribution of 1.8 $kcal.mol^{-1}$ related to the PNP and 1.5 $kcal.mol^{-1}$ with respect to the glycerol derivative (PNPG).

This result indicates that the glycophane shows axial/equatorial selectivity for these glycosides with the best selectivity for the αLFuc derivative with a $K_{axial}/K_{equatorial}$ ratio of 2.4 which corresponds to a stabilization of 0.5 kcal/mol^{-1} in favor of the axial glycoside. We think that the observed stabilization and

Table 1. Association constants (K, M^{-1}) and free energy of binding (-ΔG, Kcal.mol^{-1}) of the complexes between glycophane 1 and the p-nitrophenyl glycosides, p-nitrophenol and 1-(p-nitrophenyl)glycerol in D_2O at 303 K.

axial glycosides	K	-ΔG	$\Delta\Delta G_{PNP}$ [a]	$\Delta\Delta G_{PNPG}$ [b]	equatorial glycosides	K	-ΔG	$\Delta\Delta G_{PNP}$ [a]	$\Delta\Delta G_{PNPG}$ [b]
α-Glc	111	2.84 ± 0.01	1.3	1.0	βGlc	98	2.7 ± 0.01	1.2	0.9
α-Gal	79	2.6 ± 0.06	1.1	0.8	βGal	91	2.7 ± 0.1	1.2	0.9
α-Man	162	3.0 ± 0.06	1.5	1.2	βMan	70	2.6 ± 0.06	1.0	0.7
β-L-Ara	166	3.0 ± 0.08	1.5	1.2	αLAra	81	2.6 ± 0.1	1.1	0.8
α-L-Fuc	242	3.3 ± 0.1	1.8	1.5	βLFuc	99	2.8 ± 0.1	1.2	0.9
PNP	13	1.5 ± 0.1	-	-	βXyl	96	2.7 ± 0.1	1.1	0.8
PNPG	21	1.8 ± 0.05	0.3	-	βDFuc	81	2.6 ± 0.1	1.2	0.9

a) ΔG glycosides -ΔG p-nitrophenol in Kcal.mol^{-1}
b) ΔG glycosides -ΔG 1-(p-nitrophenyl) glycerol in Kcal.mol^{-1}

selectivity is due to a specific CARB-CARB interaction between host and guest and not the result of unspecific hydrophobic contributions, with the stereochemistry of the sugar playing an important role. A comparison of this result with those obtained for the complexes of αCD with the p-nitrophenyl glycosides (Table 2) indicated that αCD did not show additional stabilization related to their complexes with PNP (pH 3 and pH 10) or PNPG.

These differences in complexing properties of **1** and the CDs may be due to the different orientations of the interacting sugar surfaces. With the αCD, the botton face of the CARB-moiety of the guest will find either the CH_2OH groups or the OH-2 group when approaching the receptor, while with the glycophane is the α- or the β-face of the glucose units of the trehalose which will be found by the botton face of the approaching guest. This different approach makes possible the van der Waals contacts between both CARB-faces in the case of the glycophane, thus favoring desolvation of both surfaces upon binding and stabilizing the complexes. Differences in the van der Waals contacts and solvation effects between the axial and equatorial glycosides may also be the origin of the observed selectivity in the glycophane **1**.

Preliminary molecular modelling studies on the p-nitrophenyl D-mannopyranosides interacting with the glycophane **1** show slight differences in the geometry of the α- and β-derivatives, specially in the orientation of the mannose moiety with respect to the trehalose molecule. In the axial (α) glycoside this orientation places the OH-2, OH-3 and OH-6 groups of the mannose moiety in close contact with the lipophilic surfaces of the trehalose molecule, while in the equatorial (β) glycoside only the OH-2 hydroxyl group approaches the trehalose moiety.

The spatial proximity between the CARB-moieties was also confirmed in the **1**.α **Man** complex by the observation of negative intermolecular nOes between the H2 and H3 protons of the mannose moiety and the H6 of the trehalose molecule. Intermolecular nOes between the aromatic protons of both host and guest were also observed.

Table 2. Association constants (K, M^{-1}) and free energy of interaction (-ΔG, $Kcal.mol^{-1}$) of the complexes between α-CD and the p-nitrophenyl glycosides, p-nitrophenol, and 1-(p-nitrophenyl) glycerol.

axial glycosides	K	-ΔG	equatorial glycosides	K	-ΔG
α Glc	129	2,9 ± 0.07[a]	β Glc	83	2.6 ± 0.1[a]
	111	2.8 ± 0.03[b]		104	2.75 ± 0.04[b]
α Gal	120	2.8 ± 0.05[a]	β Gal	127	2.9 ± 0.1[a]
	94	2.79 ± 0.04[b]		121	2.84 ± 0.04[b]
α Man	179	3.1 ± 0.1[a]	β Man	122	2.89 ± 0.05[a]
	140	2.93 ± 0.03[b]		113	2.80 ± 0.04[b]
βLAra	198	3.2 ± 0.05[a]	αLAra	111	2.83 ± 0.01[a]
	146	2.95 ± 0.03[b]		80	2.60 ± 0.04[b]
α LFuc	137	2.9 ± 0.1[a]	βLFuc	106	2.8 ± 0.1[a]
PNP (pH3)	192	3.11 ± 0.03[b]			
PNP (pH10)	1924	4.4 ± 0.05[b]			
PNPG	122	2.9 ± 0.1[a]			

a) Binding constant calculated by NMR titration in D_2O at 303 K.
b) Binding constant calculated by calorimetry in H_2O at 298K.

The picture has not been completed as of yet and more data are necessary in order to establish the correct geometry of the sugar-sugar interaction to understand the origin of the observed stereoselectivity. Nevertheless, the structural and thermodynamic results shown here unambigously, indicate the existence of a stereospecific CARB-CARB interaction in water between the glycophane and the p-nitrophenyl glycosides. Many questions remain to be answered concerning the role played by the hydroxyl groups in this interaction. Are they all involved in hydrogen bond formation or in the solvation-desolvation processes of the lipophilic surfaces upon binding?. A systematic study with a series of different substituted p-nitrophenyl glycosides will give some insight into this problem.

It would also be of special importance to have thermodynamic parameters (ΔH°, ΔS°, ΔCp°) to evaluate the role of water in these interactions. We are presently carrying out a comparative calorimetric study with the α- and β-CD complexes and those of the glycophane **1**. The first results with the complexes of αCD indicate the enthalpic driven character of the complexes and the existence of enthalphy-entropy compensation in the binding.

In conclusion, although it is usually expected from a carbohydrate to interact *via* hydrogen bonds, the glycophane **1** interacts with the amphiphilic glycosides through *non polar forces*. Furthermore, the observed differences in the selectivity and affinity of the glycophane **1** for the glycosides, although small, imply a specific hydrophobic interaction between both host and guest (the specificity depending on the sterochemistry of the sugar surfaces).

The message that follows is: whenever dealing with CARB-interactions in aqueous solutions a carbohydrate should be considered as an amphiphilic molecule as a whole, with hydrophobic (lipophilic) and hydrophilic areas contributing to the binding. The distribution of these areas depending more on the conformation of the molecule and on the orientation of the polar and non polar groups (stereochemistry) than on the number of these groups within the molecule.

Acknowledgements

The author thanks the following persons for their work, which made this article possible: J.M. Coteron, J. Jimenez-Barbero, E. Junquera, M. Martin, J.C. Morales, S. Sharma, C. Vicent (in alphabetical order). Financial support from the DGICYT (PB87-0367) is gratefully acknowledged.

References

1. Bender, M.L. and Komiyama, M. (1977) *Cyclodextrin Chemistry*, Springer-Verlag, Berlin.
2. Diederich, F. (1991), *Cyclophanes*, The Royal Society of Chemistry, Cambridge.
3. Ciba Foundation Symposium 145 (1989) *Carbohydrate Recognition in Cellular Function*, Wiley-Interscience Publication, Chichester, UK.
4. Quiocho, F.A. (1986) Carbohydrate Binding Proteins: Tertiary structures and Protein-Sugar Interactions, *Ann. Rev. Biochem.* **55**, 287-315.
5. Hakomori, S. (1991) Carbohydrate-Carbohydrate Interaction as an Initial Step in Cell Recognition, *Pure Appl. Chem.* **63**, 473-482.
6. Lemieux, R.U., Delbaere, L.T.J., Bierbeck, H. and Spohr U. (1991) Involvement of Water in Host-Guest Interactions, in *Host-Guest Molecular Interactions: From Chemistry to Biology*, Wiley Interscience Publication, Chichester, UK.
7. Lemieux, R.U., (1989) The Origin of the Specificity in the Recognition of Oligosaccharides by Proteins, *Chem. Soc. Rev.* **18**, 347-374.
8. Quiocho, F.A. (1989) Protein-Carbohydrate Interactions: Basic Molecular Features, *Pure Appl. Chem.* **61**, 1293-1306.
9. Barone, G., Castronuovo, G., Doucas, D., Ella, V. and Mattia, C.A. (1983) Excess Enthalpies of Aqueous Solutions of Monosaccharides at 298.15 K: Pentose and 2-Deoxy Sugars, *J. Phys. Chem.* **87**, 1931-1937.
10. Coterón, J.M., Vicent, C.; Bosso, C. and Penadés, S. (1993) Glycophanes, Cyclodextrin-Cyclophane Hybrid Receptors for Apolar Binding in Aqueous Solutions, *J. Am. Chem. Soc.*, **115**, 10066-10076.

Design of Host Molecules Capable of Forming Extremely Stable Host-Guest Complexes

Barbara J. Whitlock, and Howard W. Whitlock
Department of Chemistry,
University of Wisconsin,
Madison, Wisconsin 53706
USA

Construction of extremely sticky host molecules, those capable of forming extremely stable complexes with suitable guests, is an interesting and important puzzle. The more important "design principles" for this class of compound are summarized:

• Hydrophobic complexation, as illustrated by the cyclodextrins, is historically the first and "defining" force in host-guest complex stabilization.[1] Its great importance to biological systems has caused it to be the subject of intense scrutiny. While originally ascribed to entropic effects[2] current thinking is revisionist in ascribing it primarily to enthalpic features of the hydrogen bond and differential solvation of host and guest.[3] A particularly illuminating series of papers by Diederich summarizes the arguments for this viewpoint.[4] The rather amorphous nature of hydrophobic effects is accented by Dougherty's work on the importance of ion-dipole effects in several anthracenophane-based hosts.

• The hydrogen bond has been exploited to great effect in the design of hosts capable of acting primarily in aprotic organic solvents.[5]

• Aromatic-aromatic stacking has been exploited by Rebek, Hamilton and others in building nucleoside binding hosts.[6]

• The "lock and key effect", refined as Cram's preorganization concept has received increasing consideration[7] recently in the context of immunophilin binding studies.

• Edge-face aromatic-aromatic interactions have been advanced as a stabilizing force on protein conformations. In favorable cases they exercise a positive effect on complex stability.

• Ion-ion and ion-dipole effects have proven effective in cases studied by Dougherty[8] and others.

• Computational tools are increasingly important in this area with respect to both the nature of solvophobic interactions and the origins of host-guest interactions.[9]

The above principles are not mutually exclusive; for example hydrophobic interactions frequently accompany π-stacking.[6d]

J. S. Siegel (ed.), Supramolecular Stereochemistry, 117–125.

It is instructive to consider several complexation systems in which large binding constants ($K_{assoc} > 10^5\ M^{-1}$) have been observed.

• Diederich's perylene cyclophane complex 1,[10] with a $K_{assoc} > 1.6*10^7\ M^{-1}$ in water, attests to the relationship between surface area and extent of hydrophobic interactions. These workers have demonstrated the important effect of charge interactions in aqueous complexation of this tetra-positively charged host with 2,6-toluidinonaphthalene sulfonate.

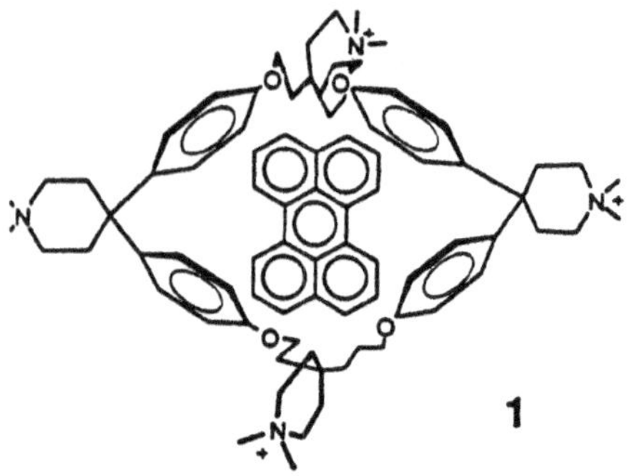

1

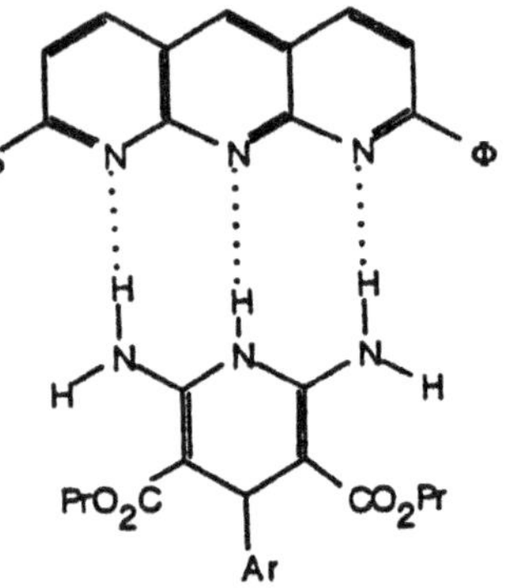

2

• Zimmerman and Murray's complex 2,[5] Kassoc > 10^7 M-1, is noteworthy for both its rational design[11] and the large binding constant for a system possessing only three hydrogen bonds.

• Hamilton's uracil binding host 3,[12] with its six hydrogen bonds, is one of the most important of the high efficiency hosts ($K_{assoc} = 1.37*10^6$ in $CDCl_3$) in that very large binding constants are attainable in nonaqueous medium. While the individual hydrogen bonds are weak (1.2 kcal/mol), they act in an additive manner. The observation that very large binding constants are attainable by additivity effects is a central and exceedingly important concept in the design of high efficiency hosts.

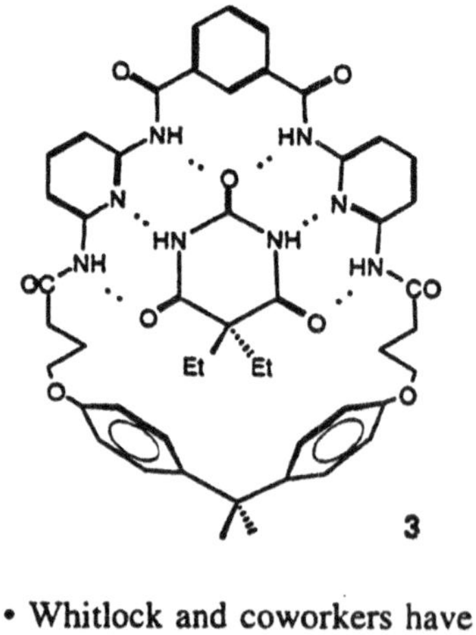

3

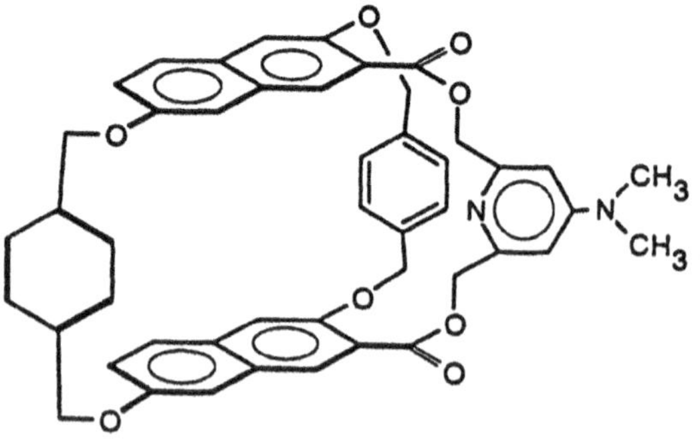

4, $K_{assoc} > 450000\ M^{-1}$

• Whitlock and coworkers have succeeded in designing host molecules such as **4** with exceedingly large association constants, despite a limited number of formal binding sites.

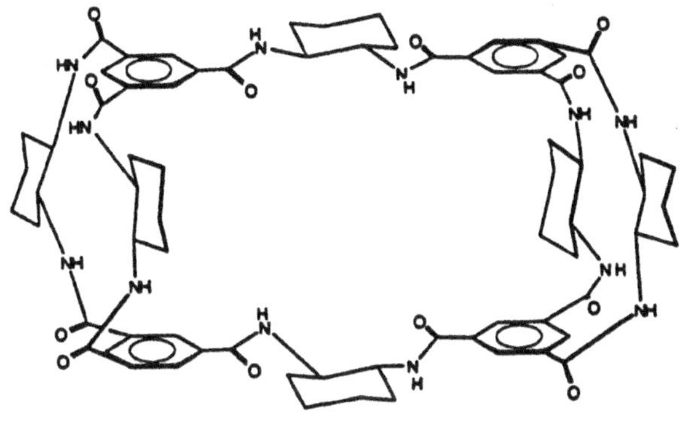

5

• The effect of additivity is clear also in the dodecaamide peptide host **5** recently described Still and coworkers. The large binding constants ($K_{assoc} > 10^6\ M^{-1}$ for tripeptides) were interpreted in terms of a model involving trans-ring hydrogen bonding by the peptide. Molecular modeling supports this idea.

We have published a series of papers devoted to the study of the above question,[13] involving structures shown in figure 1. Where are we at this time? From our work and that of others, it is clearly possible to build really sticky host molecules. It seems worthwhile at this time to inquire as to what features of a host molecule are associated with these large affinities for guests. In a statesmanlike manner we gently suggest the obvious, namely that it is either "this", "that", or a mixture of the two. The "this" is essentially entropy, while the "that" is enthalpy. At one extreme are Cram's cavitands which have moderate association constants, but exhibit bizarrely slow exchange rates as a result of the difficult passage of guests through the portals of the host. At the other extreme is Hamilton's barbiturate receptor, with six hydrogen bonds and association constants on the order of $10^6\ M^{-1}$. Clearly, there is a tradeoff between these two extremes, so we may expect that the general case of a high efficiency host will exhibit both entropic immobilization factors and enthalpically attractive sites.

The question of immobilization and cavities in the formation of stable host-guest complexes has been both puzzling and vexatious. We have flirted silently with the question of the magical nature of enzyme-substrate interactions. In its purest form this leads to the idea that the immobilization of the guest by the host in some manner conveys stability upon the host-guest complex and that this in turn is associated with the source of catalytic acceleration of reactions. However this is not supported by molecular dynamics calculations of simple host-guest models.

By itself, immobilization of the guest cannot reasonably be equated with large association constants, and in general must be related to large negative entropy changes. This general issue has been analyzed to death in the literature, leading to little that is predictively useful in design of enzyme mimics. But then why do we have the ubiquitous active site cavity-cleft-channel? Certainly it is a transcendently important feature of enzyme-substrate interaction; one never encounters enzymes with catalytic groups extending on crab-like eyestalks from the surface of the peptide ball. Molecular cavities go hand in hand with

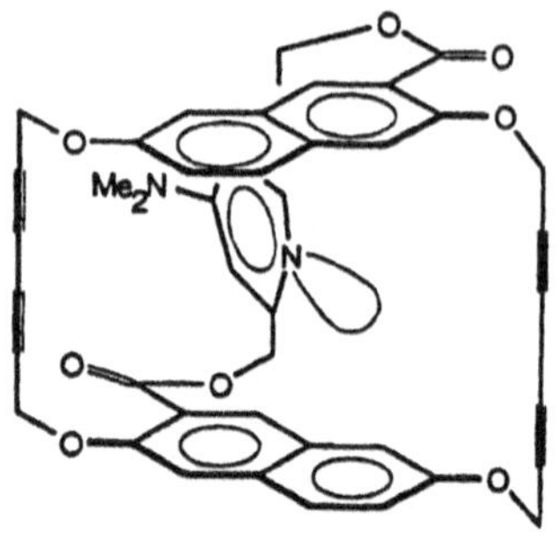

Host **H1**, 3000 M^{-1}

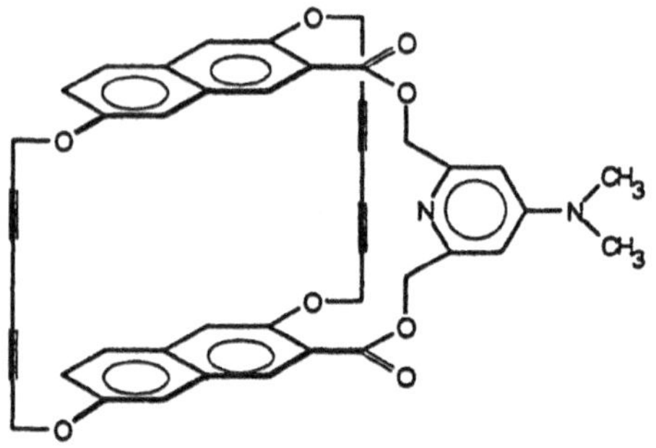

Host **H2**, 15000 M^{-1}

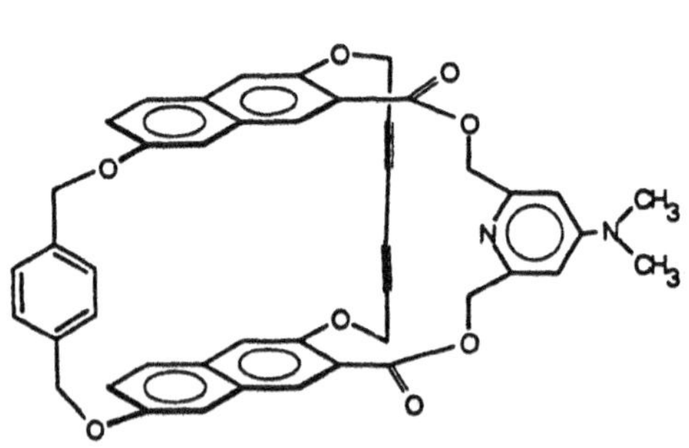

Host **H3**, 35000 M^{-1}

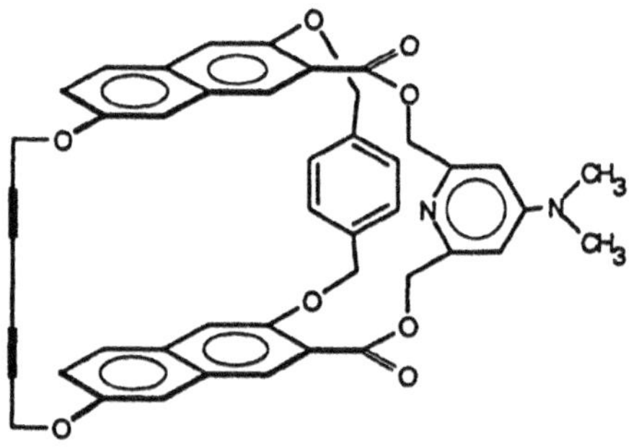

Host **H4**, ~120000 M^{-1}

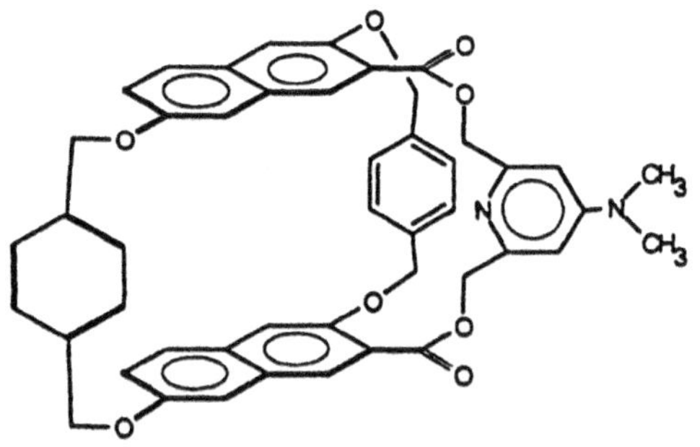

Host H5, > 450000 M^{-1}

Association constants, Guest p-Nitrophenol

Figure 1

catalytic activity. We suggest that there are two important effects of the cavity in generating large host-guest interactions in the series of host molecules which have been discussed .

The first interaction is the fruitful conformation effect arising from an exact fit of the guest in the cavity. Whether we call it the lock and key effect, conformational depopulation, effective molarity, the spatiotemporal or Thorpe-Ingold effect is immaterial; these are all manifestations of the same phenomena. What immobilization does is to hold the guest so that it has no alternative but to interact with the host's functionality. In our hosts (**1-5**, figure 1), this is a complex-stabilizing hydrogen bond accepting pyridine. It could also be a catalytic nucleophile. We personally are enamored of Menger's spatiotemporal effect because of its intuitively wonderful predictive power.

Secondly, a somewhat proletarian side effect of the close fit of host and guest is associated with immobilization of the guest by the host. If we are going to have a close fit between host cavity and guest leading to immobilization, we must also have a increased van der Waals stabilization between the two. These are small effects when taken in isolation, but they do add up. This is especially true when we consider that van der Waals effects are essentially multipliers of stability, as measured by association constants. A close-fit multiplier matters little when the nascent association is weak (10*10 = 100), but when expressed in the context of existing enthalpically stabilizing interactions it can lead to impressive boosts (10*1000 = 10000) in binding. We must not lose sight of the deep philosophical precept, presented here in the popular argot: "It's the binding constant, dummy!".

Our hosts are remarkable in exhibiting extremely large selectivity and association constants for their "natural prey p-nitrophenol, despite the absence of hydrophobic effects or multiple enthalpic sites. We are comfortable with the idea that the above two points account satisfactorily for this behavior, at least qualitatively. Every window can handle up to 1000 atoms and 1500 bonds from up to 9 different models. The Unfortunately there is a downside to this rosy picture. What we have done inadvertently is to build a perfectly normal mundane, enzyme binding site for p-nitrophenol. Unfortunately it's also unique to p-nitrophenol. We find that considerable structural modification of the cyclophane host structure is required when one changes the guest molecule. It's highly sticky and quite selective, and highly ordered. The phenolic amino acid tyrosine is bound in these systems, but weakly. The possibility of high efficiency complexation of tyrosine residues needs further work.

References

[1] (a) F. Cramer, "Einschlussverbindung", Springer-Verlag, 1954. (b) M.L. Bender and M. Komiyama, "Cyclodextrin Chemistry", Springer-Verlag, 1978.

[2] C. Tanford, "The Hydrophobic Effect: Formation of Micelles and Biological Membranes", 2nd ed., Wiley, 1980.

[3] F. Diederich, "Cyclophanes", Monographs in Supramolecular Chemistry, no 2., Chap 7., The Royal Society of Chemistry,1991.

[4] (a) Ferguson, E.M. Sanford, E.M. Seward and F. Diederich, " Cyclophane Arene Inclusion Complexation in Protic Solvents - Solvent Effects Versus Electron-Donor

Acceptor Interactions" J. Amer Chem. Soc, 1991, 13(14), 5410-5419. (b) D. B. Smithrud, T.B. Wyman and F. Diederich, " Enthalpically Driven Cyclophane Arene Inclusion Complexation - Solvent- Dependent Calorimetric Studies" J. Am. Chem. Soc, 1991,113(14), 5420-5426. (c) D. B. Smithrud, E.M. Sanford, I. Chao, S.B. Ferguson, D. R. Carcanague, J. D. Evanseck, K.N. Houk and F. Diederich,"Solvent Effects in Molecular Recognition" Pure and Applied Chemistry, 1990, 62(12), 2227-2236. (d) D.B. Smithrud and F. Diederich, "Strength of Molecular Complexation of Apolar Solutes in Water and in Organic-Solvents Is Predictable by Linear Free-Energy Relationships - A General-Model for Solvation Effects on Apolar Binding," J. Amer Chem. Soc, 1990, 112(1), 339-343. (e) F. Diederich, D.B. Smithrud, E.M. Sanford, T.B. Wyman, S.B. Ferguson, D.R. Carcanague, I.Chao and K.N. Houk, " Solvent Effects in Molecular Recognition," Acta Chem. Scand., 1992, 46(3), 205-215. (f) L. Owens C. Thilgen-C F. Diederich C.B. Knobler, "A New Helicopodand - Molecular Recognition of Dicarboxylic-Acids with High Diastereoselectivity", Helv. Chim Acta, 1993, 76, 2757-2774.

[5] (a) F.Garciatellado, S.J.Geib, S.Goswami, A.D. Hamilton, " Molecular Recognition in the Solid-State - Controlled Assembly of Hydrogen-Bonded Molecular Sheets ", J. Amer Chem. Soc., 1991, 113(24), 9265-9269. (b) S.K.Chang, D.Vanengen, E.Fan, A.Hamilton, " Hydrogen-Bonding and Molecular Recognition - Synthetic, Complexation, and Structural Studies on Barbiturate", J. Am. Chem. Soc., 1991,113(20), 7640-7645. (c) F.Garciatellado, S.Goswami, S.K.Chang, S.J.Geib, A.D.Hamilton, J.Am. Chem. Soc., 1990, 112(20), 7393-7394. (d) T.R.Kelly, G.J.Bridger, C.Zhao," Bisubstrate Reaction Templates - Examination of the Consequences of Identical Versus Different Binding-Sites", J. Am. Chem. Soc., 1990, 112(22), 8024-8034. (e) T.R.Kelly, G.J.Bridger, M.T.Bilodeau, C.Zhao, "Receptors for Uric Acids .2. A Cautionary Observation ", Tetrahedron Letters, 1989, 30(19), 2485-2489. (f) H.Vidyadhar, C.Hung, M.Puttannachetty, R.Cunningham, T.Hopfner, R.P.Thummel, "Design of Receptors for Urea Derivatives Based on the Pyrido[3,2-g] indole Subunit", J. Am. Chem. Soc, 1993, 115(3), 872-878. (g) J.C.Adrian, C.S.Wilcox, "The Chemistry of Synthetic Receptors and Functional Group Arrays .19. General Effects of Binding-Site Water Exclusion on Hydrogen-Bond Based Molecular Recognition Systems - A Closed Binding-Site Is Less Affected by Environmental-Changes Than an Open Site", J. Am. Chem. Soc, 1992, 114(4),1398-1403. (h) C.S.Wilcox, J.C.Adrian, Jr., T.H.Webb, F.J.Zawacki, "Chemistry of Synthetic Receptors and Functional-Group Arrays .18. Approaches to Quantitative Supramolecular Chemistry - Hydrogen-Bond-Based Molecular Recognition Phenomena and Sigmoidal Behavior in Multicomponent", J. Am. Chem. Soc., 1992, 114(26), 10189-10197. (i) K.Kobayashi, F.Ikeuchi, S.Inaba, Y.Aoyama, " Molecular Recognition .19. Accommodation of Polar Guests in Unimolecular Polyamine Polyhydroxy Cores - Solubilization of Sugars in Apolar Organic Media via Intramolecular Polar Microsolvation ", J. Am. Chem. Soc.,1992, 114(3), 1105-1107. (j) Tanaka-Y Aoyama-Y, "Title: Molecular Recognition .13. ^{1}H NMR Characterization of the Hydrogen-Bond Network in the Water Complex of Resorcinol-Dodecanal Cyclotetramer", Bull Chem. Soc. Jap, 1990, 63, 3343-3344. (k) Y.Tanaka, Y.Kato, Y.Aoyama," Molecular Recognition .8. 2-Point Hydrogen-Bonding Interaction - A Remarkable Chain- Length Selectivity in the Binding of Dicarboxylic-Acids with Resorcinol Aldehyde Cyclotetramer As a Multidentate Host", J. Am. Chem. Soc., 1990, 112(7), 2807-2809. (l) T.J.Murray,

S.C.Zimmerman, " New Triply Hydrogen-Bonded Complexes with Highly Variable Stabilities ", J. Am. Chem. Soc., 1992, 1 14(10), 4010-4011. (m) S. C. Zimmerman, W.M.Wu, Z.J.Zeng," Complexation of Nucleotide Bases by Molecular Tweezers with Active-Site Carboxylic-Acids - Effects of Microenvironment ", J. Am. Chem. Soc., 1991, 113(1), 196-201. (n) P.E.J.Sanderson, J.D.Kilburn, W.C.Still, " Enantioselective Complexation of Simple Amides by a C_2 Host Molecule ", J. Am. Chem. Soc., 1989,111, 8314-8315. (o) R.Liu, P.E.J.Sanderson, W.C.Still, " Enantioselective Complexation of the Alanine Dipeptide by a C-2 Host Molecule ", J. Org. Chem., 1990, 55, 5184-5186. (p) T.W.Bell, J.Liu, " Torand Synthesis by Trimerization - New Receptors for Guanidinium ", Angew. Chem., 1990,102, 931-933. (q) L.S.Flatt, V.Lynch, E.A.Anslyn, " Binding Multiple Phosphodiesters with a Polyazacleft ", Tetrahedron Letters, 1992, 33, 2785-2789. (r) E.Fan, S.A.VanArman, S.Kinkaid, A.D.Hamilton, " Molecular Recognition - Hydrogen-Bonding Receptors That Function in Highly Competitive Solvents ", J. Am. Chem. Soc., 1993, 115, 369-370. (s) S.S.Yoon, W.C.Still, " An Exceptional Synthetic Receptor for Peptides ", J. Am. Chem. Soc., 1993,115, 823-824.

[6] (a) G.Deslongchamps, A.Galan, J.Demendoza, J. Rebek, " A Synthetic Receptor for Cyclic Adenosine-Monophosphate ", Angew Chem Int Ed Engl, 1992, 31(1), 61-63. (b) A.Galan, J.Demendoza, C.Toiron, M.Bruix, G.Deslongchamps, J.Rebek, " A Synthetic Receptor for Dinucleotides ", J. Am. Chem. Soc., 1991, 113(24), 9424-9425. (c) A.Galan, D.Andreu, A.M.Echavarren, P.Prados, J.de Mendoza, " A Receptor for the Enantioselective Recognition of Phenylalanine and Tryptophan Under Neutral Conditions ", J. Am. Chem. Soc., 1992, 114(2), 1511-1512. (d) V.M.Rotello, E.A.Viani, G.Deslongchamps, B.A.Murray, J.Rebek, Jr., " Molecular Recognition in Water - New Receptors for Adenine-Derivatives ", J. Am. Chem. Soc., 1993, 115, 797-799. (e) J.Rebek, Jr., "Molecular Recognition and Model Systems", Angew Chem. Int. Ed Engl, 29, 245-255. (f) J.Rebek, Jr., D.Nemeth, P.Ballester, F-T. Lin, "Molecular Recognition Size and Shape Specificity in the Binding of Dicarboxylic acids", J. Am. Chem. Soc., 1987, 109, 3474-3475. (g) J.Rebek, Jr., B.Askew, P.Ballester, C.Buhr, S.Jones, D.Nemeth, K.Williams, " Molecular Recognition - Hydrogen-Bonding and Stacking Interactions Stabilize a Model for Nucleic-Acid Structure", J. Am. Chem. Soc., 1987, 109, 5033-5035. (h) A.D.Hamilton, D.VanEngen, " Induced Fit in Synthetic Receptors - Nucleotide Base Recognition by a Molecular Hinge ", J. Am. Chem. Soc., 1987,109, 5035-5036. (i) A.Galan, J.deMendoza, C.Toiron, M.Bruix, G.Deslongchamps. J.Rebek, Jr, " A Synthetic Receptor for Dinucleotides ", . J. Am. Chem. Soc., 1991,113, 9424-9425. (j) S.C.Zimmerman, C.M.VanZyl, " Rigid Molecular Tweezers - Synthesis, Characterization, and Complexation Chemistry of a Diacridine ", J. Am. Chem. Soc., 1987,109, 7894-7896. (k) S.Shinakai, G-X.He, T.Matsuda, A.D.Hamilton, H.S.Rosenzweig, " Specific-Inhibition of Flavin Catalyzes by a Molecular Hinge ", Tetrahedron Letters, 1989, 30, 5895-5898. (l) S.C.Zimmerman, M.Milan, M.Baloga, " Highly Efficient Complexation of a Pi-Acceptor by a Molecular Tweezer Containing Pi-Donors - The Role of Preorganization ", J. Am. Chem. Soc., 1989,1 1 1, 8528-8530. (m) D.J.Cram, H-J.Choi, J.A.Bryant, C.B.Knobler, " Host-Guest Complexation .62. Solvophobic and Entropic Driving Forces for Forming Velcraplexes, Which Are 4-Fold, Lock-Key Dimers in Organic Media ", J. Am. Chem. Soc., 1992, 114, 7748-7765. (n) M.W.Hosseini, A.J.Blacker, J-M. Lehn, "Multiple Molecular

Recognition and Catalysis: Nucleotide Binding and ATP Hydrolysis by a Receptor Molecule Bearing an Anion Binding Site, an Intercalator Group, and a Catalytic Site", J. Chem. Soc. Chem. Commun., 1988, 596.

[7] (a) E. Fischer, "Einfluss der Configuration auf die Wirkung der Enzyme", Ber. Dtsch. Chem. Ges., 1894, 27, 2985. (b) D.J. Cram, " Preorganization - From Solvents to Spherands ", Angew. Chem. Int. Ed. Engl., 1986, 25,1039-1057. (c) D.Yang, M.K.Rosen, S.L.Schreiber, "A Composite Fkbp2-FK506 Surface That Contacts Calcineurin", J. Am. Chem. Soc., 1993, 115, 819-820. (d) M.K.Rosen, D.Yang, P.K.Martin, S.L.Schreiber, "Activation of Inactive Immunophilin by Mutagenesis", J. Am. Chem. Soc., 1993, 115, 821-822. (e) W.L.Jorgensen, " Rusting of the Lock and Key Model for Protein-Ligand Binding ", Science, 1991, 254, 954-955.

[8] (a) Dougherty, D.A.Stauffer, " Acetylcholine Binding by a Synthetic Receptor - Implications for Biological Recognition ", Science, 1990, 250, 1558-1560. (b) D.A.Stauffer, R.E.Barrans, D.A.Dougherty, " Concerning the Thermodynamics of Molecular Recognition in Aqueous and Organic Media - Evidence for Significant Heat-Capacity Effects ", J. Org Chem., 1990, 55, 2762-2767. (c) H.J.Schneider, D.Guttes, U.Schneider," Host Guest Chemistry .15. Host Guest Complexes with Water-Soluble Macrocyclic Polyphenolates Including Induced Fit and Simple Elements of a Proton Pump ", J. Am. Chem. Soc, 1988,110, 6449.

[9] (a) W.L.Jorgensen, T.B.Nguyen, E.M.Sanford,I.Chao, K.N.Houk, F.Diederich, " Enhanced View of Structure and Binding for Cyclophane Arene Complexes Through Joint Theoretical and Experimental-Study ",J. Am. Chem. Soc., 1992,114(10), 4003-4004. (b) J.Pranata, W.L.Jorgensen, " Computational Studies on FK506 - Conformational Search and Molecular-Dynamics Simulation n Water ",J. Am. Chem. Soc., 1991,113(25), 9483-9493. (c) J.Pranata, S.G.Wierschke, W.L.Jorgensen, " Opls Potential Functions for Nucleotide Bases - Relative Association Constants of Hydrogen-Bonded Base-Pairs in Chloroform ", J. Am. Chem. Soc., 1991,113, 2810-2819.

[10] F. Diederich and D. Griebel, "^{1}H NMR Investigation of Host-Guest Complexation Between a Macrocyclic Host of the Cyclophane Type and Aromatic Guests in Aqueous Solutions", J. Am. Chem. Soc., 1984,106, 8037.

[11] W. L. Jorgensen and J. Pranata, " Importance of Secondary Interactions in Triply Hydrogen-Bonded Complexes - Guanine-Cytosine vs Uracil-2,6-Diaminopyridine ", J. Am. Chem. Soc., 1990,112, 2008-2010.

[12] S-K Chang and A. D. Hamilton, " Molecular Recognition of Biologically Interesting Substrates - Synthesis of an Artificial Receptor for Barbiturates Employing 6 Hydrogen-Bonds ", J. Am. Chem. Soc., 1988,110,1318.

[13] (a) C. W. Chen and H. W. Whitlock, J. Am. Chem. Soc. 1978,100, 4921. (b) E.T. Jarvi and H. W. Whitlock, J. Am. Chem. Soc. 1980,102, 657; and J. Am. Chem. Soc., 1982,104, 7196-7204. (c) S.P.Adams and H. W. Whitlock, J. Org. Chem. 1981, 46, 3474-8; and J. Am. Chem. Soc. 1982, 104,1602-1611. (d) B.J. Whitlock, E. T. Jarvi and H W.Whitlock, J. Org. Chem. 1981, 46,1832. (e) B. J. Whitlock and H.W.Whitlock, J. Am. Chem. Soc. 1983,105, 838-844. (f) B. J. Whitlock and H.W.Whitlock, J. Am. Chem. Soc. 1985,107,1325-1329. (g) B. J. Whitlock and H. W. Whitlock, Tetrahedron Lett. 1988, 6047. (h) A.B.Brown, K. J. Haller and H.W.Whitlock, Tetrahedron Lett. 1982, 23(33), 3311-3314;. (i) H.W.Whitlock and A. B. Brown, J. Am. Chem. Soc. 1986,108, 3509-3510;. (j) A. B. Brown and H. W. Whitlock, J. Am. Chem. Soc. 1989, 111, 3640-3651. (k)

H.W.Whitlock and S.P.Miller, J. Am. Chem. Soc. 1984,106,1492-1494. (l) M. E. Haeg, B. J. Whitlock and H. W. Whitlock, J. Am. Chem. Soc. 1989, 111, 692-696. (m) B.P.Friedrichsen and H. W. Whitlock, J. Am. Chem. Soc., 1989, 111, 9132-9134. (n) R.E. Sheridan and H.W.Whitlock, J. Am. Chem. Soc. 1986,108, 7210-7221;. (o) R. E. Sheridan and H. W. Whitlock, J. Am. Chem. Soc. 1988, 110, 4071-4073. (p) K. M. Neder and H. W. Whitlock, J. Am. Chem. Soc. 1990,112(25), 9412-9414. (q) B.J. Whitlock and H.W. Whitlock, J. Am. Chem. Soc. 1990, 112(11), 3910-3915. (r) J.E.Cochran, T.J.Parrott, B.J.Whitlock and H.W.Whitlock, J. Am. Chem. Soc., 1992, 114, 2669-2670. (s) B.J.Whitlock and H.W.Whitlock, J. Am.Chem.Soc., 1994, 116, 2301-2312.

STRUCTURAL CONTROL OVER THE FORMATION OF CALCIUM CARBONATE MINERAL PHASES IN BIOMINERALIZATION

L. ADDADI*, J. AIZENBERG*, S. ALBECK*, G. FALINI† AND S. WEINER*
**Dept. of Structural Biology, Weizmann Institute of Science, Rehovot, Israel*; †*Dept. of Chemistry, University of Bologna, Italy*

1. Introduction

Calcite is the most common of the $CaCO_3$ mineral phases. It has a very simple structure (Fig. 1a), that can essentially be rationalized as a deformed NaCl structure[1]. Because of this relation to the halite structure, this was one of the first structures solved by Bragg[2]. Aragonite is thermodynamically slightly less stable than calcite at ambient temperatures and pressures. Its calcium ions in the (001) calcium layer are almost in exactly the same positions as those of calcite, but it differs from calcite in terms of the carbonate ion organization (Fig. 1b). Vaterite is the least stable of these three polymorphs. It differs significantly from the other two, in that the planes of the carbonate molecules are perpendicular to the calcium ion planes (Fig. 1c).

Despite the simplicity of these structures, and the rather small differences that exist between them, most organisms that precipitate $CaCO_3$, completely control the polymorph type[3]. Furthermore, some organisms are even able to produce stable amorphous $CaCO_3$. This phase, to our knowledge, is never stable abiologically under ambient conditions.

The abilities of organisms to control these mineralization processes is well illustrated by qualitatively comparing the abundance distributions of the $CaCO_3$ mineral phases produced biologically, with those produced abiologically (Table 1). The dichotomy between these two "worlds", illustrates well the degree of control that organisms have over determining the phase to be precipitated[3].

TABLE 1. Qualitative comparison of the distributions of biologically- and abiologically-produced $CaCO_3$ mineral phases

	Biologic	Abiologic
Calcite	Very common	Very common
Aragonite	Very common	Rare
Vaterite	Rare	Very rare
Dolomite	Very rare	Common
Amorphous $CaCO_3$	Rare	Non-existent

J. S. Siegel (ed.), Supramolecular Stereochemistry, 127–139.

One of the most puzzling aspects of this subject relates to the possible functional advantages that one polymorph may afford over another. They do differ in terms of cleavage, specific gravity, solubility and their abilities to incorporate other ions such as Mg^{++} and Sr^{++} [1]. Examination, however, of the distributions of polymorphs in biological structures, cannot be usually reconciled with exploitation of these physical differences for functional purposes.

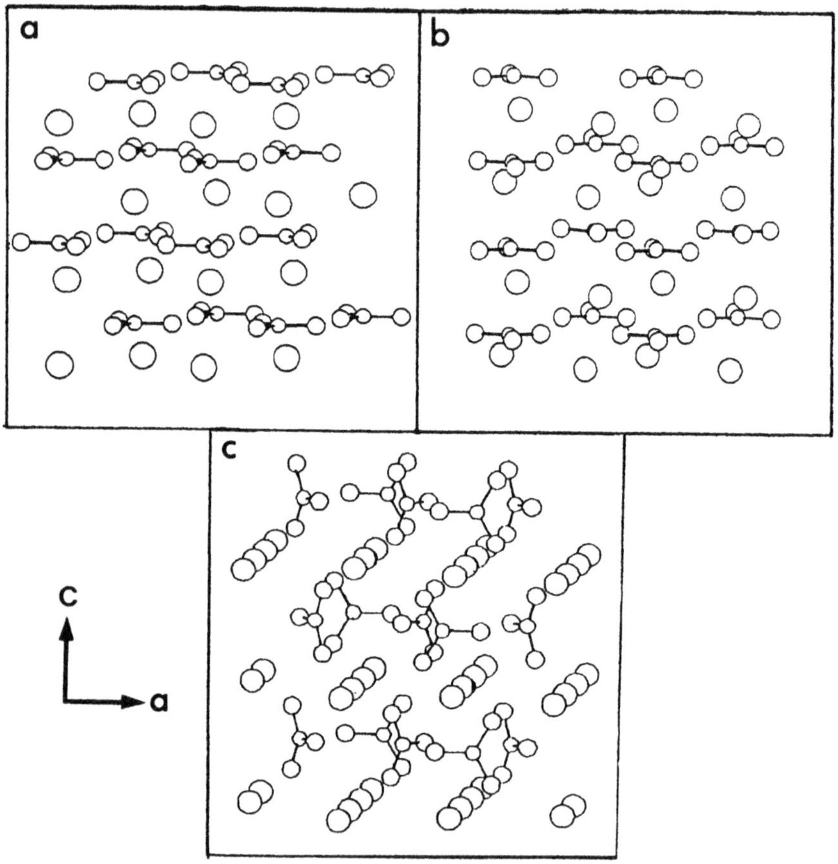

Figure 1. Packing arrangements of the three main polymorphs of calcium carbonate, viewed along a direction close to the *b** axis. The *c* and *a* axes have been tilted out of plane by 5°, to achieve a better view of the structure. a) Calcite; b) aragonite; c) vaterite.

A common, well-studied example, is the mollusk shell. They are composed of either calcite, aragonite or both. The polymorph distribution follows strict taxonomic lines, and even in cases where both polymorphs are found in a single shell, they are never mixed within the same layer. For example, many bivalve shells contain a thin layer that is laid down by the adductor muscle cells (the myostracum). It is always composed of aragonite[4], irrespective of the mineralogy of the bulk of the shell (Fig. 2a). In many other shells, aragonite is used widely to line the inner portion of the shell, and its unique uniformly thick crystal layers, which are separated by organic matrix layers, give rise to the well-known pearly luster. The outer layers of many shells are composed of crystals in the shapes of prisms. In some taxonomic groups these are composed of aragonite, whereas in others they are calcite[5]. It is obviously difficult to reconcile all these facts with any obvious functional advantage.

A variety of different organisms form vaterite, the most unstable of the crystalline polymorphs[3]. Furthermore, the biological vaterite is stable during the life-time of the organism, and in some cases, for years after extraction from the organism as well[6]. The vaterite polymorph is usually used for structural purposes. One of the most picturesque examples, is the multicrystalline spines of the ascidian, *Herdmania* (Fig. 2b), which apparently provide protection to the organism[6,7]. Vaterite is also formed in the smallest of the three otic chambers of most fish (Fig. 2c), whereas aragonite is generally formed in the other two chambers[3,8]. The otic chambers function mainly in sound reception, and it is not known why vaterite is consistently used in this one chamber.

Amorphous $CaCO_3$ is formed by several organisms in widely divergent taxa[3]. In most cases, the function is clear, and can even be rationalized. The mineral functions as a temporary storage site for calcium and carbonate, and presumably its amorphous nature allows for easy redissolution[9]. Strangely, however, amorphous $CaCO_3$ is also used by a few organisms for structural/protective purposes. Certain of the spicules formed by the ascidian *Pyura*, are composed of amorphous $CaCO_3$ (Fig. 2d). Other spicules formed by the Pyuridae are composed of vaterite, calcite, amorphous calcium phosphate, carbonate apatite (dahllite) and amorphous fluorite[7]. It is not known, what, if any, advantages each of these minerals affords these organisms.

The absence of data on $CaCO_3$ mineral phase functions, provides ample opportunity for speculation. The amorphous phase should be easy to mold into irregular shapes, it is readily redissolved and has isotropic mineral properties, which could, if hardness is not important, be well exploited. Recent unpublished observations have shown that the spicules of certain calcisponges are composed of both crystalline calcite and amorphous calcium carbonate; a situation that is technically not easy to detect. We suspect that amorphous $CaCO_3$ may well be a lot more common, and perhaps the apparent advantages that amorphous phases offer, may be more widely exploited by organisms.

Crystalline phases are more stable and harder than amorphous phases, and in this respect are better materials for structural purposes. In the case of calcite, the well developed cleavage planes are a distinct disadvantage. Shaping a crystalline phase is much more difficult because of its anisotropic nature. It can be either used "as is", or has to be actively shaped. Most organisms are clearly capable of the latter.

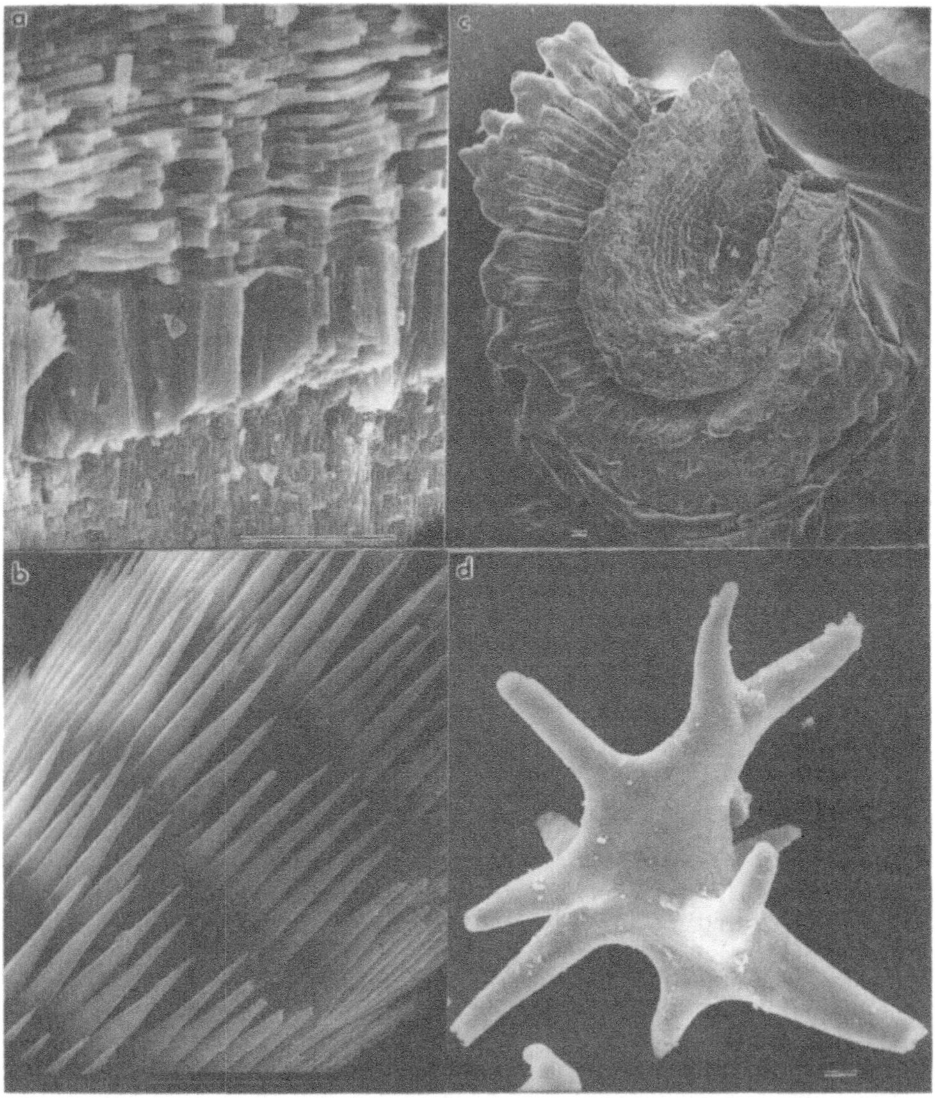

Figure 2. a) Broken section of the shell of the bivalve mollusk *Mytilus californianus*. The nacreous layer (top) is composed of aragonite, as is the myostracum layer (middle). The inner prismatic layer is composed of calcite (bottom). Scale bar=10 μm. b) Portion of the spine of the ascidian *Herdmania momus*. The mineral is vaterite. Scale bar=10 μm. c) Otolith from the small otic chamber (lagena) of the fish *Ptychecheilus lucius*. The mineral is vaterite. Scale bar=10 μm. d) Spicule from the ascidian *Pyura sacciformis*, composed of amorphous calcium carbonate. Scale bar=10 μm.

There is a great deal of descriptive evidence that supports the notion that organisms are able to control the $CaCO_3$ mineral phase precipitated, the orientation of the crystals (if a crystalline phase is opted for), the shape of the crystals and their texture. How they achieve all this is also still largely "terra incognita". Some progress has been made, however, and the evidence available to date, which forms the subject of the rest of this review, points strongly towards stereochemical control mechanisms. We address these various control "steps" in a roughly consecutive sequence.

2. Step 1: Crystalline or Amorphous

Amorphous calcium carbonate is extremely unstable under normal conditions[10]. It may form *in vitro* as a result of excessive supersaturation in the solution, but is rapidly transformed into one of the crystalline phases, following dehydration. Unlike amorphous silica, amorphous calcium carbonate is also kinetically unstable. Thus an organism, having "made the choice" of amorphous calcium carbonate as a structural material, needs to provide a permanent means to stabilize a phase that is both thermodynamically and kinetically unstable. Conceivably, the problem could be solved with the addition of additives, either other ions or macromolecules, that stabilize the amorphous phase by inhibiting the formation of nuclei of crystallization and hence preventing them from growing into stable crystals. Addition of foreign ions would presumably require a relatively high concentration, because the inhibiting action must be active at a microscopic level in the entire bulk phase. Indeed, the examples reported in the literature usually report an association between amorphous calcium carbonate and phosphate, although whether or not phosphate really functions in this manner, remains to be proved.

The antler spicules of the ascidian *Pyura pachidermatina* are composed of phosphate-containing amorphous calcium carbonate[11]. They do, however, also contain protein. The amino acid composition of the protein shows it to be glutamic acid, glycine and serine-rich in approximately the same proportions (18 mole %). We note that in crystalline calcium carbonate biomineralized tissues, the major associated acidic proteins are typically aspartic acid-rich, rather than glutamic acid-rich[12]. Crystallization *in vitro* of calcium carbonate, under conditions leading to abundant formation of calcite, was totally inhibited in the presence of the protein extracted from these spicules, and in the absence of phosphate ions (unpublished). An additional indication of the possible function of Glu, Gly, Ser-rich proteins in the stabilization of amorphous calcium carbonate is provided by the triradiate spicules of the calcareous sponge *Clathrina*. These spicules are composed of calcite and amorphous calcium carbonate in separate domains (unpublished)[13]. When the amorphous part was differentially solubilized, it was found to contain a protein component of amino acid composition very similar to the *Pyura* antler spicules. The proteins occluded in the calcitic fraction are, in contrast, Asp-rich, as is typical of many other calcitic phases. This observation is not yet mechanistically justified. We can only speculate about the possibilities of different conformations or of a greater degree of side-chain conformational freedom, as well as on the possible association of the protein with saccharides. Clearly, much more work has to be performed.

3. Step 2: Determination of the Polymorph of Choice

Calcite and aragonite have very similar thermodynamical stabilities at ambient temperatures and pressures. Vaterite is less stable, but can easily be obtained *in vitro* by very rapid crystallization resulting in the formation of spherulites. Calcite is obtained from aqueous solutions under ambient conditions where calcium is the only cation present. Aragonite can be precipitated, however, from solutions containing magnesium in a Mg^{++}/Ca^{++} ratio of at least 4:1 [14]. The role of magnesium in this case involves selectively inhibiting calcite crystallization. The selectivity is attributed to the higher density of aragonite, that prevents adsorption on the crystal surface of the hydrated Mg^{++} ions[1]. As sea water and cell cytoplasm contain high levels of Mg^{++}, a plausible mechanism for regulating aragonite/calcite polymorphism in organisms could involve the controlled introduction (or removal) of magnesium and other ions from the crystallization sites.

A recent series of experiments performed in our laboratory led, however, to a surprising discovery (unpublished). Calcium carbonate crystallizes *in vitro* in the form of calcite inside an assembly of β-chitin and silk, onto which acidic proteins extracted from the calcitic layers of the shells of *Atrina rigida* or *Mytilus californianus* were adsorbed. It crystallizes in the form of aragonite in the same assembly except in this case the adsorbed proteins were extracted from the aragonitic layer of *Mytilus californianus*. In the absence of adsorbed acidic proteins, it crystallizes in the form of vaterite spherulites deposited on the outer surface layers of the chitin matrix. In these experiments neither the chitin (from squid pen) nor the silk (from cocoons of the silkworm) were previously associated with calcium carbonate crystals. No magnesium or other ions were present in solution. The acidic proteins were used from solution, after extensive decalcification, dialysis and filtration steps, such that no nuclei of the respective crystalline phases could be present. Furthermore, the same proteins, adsorbed on a polystyrene surface, did not induce polymorph-specific crystallization. We conclude that the protein alone, present in the appropriate environment, is sufficient to determine the polymorphic form of the calcium carbonate of choice. We note that the protein activity must involve an active induction of the required phase, rather than inhibition of the undesired ones, because in the absence of protein the less stable polymorph is formed, and then only on the outer surface of the chitin layer.

Once more, these preliminary results are at the observational level, and are still far from having a mechanistic explanation. They are particularly puzzling, because the calcite and aragonite structures are so similar. Active nucleation must involve an impressive level of stereochemical match between the protein surface and the nucleus surface. It cannot involve the common calcium (001) layer, because this is practically identical in the two structures. It is easier to conceive of nucleation from a different crystallographic plane, where the differences between the two structures are more pronounced. Alternatively, a three-dimensional nucleation site could be invoked, with protein moieties fixing carbonate ions in place, in addition to calcium. Only when the structures of the nucleating proteins are known, shall we be able to propose a mechanism.

4. Step 3: Oriented Nucleation

In most cases of biological calcium carbonate formation, crystallization is induced from a specific crystallographic plane. In many aragonitic and calcitic shell layers of bivalve mollusk shells, all the crystals are oriented such that their *c* axes are parallel to each other, and perpendicular to the shell matrix[3]. In all the different types of calcareous sponge spicules that we have examined, in larval spicules and adult sea urchin spines, and in many other examples of calcite single crystal spicules, the orientation of the crystallographic axes is fixed relative to the skeletal element's morphology, even though this can vary widely from element to element. These phenomena can most easily be explained by assuming oriented nucleation of the crystal from a defined site.

The idea of oriented nucleation is logically consistent with proteins that actively induce crystallization by virtue of a match to the structure of the crystal on a certain plane. This hypothesis is supported by some observations of a spatial relation existing between matrix and mineral *in vivo*[15]. It is also supported by experimental *in vitro* evidence from different systems. Oriented nucleation of calcite from the (001) plane was demonstrated on acidic proteins extracted from the nacreous or prismatic layers of bivalve mollusk shells, adsorbed on polystyrene dishes[16]. The same effect was induced when polyaspartate was adsorbed on sulphonated polystyrene surfaces, suggesting a cooperative activity of moieties concentrating cations, and organized motifs inducing a specific two-dimensional organization[17]. Calcite was also crystallized from monolayers of long-chain fatty acids at the air-water interface[18]. Even more surprising is the oriented nucleation of ice crystals from monolayers of long-chain alcohols, where it was found that a small difference in the tilt angle of the chain to the water surface is sufficient to influence the resulting nucleation effect[19]. The tilt angle in turn is determined by the number of carbon atoms in the chain. It is thus evident that small differences in the stereochemistry of the nucleating layer can result in very substantial differences in the nucleated crystal. Proteins, with their wide variability in side chain group distributions and conformations, can be the source of an infinite number of possible structural motifs, each conceivably matched to a different crystal surface.

5. Step 4: Modulating Crystal Growth

While nucleation involves a single intervention event, modulation of crystal shape requires continuous intervention during growth, when a specific morphology, different from the one normally expressed by the crystal, is produced. The same hypothetical match between protein and crystal surfaces may, however, be invoked in modulation of crystal growth, as in nucleation. The difference between the two processes is a reversal in the physical state of the two interacting partners. For oriented crystal nucleation, the protein needs to be adsorbed and rigidized on a solid surface, and the prenucleation crystal aggregates form from solution. For growth modulation to occur, the protein is adsorbed from solution onto specific surfaces of a growing crystal. As a consequence of adsorption, the rate of growth of the crystal decreases in some directions relative to others, resulting in modified crystal morphologies. The adsorbed protein is consequently occluded inside the growing crystal by overgrowth.

The presence of intracrystalline proteins has been extensively documented, especially in biogenic single crystals of calcite[20,21]. Each of these skeletal elements has a specific morphology, that is determined both by the space in which it is grown, as well as by specialized proteins secreted during crystal growth. An example that we have recently studied involved the calcitic prisms of the external layer of *Atrina rigida* bivalve shells[22]. Various intracrystalline proteins were extracted and separated after dissolution of cleaned dispersed prisms. When calcite was crystallized from solutions containing the protein mixture, non-specific inhibition of crystal growth ensued. Two protein fractions were, however, separated and tested individually. The major protein component has an amino acid composition consisting of more than 50 mole % aspartic acid.

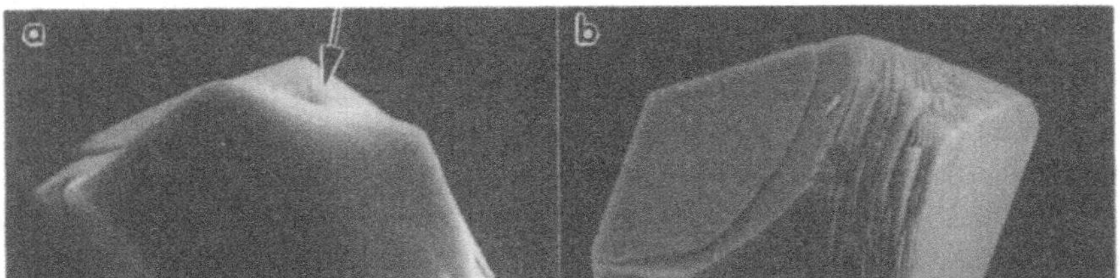

Figure 3. Crystals of calcite grown *in vitro* in the presence of two intracrystalline protein fractions isolated from the prismatic layer of the bivalve *Atrina rigida*. a) Calcite crystal grown in the presence of the abundant aspartic acid-rich fraction. The (001) face (arrow) is developed. b) Calcite crystal grown in the presence of the minor, less acidic fraction, associated with saccharide. The rough faces are indexed as $\{01\ell\}$.

When present in solutions where calcite is crystallized, it adsorbed selectively on the calcite (001) plane, resulting in crystals displaying well formed (001) faces perpendicular to the *c* axis (Fig. 3a, Fig. 4). In contrast, a minor protein component, poorer in aspartic acid and associated with saccharide, induces the formation of large {01ℓ} faces, slightly inclined relative to the *c* axis (Fig. 3b, Fig. 4). Clearly, the biogenic crystals are formed in the presence of both macromolecular types, whose introduction into the crystal growth environment and subsequent effect on crystal growth are accurately orchestrated by the cells. Again, a high degree of stereochemical recognition must be involved. The isolation from the same crystal of two protein-containing fractions having different compositions, demonstrates just how much additional information on the interactions between crystal and macromolecule at the molecular level may be obtained in this way.

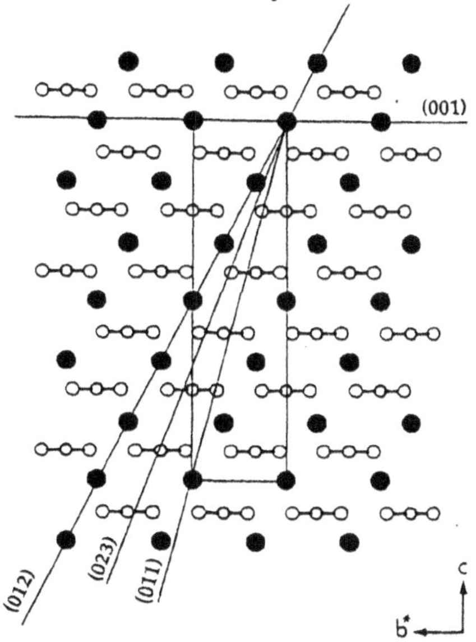

Figure 4. Packing arrangement of calcite viewed in the *cb** plane. The (001), (023), (011) and (012) planes are all viewed edge-on. The faces indexed as {01ℓ} are between (011) and (023).

The presence of the macromolecules inside the biogenic crystals alters crystal texture. This can be monitored by high resolution X-ray diffraction experiments[20,23]. The macromolecule, being much larger than a crystal unit cell, leaves a permanent imprint in the crystal bulk, in the form of a dislocation, or domain boundary. The average shape, size and degree of alignment of the perfect microscopic domains composing each single crystal can be mapped measuring the appropriate parameters from the width of diffraction peak line shapes[24]. Each diffraction peak, corresponding to a particular direction inside the crystal, can be monitored separately. The reconstructed domain shape can then be compared and correlated to the existing information on the protein effect, deduced from *in vitro* experiments[25].

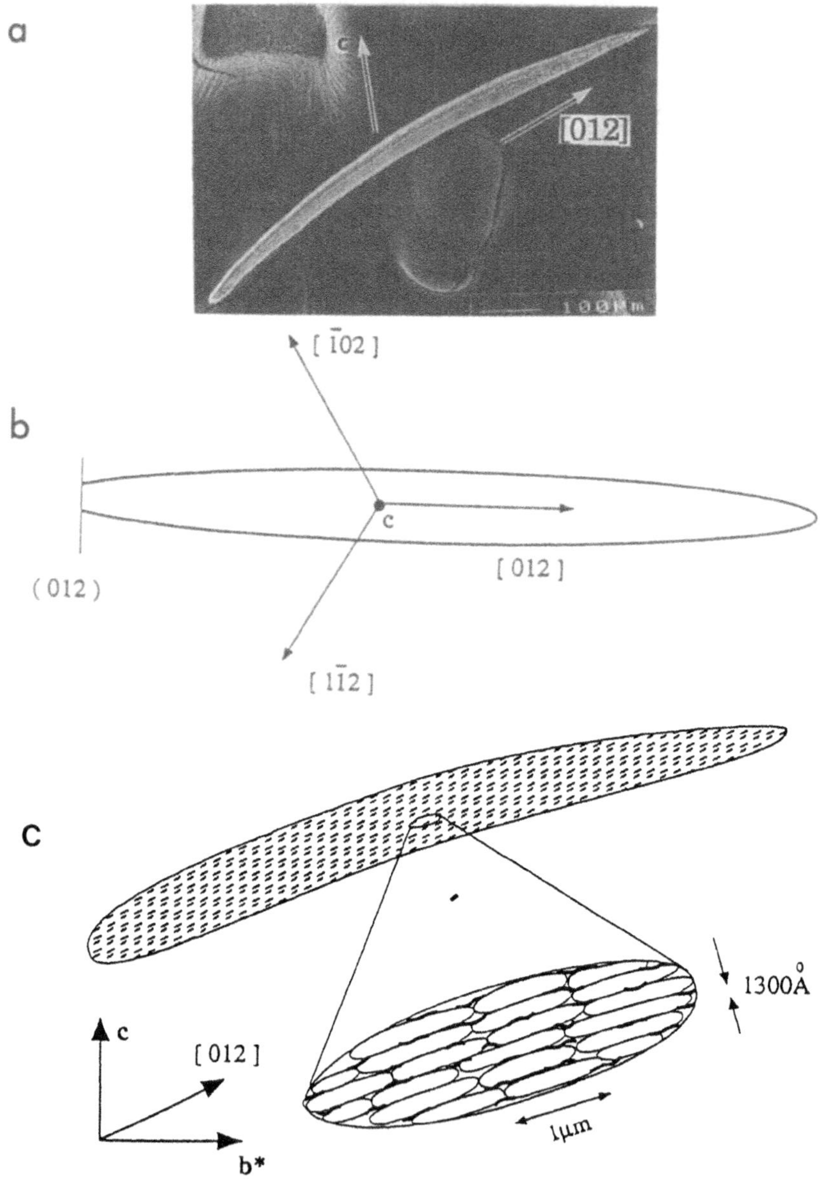

Figure 5. a) Curved monaxon from the calcareous sponge *Sycon sp.*
b) Schematic representation of the orientation of the vectors of the symmetry-related {012} planes relative to the spicule morphology. The family {012} includes the identical planes (012), ($\bar{1}$02) and (1$\bar{1}$2). Their vector directions are shown. c) Schematic representation of the microscopic domain structure of the spicule. The adsorbed proteins (~) located along domain boundaries, are elongated in the direction of the spine axis.

The curved monaxons from the calcareous sponge *Sycon* were recently investigated, and provide an astounding example of control of organisms over crystal growth (Fig. 5a). The spicule is composed of one single crystal of calcite[26]. The 'arc-shaped' spicule morphology does not relate in any way to the crystal symmetry. The hexagonal *c* axis forms an angle of approximately 20° to the radius of the arc, and its orientation is fixed. The predominant elongation direction of the spicule corresponds to [012] (Fig. 5b). The shape of the perfect microscopic domains, reconstructed as described above, reflects, to a remarkable degree, the macroscopic shape of the spicule (Fig. 5c). This implies that proteins are introduced within the growing crystal at the desired locations irrespective of the crystallographic direction. In particular, the organism is capable of differentially affecting symmetry-related directions. For example, there are three crystallographic planes of the family {012} that are structurally identical to each other (Fig. 5b). When a crystal is grown in an isotropic environment, the rate and mode of its growth along the corresponding directions will inevitably be identical. The organism's ability to differentiate between those directions leads to effective reduction in the morphological and textural symmetry of the crystal, although the structure of the bulk crystal still preserves its intrinsic symmetry.

There are at least two scenarios that we can envisage, that can give rise to the observed reduction in symmetry. Both involve oriented nucleation of the crystal from a 'general' plane (different from {001} or {100}). Oriented nucleation is required because the crystal axes are uniquely oriented relative to the spicule shape. We speculate that the nucleation plane could be (012), that is perpendicular to the spicule's morphological axis (Fig. 5b). This plane is composed of layers of Ca^{++} juxtaposed to layers of carbonate ions inclined to the layer in a unique geometry (Fig. 4). Once oriented nucleation occurs, restricted growth can be achieved by the secretion of proteins into the space where the crystal grows, that selectively recognize the {012} and {001} planes. As the main [012] direction is blocked after nucleation, the resulting protein interference with growth would not affect it, resulting in spicule elongation in this direction. Alternatively, the proteins could be targeted to the growth sites by some directed secretion mechanism that is totally independent of crystal structure. Irrespective of the mechanism, these observations show that organisms can, once a nucleation mode is established, manipulate crystal growth "at will", follow and exploit the intrinsic properties of the crystal if appropriate, or overrule them completely if they do not fit the final required shape.

6. Conclusions

Organisms can either induce an amorphous $CaCO_3$ phase to form, or allow crystallization to proceed. If the latter is the chosen path, then the polymorphic form is dictated apparently by the nucleation conditions. The nucleated crystals can be oriented, and modulation during growth can determine both shape and material properties. The latter are achieved by the accurate orchestration of space delineation, solution chemistry and the introduction of specialized macromolecules. It appears that the intervention mode of choice is to always actively direct the process, rather than indirectly inhibiting other undesired processes. Although we know very little about the manner in which these active interventions occur, we have to assume that they involve most exquisite stereochemical recognition at the molecular and supramolecular level.

7. Acknowledgments

We dedicate this paper to the memory of Heinz A. Lowenstam. The micrographs in figure 2 were obtained from his library. Supported in part by a grant from the US-Israel Binational Science Foundation. S.W. is the incumbent of the I.W. Abel Professorial Chair of Structural Biology, and L.A. is the incumbent of the Patrick A. Gorman Professorial Chair of Biological Ultrastructure.

8. References

1. Lippman, F. (1973) *Sedimentary Carbonate Minerals*, Springer Verlag, Berlin.
2. Bragg, W.L. (1914) The analysis of crystals by the X-ray spectrometer, Proc. Roy. Soc. London, Ser. A **89**, 468-489.
3. Lowenstam, H.A. and Weiner, S. (1989) *On Biomineralization*, Oxford University Press, New York.
4. Kennedy, W.J., Taylor, J.D. and Hall, A. (1969) Environmental and biological controls on bivalve shell mineralogy, *Biol. Rev.* **44**, 499-530.
5. Carter, J.G. (1990) Shell microstructural data for the bivalvia, J.G. Carter (ed), *Skeletal biomineralization: Patterns, Processes and Evolutionary Trends*, Vol. 1, Van Rostrand Reinhold, New York, pp. 297-412.
6. Lowenstam, H.A. and Abbott, D.P. (1975) Vaterite: a mineralization product of the hard tissues of a marine organism (Ascidiacea), *Science* **188**, 363-365.
7. Lowenstam, H.A. (1989) Spicular morphology and mineralogy in some Pyuridae (Ascidiacea), *Bull. Mar. Sci.* **45**, 243-252.
8. Lowenstam, H.A. and Fitch, J.E. (1978) Morphology, mineral diversity and ontogenetic minral charges in extant elasmobranch otoconia, *Geol. Soc. Am. Abstr.* **10**, 114.
9. Taylor, M.G., Simkiss, K., Greaves, G.N., Okazaki, M. and Mann, S. (1993) An X-ray absorption study of the structure and transformation of amorphous calcium carbonate from plant cystoliths, *Proc. R. Soc. Lond.* B **252**, 75-80.
10. Clarkson, J.R., Price, T.J. and Adams, C.J. (1992) Role of metastable phases in the spontaneous precipitation of calcium carbonate, *J. Chem. Soc. Faraday Trans.* **88**, 243-249.
11. Lambert, G. Lambert, C.C. and Lowenstam, H.A. (1990) Protochordate biomineralization, J.G. Carter (ed), *Skeletal Biomineralization: Patterns, Processes and Evolutinary Trends*, Vol. 1, Van Rostrand Reinhold, New York, pp. 461-469.
12. Weiner, S. and Addadi, L. (1991) Acidic macromolecules of mineralized tissues: the controllers of crystal formation, *Trends in Biochem. Sci.* **16**, 252-256.
13. Minchin, E.A. (1908) Materials for a monograph of the *Ascons*. I. On the origin and growth of the triradiate and quadriradiate spicules in the family, *Clathrinidae Quart. J. Micr. Sci.* **40**, 469-587.
14. Kitano, Y. (1962) The behavior of various inorganic ions in the separation of calcium carbonate from a bi carbonate solution, *Bull. Chem. Soc. Japan* **35**, 1973-1980.

15. Weiner, S. and Traub, W. (1984) Macromolecules in mollusk shells and their functions in biomineralization, *Phil. Trans. R. Soc. Lond. B*, **304**, 421-438.
16. Addadi, L. and Weiner, S. (1985) Interactions between acidic proteins and crystals: stereochemical requirements in biomineralization, *Proc. Natl. Acad. Sci. U.S.A.* **82**, 4110-4114.
17. Addadi, L., Moradian, J., Shay, E., Maroudas, N.G. and Weiner, S. (1987) A chemical model for the cooperation of sulfates and carboxylates in calcite crystal nucleation. Relevance to biomineralization, *Proc. Natl. Acad. Sci. U.S.A.* **84**, 2732-2736.
18. Mann, S., Heywood, B.R., Rajam, S. and Walker, J.B.A. (1991) Structural and stereochemical relationships between Langmuir monolayers and calcium carbonate nucleation, *J. Phys. D: Appl. Phys.* **24**, 154-164.
19. Popovitz-Biro, R., Wang, J.L., Majewski, J., Shavit, E., Leiserowitz, L. and Lahav, M. (1994) Induced freezing of supercooled water into ice by self-assembled crystalline monolayers of amphiphilic alcohols at the air-water interface, *J. Am. Chem. Soc.* **116**, 1179-1191.
20. Berman, A., Hanson, J., Leiserowitz, L., Koetzle, T.F., Weiner. S. and Addadi, L. (1993) Biological control of crystal texture: a widespread strategy for adapting crystal properties to function, *Science* **259**, 776-779.
21. Curry, G.B. Cusack, M., Walton, D., Endo, K., Clegg, H., Abbott, G. and Armstrong. H. (1991) Biochemistry of brachiopod intracrystalline molecules, *Phil. Trans. R. Soc. Lond.* B **333**, 359-366.
22. Albeck, S., Aizenberg, J. Addadi, L. and Weiner, S. (1993) Interactions of various skeletal intracrystalline components with calcite crystals, *J. Am. Chem. Soc.* **115**, 11691-11697.
23. Berman, A., Hanson, J., Leiserowitz, L., Koetzle, T.F., Weiner, S. and Addadi, L. (1993) Crystal-protein interactions: controlled anisotropic changes in crystal microtexture, *J. Phys. Chem.* **97**, 5162-5170.
24. Aizenberg, J., Hanson, J., Leiserowitz, L., Koetzle, T.F., Addadi. L. and Weiner, S. (1994) Morphogenesis of calcitic sponge spicules: a role for specialized proteins interactiong with growing crystals, *FASEB J.*, submitted for publication.
25. Addadi, L., Aizenberg, J., Albeck, S., Berman, A., Leiserowitz, L. and Weiner, S. (1994) Controlled occlusion of proteins: a tool for modulating the properties of skeletal elements, *Mol. Cryst. Liquid Cryst.* **248**, 185-198.
26. Jones, W.C. (1970) The composition, development, form and orientation of calcareous sponge spicules, *Symp. Zool. Soc. Lond.* **25**, 91-123.

HYDROGEN BONDING CONTROL OF MOLECULAR SELF-ASSEMBLY

JI YANG, JEAN-LUC MARENDEZ, ABDULLAH ZAFAR, STEVEN J. GEIB AND ANDREW D. HAMILTON
Department of Chemistry, University of Pittsburgh, Pittsburgh, PA, 15260, USA

1. Introduction

Molecular organization is at the center of all biological processes. The non-covalent interaction of two or more molecular subunits can lead to the formation of large, well-defined and functional molecular aggregates. In Nature this strategy of self-assembly has several key advantages[1].

a) The use of a few repeating subunits to create large structures reduces the amount of genetic information needed, compared to a single biopolymer chain.

b) The weak aggregation of multiple substrates means that association and dissociation can be fast under different conditions leading to a greater control and potential switching of structure.

c) The strict structural requirements imposed by the use of many small units on the final structure means that any errors in subunits will be rejected as incompatible and so lead to an inbuilt error checking system.

The enormous complexity inherent and necessary in biological structures cannot be reached by individual biosynthetically derived units but requires the association of many components. The size, shape and makeup of the resulting aggregates will be controlled by the nature and orientation of the binding groups on the surface of the subunits. The arrangement of these binding regions represents chemical information that is imposed at the translation and folding stages but that is nonetheless sufficient to direct the non-covalent construction of the aggregate[2,3].

2. Synthetic Approaches to Self-Assembling Molecules

Our interest lies in the design and construction of synthetic molecules that self organize with some of the features seen in the biological structures. Such study should give insights into the mechanisms of self-assembly, into the control of the size and number of an aggregate, into the balance of enthalpic and entropic factors and how they influence the stability of an aggregated structure, and into the role of the unfavorable entropy term when multiple subunits form a well-defined aggregate[4].

There have been few examples of designed synthetic molecules that self-organize into larger aggregates with defined structure and function[1]. Whitesides[5]

J. S. Siegel (ed.), Supramolecular Stereochemistry, 141–146.

has elegantly exploited the melamine-cyanuric acid lattice to form a number of well defined aggregates. Zimmerman has also developed cyclic aggregates as in **1**, based on multiple hydrogen bonding subunits[6]. In a related approach, Wuest has demonstrated that bis-pyridones can be made to change their aggregation pattern by varying the linking position[7]. Our own work has demonstrated both in solution and in the solid state, the formation of a cyclic 2+2 complex[8]. These are all relatively compact structures with no central cavity that might permit the binding of ions or the complexation of substrates. Such features are essential for the next generation of self-organizing structures that will allow more detailed comparison to biological systems and also imposition of switch or template functions. In an elegant recent study Rebek[9] has shown that carefully designed tetra-lactam derivatives can be induced to dimerize into a tennis ball-like structure, as in **2**. Lehn [10] has also formed 2+2 aggregates based on porphyrin derivatives.

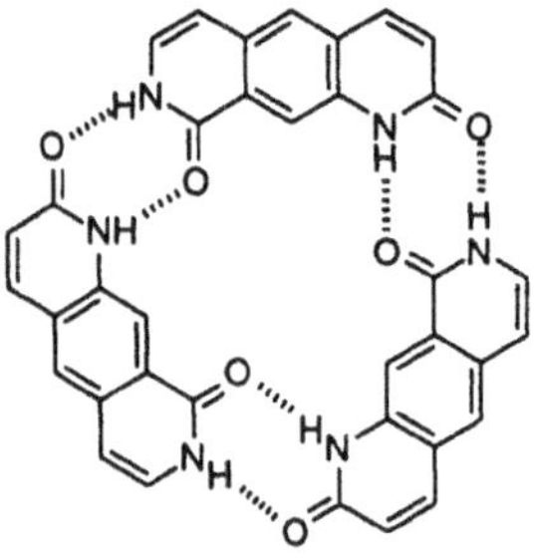

1

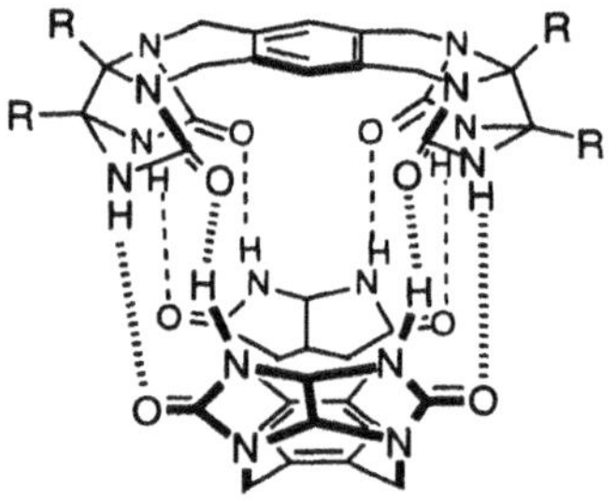

2

3. Isophthalic Acid Approach to Self-Assembly

Some insight into the design of self-organizing molecules can come from consideration of the crystal structure packing patterns of hydrogen bonding molecules[11]. For example, the X-ray structure of trimesic acid shows a infinite sheet arrangement in which each molecule forms six hydrogen bonds to three neighboring residues (Figure 1)[12]. The resulting hexagonal-lattice structure leads to large cavities each bounded by six trimesic acid molecules. Isophthalic acid derivatives would be expected to form only partial elements of the sheet structure, such as the linear or ring motifs indicated in Figure 1. Indeed, isophthalic acid itself forms the ribbon motif with a planar arrangement of phenyl and carboxylic acid groups (Figure 2). The ribbon motif allows both optimal formation of the bidentate hydrogen bonds and efficient packing interactions between the ribbons in vertical and horizontal directions.

In order to direct formation of the cyclic motif in the solid state it is necessary to disrupt the linear packing arrangement. We reasoned that this might be achieved by placing a bulky substituent in the 5-position of isophthalic acid, thus preventing easy alignment of the hydrogen bonded ribbons. The X-ray structure of 5-decyloxyisophthalic acid **3** (crystallized from THF/hexane; Figure 2a) shows the formation of a remarkable hexameric aggregate corresponding to the ring motif in Figure 1a. The six isophthalic acid molecules define a macrocyclic cavity that is 14 Å in diameter (from opposite isophthaloyl-2H sites) and is stabilized by 12 hydrogen

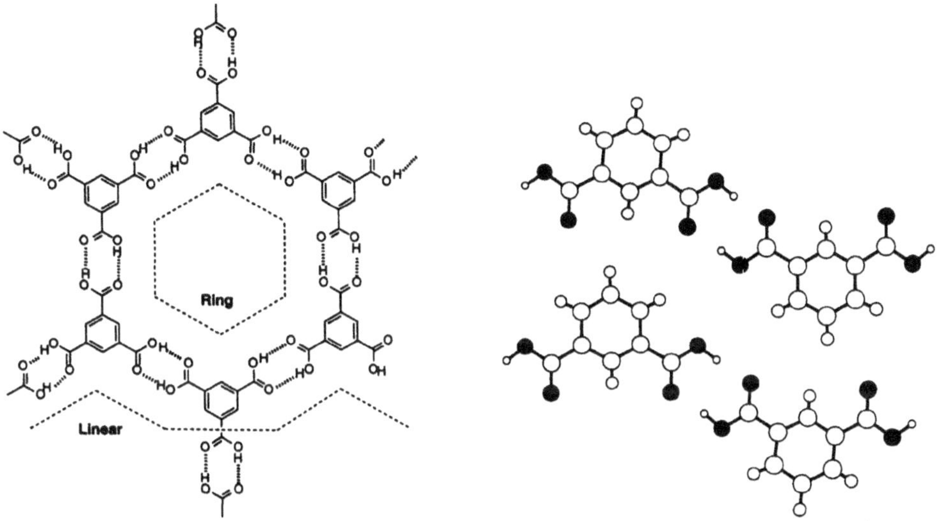

Figure 1. X-ray structure of a) trimesic acid b) isophthalic acid (linear aggregate).

bonds (O-O distance, 2.65 Å), as in Figure 2b. The isophthalate core of the structure is planar (4° angle between phenyl and carboxylate planes) while the decyloxy chains maintain an alternating up-down arrangement.

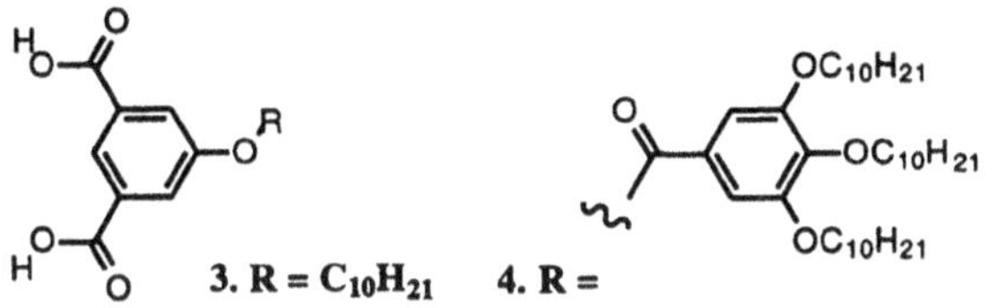

3. R = $C_{10}H_{21}$ **4. R =**

A critical question is whether these cyclic aggregates form in solution and what would be the relative stability of the linear vs. cyclic motifs. Despite the absence of crystal packing effects, the cyclic aggregate should still be favored entropically in solution since it allows the formation of 12 hydrogen bonds by six particles rather than the seven that would be necessary in a linear aggregate. We have investigated the solution aggregation properties of a more soluble 5-substituted isophthalic acid derivative **4** using vapor phase osmometry and ^1H NMR. VPO measurements were carried out in toluene at 40°C and gave molecular weights for the aggregate of 4600 (against a benzil standard) and 4900 (against a polystyrene standard) over a concentration range of 12-35 mM. The calculated MW for hexameric **4** is 4531.

Diacid **4** is very soluble in benzene and toluene (up to 100 mM). The ^1H NMR spectrum of **4** in C_6D_6 (25 mM) at 55°C shows three sharp resonances at 8.89, 8.21 and 7.51 ppm from the isophthaloyl-2H and -4/6H and benzoyl-H, respectively. The position of the isophthaloyl-2H resonance is strongly dependent on concentration,

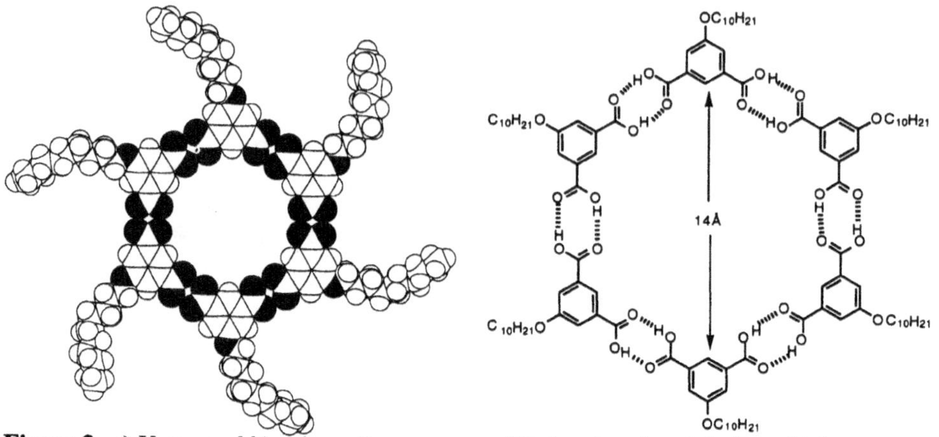

Figure 2. a) X-ray and b) schematic structure of 5-decyloxyisophthalic acid **3**

shifting upfield by 0.26 ppm over a 15-0.1 mM range (Figure 3). Above 15 mM, however, there is little change, reflecting the formation of a discrete and stable aggregate at higher concentrations.

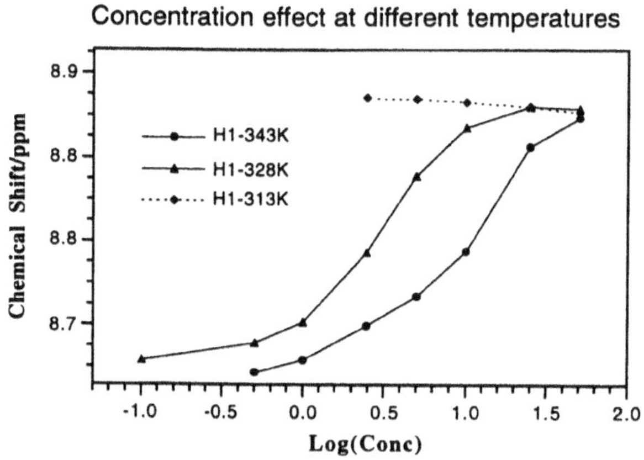

Figure 3. Plot of δ vs. log[**4**] at different temperatures.

4. Future Directions

The above results point to a clear dissociation of the hexameric complex, particularly at low concentrations. This instability is due in large part to the significant unfavorable entropy involved in bringing six independent particles together in a complex. We are currently investigating approaches to increase the stability of the aggregate. In theory this problem can be tackled in two ways; a) by increasing the enthalpy of aggregation to overcome the unfavorable entropy or b) by reducing the unfavorable entropy. The enthalpy of association can be enhanced by increasing the strengths of the hydrogen bonding interactions involved. We have previously shown that ureas and thioureas bind strongly to carboxylates in polar organic solvents[14]. Incorporating urea and carboxylate groups into the same molecular component should lead to a highly stable aggregate of the type **5**. A second approach would be to incorporate into the aggregate

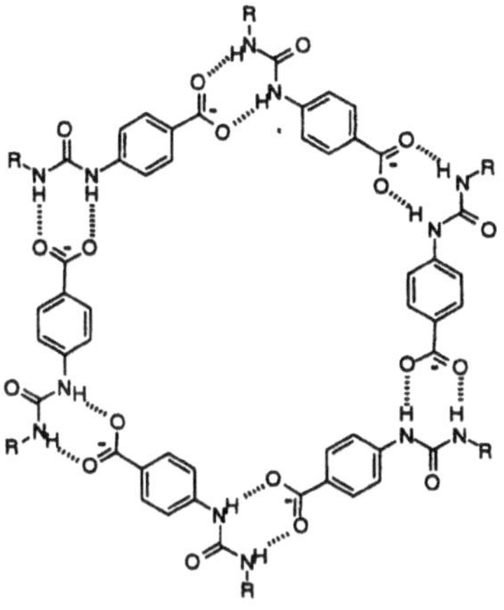

5

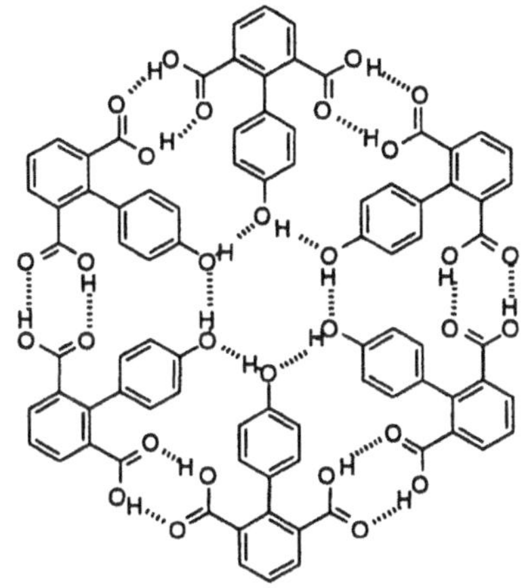

6

additional electrostatic interactions to augment the twelve hydrogen bonds. One possible design is shown in **6** where 4-hydroxyphenyl substituents in the 2-position of the isophthalic acid project into the center of the aggregate forming an additional cyclic array of six hydrogen bonds.

The primary barrier to the formation of these aggregates is the unfavorable entropy of association involved in bringing together six particles. An additional strategy to stabilizing the structure would be to maintain the six components but to increase the number of hydrogen bonds per component. This can be achieved by linking two isophthalic acid groups through a suitable spacer. The resulting monomer can now form a hexameric aggregate composed of two stacked disc structures and stabilized by twenty four hydrogen bonds, as in **7**.

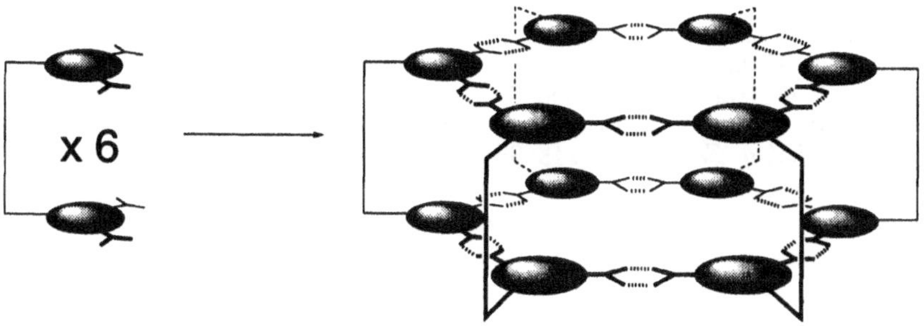

7

Acknowledgment

We thank the National Science Foundation (CHE 9213937) for financial support of this work and the Swiss National Science Foundation for a Post-doctoral Fellowship to J-L. M.

References

1. Lindsey, J. S. (1991) Self-Assembly in Synthetic Routes to Molecular Devices *New J. Chem.* **15,** 153.
2. Whitesides, G. M.; Mathias, J. P.; Seto, C. T. (1992) Molecular Self-Assembly and Nanochemistry: A Chemical Strategy for the Synthesis of Nanostrucures, *Science (Washington, D.C.)* **254** 1312.
3 Goodsell, D. S.; Olson, A. J. (1993) Soluble Proteins: Size Shape and Function, *Trends in Biochem. Sci.* **18,** 65.
4. Williams, D. H. (1991) The Basis of Biological Order, *Aldrichimica Acta*, **24,** 71.
5. Mathias, J. P.; Seto, C. T.; Whitesides, G. M. (1993) Molecular Self-Assembly through Hydrogen Bonding: Supramolecular Aggregates Based on the Cyanuric Acid•Melamine Lattice, *J. Am. Chem. Soc.* **115,** 905.
6. Zimmerman, S. C.; Duerr, B. F. (1992) Controlled Molecular Aggregation: Cyclic Trimerization via Hydrogen Bonding, *J. Org. Chem.* **57,** 2215.
7. Gallant, M.; Viet, M. T. P.; Wuest, J. D. (1991) Use of Hydrogen Bonds to Control Molecular Aggregation; Association of Dipyridones Joined by Flexible Spacer, *J. Org. Chem.* **56,** 2284.
8. Yang, J.; Fan, E.; Geib, S. J.; Hamilton, A. D. (1993) Hydrogen Bonding Control of Molecular Self-Assembly: Formation of a 2 + 2 Complex in Solution and in the Solid State, *J. Am. Chem. Soc.* **115,** 5314.
9. Wyler, R.; de Mendoza, J.; Rebek, J. Jr. (1993) A Synthetic Cavity Assembles Through Self-Complementary Hydrogen Bonds, *Angew. Chem. Int. Ed.* **32,** 1699.
10. Drain, C. M.; Fischer, R.; Nolen, E. G.; Lehn, J. M. (1993) Self-Assembly of a Bis-Porphyrin Supramolecular Cage Induced by Molecular Recognition Between Complementary Hydrogen Bonding Sites, *J. Chem. Soc. Chem. Commun.* 243.
11. Etter, M. C. (1990) Encoding and Decoding Hydrogen Bond Patterns of Organic Compounds, *Acc. Chem Res.* **23,** 120.
12. Duchamp, D. J.; Marsh, R. E. (1969) The Crystal Structure of Trimesic Acid, *Acta Cryst.* **B25,** 5.
13. Alcala, R.; Martinez-Carrera, S. (1972) The Crystal Structure of Isophthalic Acid, *Acta Cryst.* **B28,** 1671.
14. Fan, E.; Van Arman, S.; Kincaid, S.; Hamilton, A. D. (1993) Molecular Recognition: Hydrogen Bonding Receptors that Function in Highly Competitive Solvents, *J. Am. Chem. Soc.* **115,** 369.

SUPRAMOLECULAR STEREOCHEMISTRY OF SUPERCHARGED POLYCYCLIC AROMATIC HYDROCARBONS

LAWRENCE T. SCOTT
Merkert Chemistry Center, Department of Chemistry
Boston College, Chestnut Hill, Massachusetts 02167-3860
USA

1. Abstract

Reduction of corannulene and its simple alkyl derivatives with lithium metal gives tetraanions that come together to form ten-component aggregates with four lithium ions sandwiched between two decks of supercharged polycyclic aromatic hydrocarbons. Analogous species are not formed on reduction with sodium or potassium. Stereochemical and dynamic aspects of the new sandwich compounds have been studied by NMR spectroscopy (^{1}H, ^{13}C, and ^{7}Li), and a simple picture for the "supramolecular bonding" is presented.

2. Background: The Tetraanion of Corannulene

Corannulene, the minimum molecular subunit of C_{60} that retains a curved polycyclic network of trigonal carbon atoms, is a unique, bowl-shaped, aromatic hydrocarbon. Though first synthesized in the 1960s [1], this intriguing molecule has become available in quantities sufficient for extensive chemical investigations only recently as a result of new preparative methods [2].

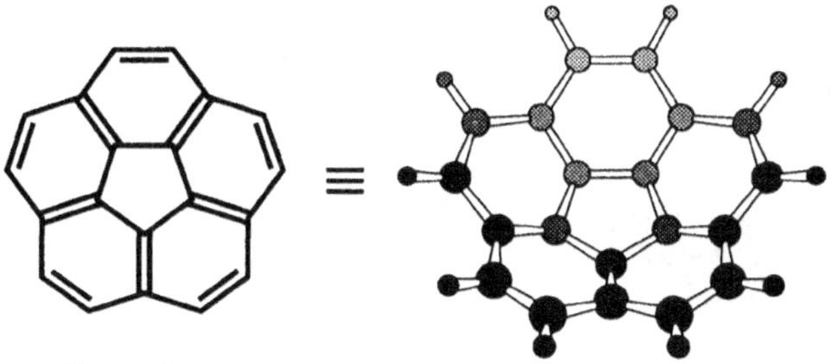

Corannulene

Like the fullerenes, corannulene has a LUMO (lowest unoccupied molecular orbital) that lies uncommonly low in energy. The LUMO of corannulene is also doubly degenerate, which should give the π system a capacity for four extra electrons.

J. S. Siegel (ed.), Supramolecular Stereochemistry, 147–150.

Indeed, we have already succeeded in preparing a stable tetraanion of corannulene by reduction of the neutral hydrocarbon with excess lithium metal in THF-d_8 [3]. A structural representation with concentric anions probably best describes this super-charged species, although the ^{13}C NMR spectrum and molecular orbital calculations [4] suggest some deviation from a perfectly uniform charge distribution (20% charge per carbon atom).

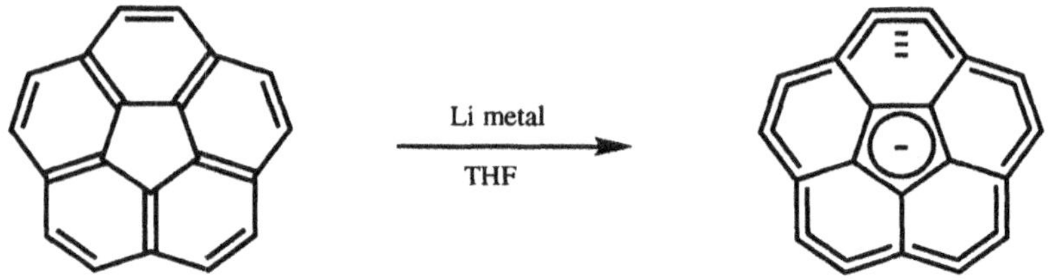

3. Stacked-Bowl Dimers Revealed by Stereoisomerism

When we extended the lithium reduction studies to simple alkyl derivatives of corannulene, *e.g.*, *t*-butylcorannulene, we were surprised to see NMR spectra (both ^1H and ^{13}C) indicative of not one but *two* tetraanions formed in nearly equal amounts. ^7Li NMR spectroscopy showed only a single peak for the four lithium ions at room temperature, so the structural difference between the two new species could not be attributed to different spatial arrangements of the counterions.

Eventually we realized that aggregation of two tetraanions into sandwich compounds with lithium ions between the decks (**1**) could explain the facts [5]. Two stereoisomers can be formed in such multi-component sandwich compounds, one meso (**2**) and one racemic (**3**). We would expect in-plane rotation of the decks to be facile in these dimers, but such a process will not interconvert **2** with **3**.

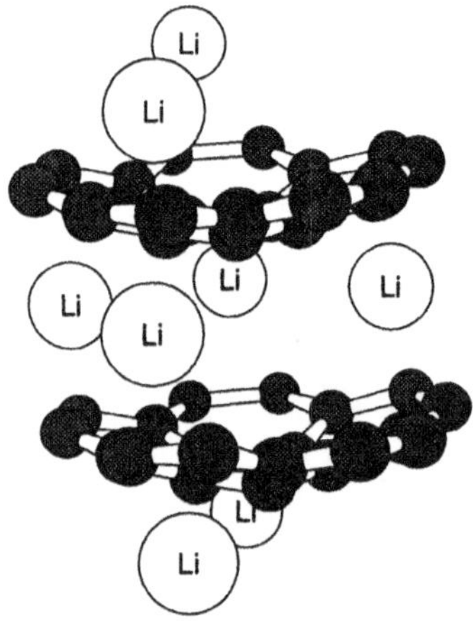

1

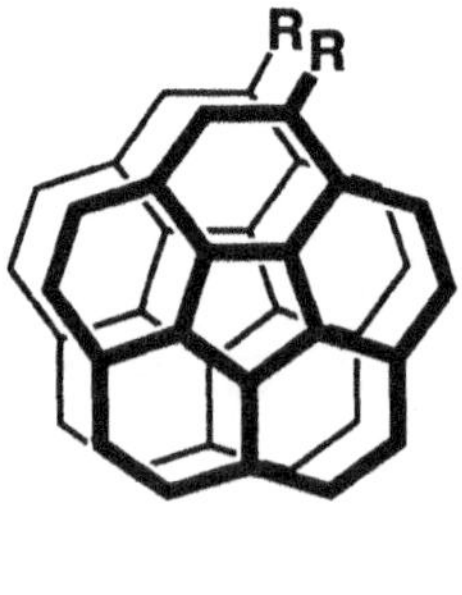

2

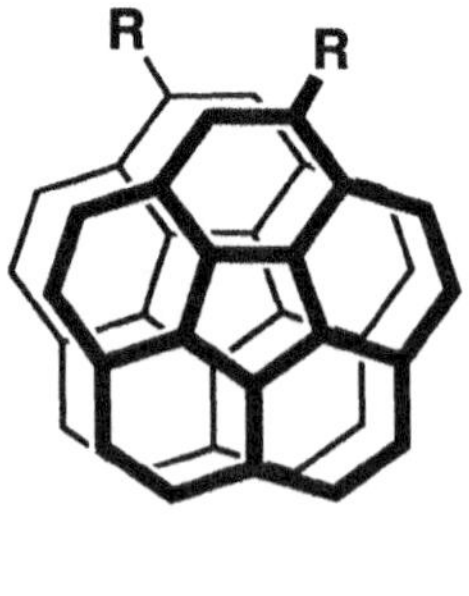

3

Strong evidence in support of this picture comes from the successful preparation of a mixed dimer in which one hydrocarbon deck bears a *t*-butyl group while the other has none [5]. NMR peaks for this mixed dimer stand out prominently in the spectrum taken after reduction of a 5:1 mixture of unsubstituted corannulene and *t*-butylcorannulene; under the conditions of the experiment, the self dimer of corannulene gives rise to only one large singlet, whereas the self dimers of *t*-butylcorannulene (**2** and **3**) are stoichiometrically disfavored relative to the mixed dimer.

Where are the lithium ions in these dimers? At room temperature, they seem to exchange positions rapidly with one another on the NMR time scale (one signal in the ^{7}Li NMR spectrum), *even though the stereoisomeric dimers remain intact and do not interconvert; they do not dissociate on the NMR time scale.* As the temperature is lowered, the ^{7}Li NMR signal splits into two peaks of equal intensity (coalescence temperature 265K). An exceptionally high field signal at -11.7 ppm can be attributed to the four interior lithium ions, all strongly shielded by the large diamagnetic ring currents of the two aromatic supercharged decks, whereas the signal at -4.5 ppm comes from the four exterior lithiums. Molecular orbital calculations (MNDO) predict a very low barrier for double bowl-to-bowl inversion in **1**, assuming partial solvation of the exterior lithium ions.

4. Supramolecular Bonding: The Special Role of Lithium

Reduction of corannulene with sodium or potassium in THF does not give the analogous sandwich compounds. Only lithium gives a stable tetraanion with corannulene and its simple alkyl derivatives. The larger ionic radii of sodium and potassium presumably render these ions less able to fit inside a sandwich of type **1**, and other factors undoubtedly also contribute to their failure to behave like lithium.

Aggregation is not a new phenomenon in organolithium chemistry; however, high-order aggregates with four lithium ions sandwiched between two supercharged polycyclic aromatic decks are unprecedented. What holds them together?

We view the supramolecular bonding in these sandwich compounds as analogous to simple covalent bonding between atoms. Thus, for example, two chlorine atoms can come together and bond to one another by sharing a pair of electrons, one electron having been contributed by each partner (Figure 1, top). The energetic benefit of having the shared electrons close to both nuclei outweighs the coulombic repulsion of the electrons for each other and the coulombic repulsion of the nuclei for each other. Similarly, two corannulene tetraanions can come together and "bond" to one another by sharing two pairs of lithium ions, one pair of lithiums having been contributed by each partner (Figure 1, bottom). The energetic benefit of having four lithium cations close to both tetraanions apparently outweighs the coulombic repulsion of the lithiums for each other and the coulombic repulsion of the two decks for each other.

One fact that this picture does not explain is the absence of triple-decker (or higher order) sandwiches. The experimental evidence, however, strongly argues against the existence of triple-decker sandwiches, either as stable species or as kinetically accessible transient intermediates. Triple-decker sandwiches of monosubstituted corannulenes would be expected to exist in *three* stereoisomeric forms of comparable energy, two meso and one racemic, but we see only two. An equilibration between double-decker and triple-decker sandwiches can likewise be ruled out, since such a process would interconvert **2** with **3**, and these dimers are kinetically stable on the NMR time scale.

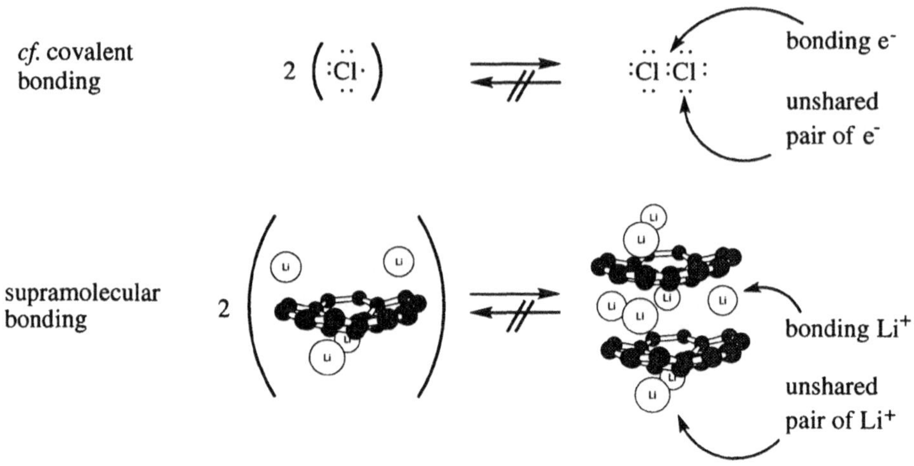

Figure 1. The parallel between covalent bonding between two atoms and supramolecular bonding between two tetralithio corannulene tetraanions.

It should be noted that there is nothing in our description of this supramolecular bonding that requires the hydrocarbon polyanions to be bowl-shaped. One could just as easily imagine planar polyanions aggregating in an analogous fashion, and we have acquired preliminary evidence for such species. We have also seen evidence for "mixed-valence" sandwiches of this type in which the two decks have different numbers of electrons. Details of these and related studies will be reported in due course.

5. References and Notes

[1] Barth, W. E. and Lawton, R. G. (1966) *J. Am. Chem. Soc.* **88**, 380-381. Barth, W. E. and Lawton, R. G. (1971) *J. Am. Chem. Soc.* **93**, 1730-1745.

[2] Scott, L. T., Hashemi, M. M., Meyer, D. T., and Warren, H. B. (1991) *J. Am. Chem. Soc.* **113**, 7082-7084. Borchardt, A., Fuchicello, A., Kilway, K. V., Baldridge, K. K., and Siegel, J. S. (1992) *J. Am. Chem. Soc.* **114**, 1921-1923.

[3] Ayalon, A., Rabinovitz, M., Cheng, P.-C., and Scott, L. T. (1992) *Angew. Chem.* **104**, 1691-1692; *Angew. Chem. Internat. Ed. Engl.* **31**, 1636-1637.

[4] Rabideau, P. W., Marcinow, Z., Sygula, R., and Sygula, A. (1993) *Tetrahedron Lett.* **34**, 6351-6354.

[5] Ayalon, A., Sygula, A., Cheng, P.-C., Rabinovitz, M., Rabideau, P., and Scott, L. T. (1994) *Science*, **265**, 1065-1067.

[6] The major breakthroughs in our understanding of these supra-molecular sandwich compounds came in 1992-93 while Professor Mordecai Rabinovitz (Hebrew University, Jerusalem) was in my laboratory on sabbatical. I am indebted to Professor Rabinovitz and his students, as well as to Professor Peter Rabideau (Louisiana State University) and his students, for a stimulating collaboration. Financial support from the U.S. Department of Energy, the U.S. National Science Foundation, and the U.S.-Israel Binational Science Foundation is gratefully acknowledged.

ENHANCEMENT OF ELECTROSTATIC BINDING THROUGH COOPERATIVE INTERACTIONS: ENTHALPY/ENTROPY COMPENSATION AND PEPTIDE—PEPTIDE RECOGNITION

MARK S. SEARLE, MARTIN S. WESTWELL, PATRICK GROVES, GARY J. SHARMAN & DUDLEY H. WILLIAMS
Cambridge Centre for Molecular Recognition
University Chemical Laboratories, Lensfield Road, Cambridge, CB2 1EW

1. Introduction

In recent work we have established a fundamental relationship between the cost in entropy of an association between two entities **A** and **B**, involving one- and two-point interactions in the gas phase, or in non-polar solvents, to give a complex **A.B**, as a function of the exothermicity of the interaction between them (1). While there is a limit to the adverse entropy of a bimolecular association in terms of the loss of translational and overall rotational freedom (ca 50 to 60 kJ/mol for $T\Delta S°$ in solution at 298 K) (2), the exothermicity of an interaction can increase far beyond the exothermicity at which the limiting cost in entropy is approached. We have proposed, and subsequently justified on the basis of a large body of experimental data from many laboratories (for example, see Figure 1A), that the enthalpic benefit versus entropic cost of associations of the type stipulated above have the general form shown in Figure 1B (1). We emphasise the approximate form of the curve; the precise entropic cost of an association is dependent on many variables such as mass, density of vibrational states and the shape of the potential energy well in which the associated species lies. In complex systems where the possibility arises that internal rotations are also restricted within the associating species, then the limiting entropic cost will be higher with the curve displaced to the right. The loss of entropy illustrated by Figures 1A and 1B reflects only loss of translational and overall rotational freedom upon association. The data clearly show that associations with small exothermicities can result in remarkably small adverse entropy changes.

In this paper we illustrate the importance of the variable slope of the enthalpy/entropy plot in providing a basis for cooperativity in binding phenomena. We use the plot to present a model illustrating how two weak interactions, acting to support the binding of one molecule to a receptor, produce a net binding free energy $\Delta G°$ which is greater than the sum of

J. S. Siegel (ed.), Supramolecular Stereochemistry, 151–159.

the parts. In this description we divide this mutual aiding of the two interactions into two distinct parts; one is entropic in origin and related to the entropic chelate enhancement of binding described by Jencks (3,4), and the other is an enthalpic benefit (5).

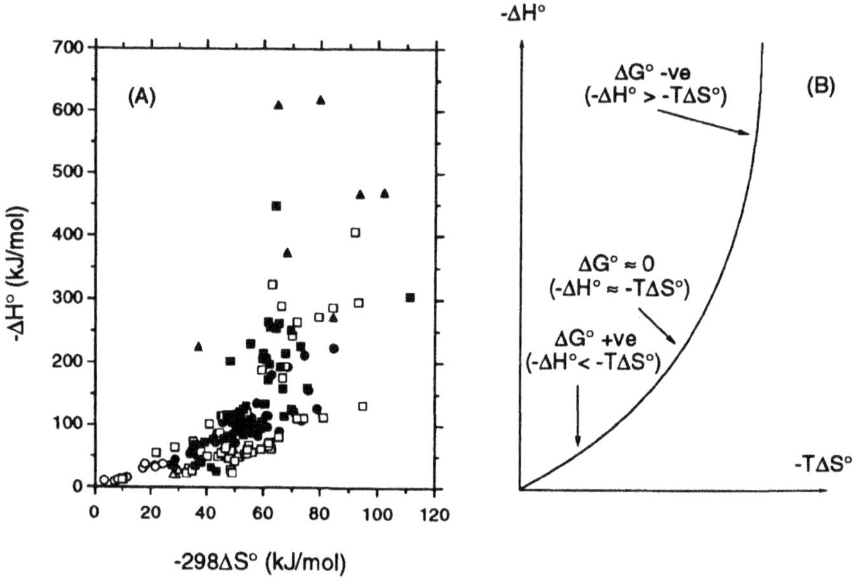

Figure 1. (A) Enthalpy ΔH° versus entropy 298ΔS° plot for gas phase associations and sublimation data for compounds involving small or negligible changes in internal rotations upon association (1). (B) The general form of the extent of the exothermicity of association as a function of entropic cost

2. Cooperativity in Binding

We consider a model for binding based upon the consideration of an interaction of a polar functional group in a molecule **X** with a complementary polar group on a receptor, and of a polar functional group **Y** with another polar group on the same receptor as they would occur in a gas phase environment. These interactions are considered when **X** and **Y** interact separately with the receptor (Figure 2A), or simultaneously by virtue of being covalently connected together (Figure 2B) such that **X** and **Y** are able to bind into their respective binding sites without strain and without any adverse entropy change needed to conformationally restrict the connection. We again emphasize that the entropy/enthalpy plot allows a crude guideline to the entropic cost as a function of the strength of the electrostatic interaction formed, and that the overall curvature of the plot allows us to make some advances from the analysis of Jencks of the chelate effect (3,4). Two cases are now considered, the first is idealised, while the second is a more realistic description of binding cooperativity that is subsequently justified on the basis of experimental observation.

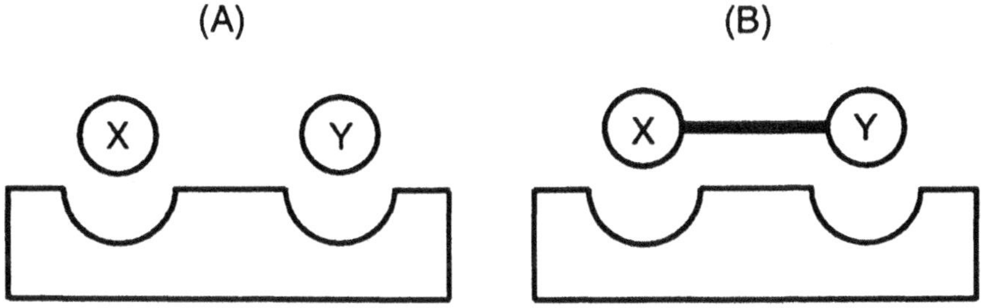

Figure 2. (A) Binding of **X** and **Y** separately to a receptor, and (B) simultaneously as **X-Y** when covalently linked.

2.1. THE EXOTHERMICITY OF BINDING X-Y IS APPROXIMATELY THE SUM OF THE EXOTHERMICITIES OF X AND Y

Suppose molecule **X** alone binds to the receptor with an exothermicity $\Delta H°_X$, and **Y** with an exothermicity $\Delta H°_Y$; on the basis of Figure 1B, the respective adverse entropy terms would be $T\Delta S°_X$ and $T\Delta S°_Y$. If the initial assumption is made that **X** and **Y** bind with the same exothermicities when separate as when binding to the receptor jointly as **X-Y**, then the exothermicity of binding **X-Y** is taken as $\Delta H°_{X\text{-}Y}$ and the cost due to loss of translational and rotational entropy ($T\Delta S°_{T+R}$) is $T\Delta S°_{X\text{-}Y}$. Thus, in the case of **X** binding with a relatively small exothermicity, and **Y** binding with a moderate exothermicity, the enhancement in binding free energy for **X-Y** relative to the sum of the separate components is given by the difference in the entropic cost to binding $T\Delta S°_{X\text{-}Y} - (T\Delta S°_X + T\Delta S°_Y)$. In Figure 3A, each binding interaction (**X**, **Y** and **X-Y**) is represented by a vector from the origin to the relevant point on the entropy/enthalpy curve, with dotted lines identifying the pertinent values for $\Delta H°$ and $T\Delta S°$ for each interaction. In Figure 3B, the vectors for binding **X** and **Y** separately are summed such that overall $\Delta H°_X + \Delta H°_Y = \Delta H°_{X\text{-}Y}$, but the difference in $T\Delta S°$ for binding **X-Y** versus the sum for **X** and **Y** alone represents the entropic chelate enhancement ($T\Delta S°_{chelate}$). Whatever a more precise form of Figure 1B may be, given only the generality that more exothermic interactions approach a limiting cost in entropy, it is seen that this expression of cooperativity is greatest where each of the associations of **X** and **Y** are sufficiently exothermic to approach the limiting cost in entropy (Figure 3C).

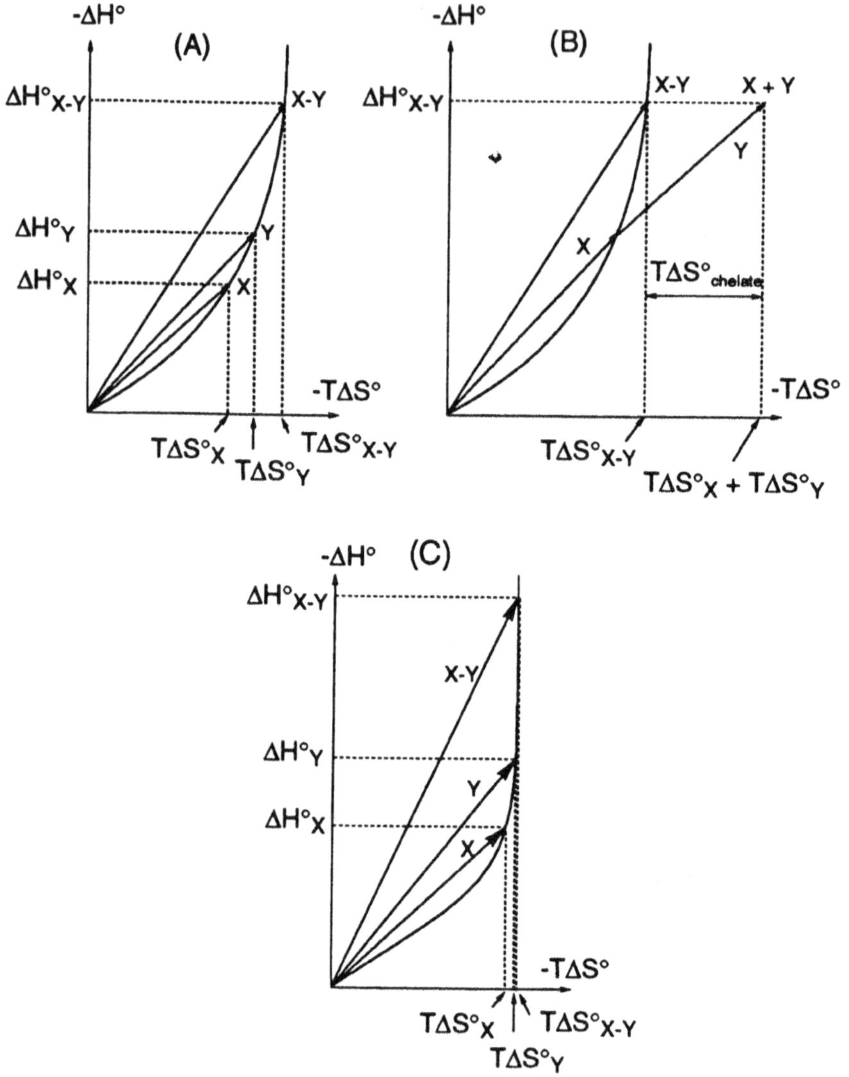

Figure 3. Vector diagrams describing (A) the gas phase enthalpies and entropies of binding **X**, **Y** and **X-Y** to a receptor, where the exothermicity of binding **X-Y** is taken as the sum of the exothermicities of **X** and **Y** when bound alone. Each binding event is represented by a vector from the origin to the relevant point on the enthaply/entropy curve. In (B), the summation of the **X** and **Y** vectors is visualized. The difference in the entropic terms, corresponding to the binding of **X-Y** and the sum of the terms for **X** and **Y**, represents the chelate enhancement of binding ($T\Delta S°_{chelate}$). The vectors corresponding to the maximum possible chelate enhancement of binding, due to **X** and **Y** binding with large exothermicities and limiting adverse entropies, are represented in (C).

There are many intramolecular model reactions that exhibit rate increases, compared with the corresponding bimolecular reactions, of between $\approx 10^5$ and $\approx 10^8$ M (3,4), where the reacting groups in the bimolecular association have the freedom to undergo independent translation and overall rotation. Here, of course, the large adverse entropy involved in forming the transition state in a bimolecular association is largely negated in the intramolecular reaction by a strongly exothermic interaction between the two reacting species that holds them in close proximity, namely a covalent bond. Conversely, when **X** and **Y** bind with very small exothermicities, then the chelate effect may be small because the entropic cost of binding **X-Y** will not be much less than the sum of the entropic cost of binding **X** and **Y** separately. This effect arises because of the large change in slope of the curve in Figure 1B on passing from small to large exothermicities.

We next consider the case where cooperativity is also expressed as an enthalpic benefit when **X** and **Y** are tethered together in **X-Y**. This effect operates in addition to the entropic chelate effect just described.

2.2. THE EXOTHERMICITY OF BINDING X-Y IS GREATER THAN THE SUM OF THE EXOTHERMICITIES FOR BINDING X AND Y SEPARATELY

For each of the interactions of **X** and **Y** with the receptor, in general only part of the theoretical maximum translational and rotational entropy which could be lost is lost. In a weak interaction the theoretical maximum bonding (which could only be expressed at 0 K) is not attained at room temperature due to the opposing entropic advantage of residual motion. The result is that each molecule binds and occupies an average position in the enthalpic well corresponding to ΔH°_X when **X** binds alone, and ΔH°_Y when **Y** binds alone. However, if the restriction of motion is aided by a neighbouring exothermic interaction, then the average position in the enthalpic well will correspond to some larger exothermicity. Thus, the binding of the **X** part of **X-Y** is aided by the binding of the **Y** part of **X-Y**, and *vice versa*, as a consequence of an increased barrier to dissociation of **X** when **X** is directly bonded to **Y**. Therefore, in the general case it is to be expected that the exothermicity of binding **X-Y** will be greater than the sum of the exothermicities of **X** and **Y** alone. The effect of this second cooperative factor, which is quite distinct from the entropic chelate effect described above, is illustrated in Figure 4. The dotted lines represent the enthalpies of association of **X**, **Y** and **X-Y**, as shown in Figure 3 (ΔH°_X, ΔH°_Y and $\Delta H^\circ_{X\text{-}Y}$, respectively), which were analysed above with the assumption that $\Delta H^\circ_{X\text{-}Y} = \Delta H^\circ_X + \Delta H^\circ_Y$. The dashed lines indicate the corresponding analysis for which **X-Y** binds more exothermically than the sum of its parts. Thus, $\Delta H^{\circ\prime}_X$ and $\Delta H^{\circ\prime}_Y$ represent the exothermicity of binding **X** and **Y**, respectively, when each is tethered to the other, and $\Delta H^{\circ\prime}_{X\text{-}Y}$ represents the total exothermicity of binding **X-Y** ($\Delta H^{\circ\prime}_X$ + $\Delta H^{\circ\prime}_Y$). Hence, the cooperativity expressed as an increase in electrostatic bonding ($\Delta H^{\circ}_{chelate}$) is distinct from, and expressed in addition to, the entropic chelate effect, $T\Delta S^{\circ}_{chelate}$. The implications of this analysis are now illustrated experimentally.

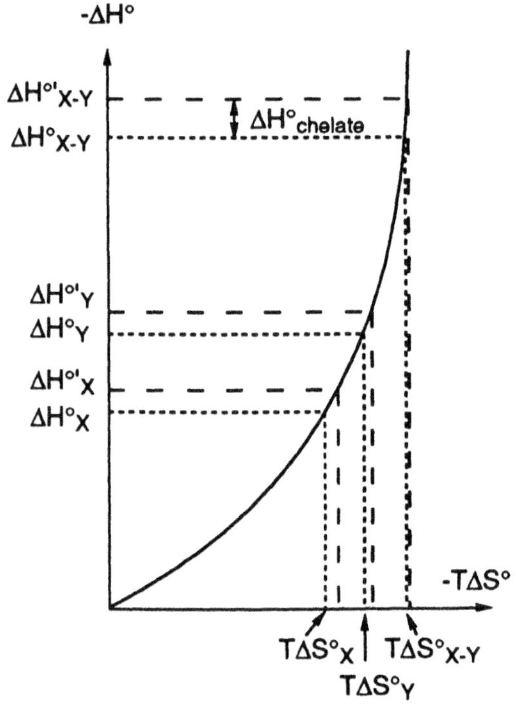

Figure 4. Enthalpy/entropy curve illustrating cooperativity through the enthalpic benefit to binding when the electrostatic interactions of **X** are aided by **Y**, and **Y** by **X**, when the two are tethered together in **X-Y**. Dotted lines represent the enthalpies and entropies of binding **X**, **Y** and **X-Y** when the exothermicity of **X-Y** is taken to equal the sum of the exothermicities of the constituent parts, while the terms $\Delta H^{\circ\prime}_X$ and $\Delta H^{\circ\prime}_Y$ represent the numerically larger exothermicities of binding **X** and **Y**, respectively, when the binding of **X** and **Y** are mutually enhanced as part of **X-Y**. Thus, $\Delta H^{\circ}_{chelate}$ represents cooperativity expressed as an enhancement of electrostatic bonding.

3. Binding of the carboxylate group of cell wall analogues to glycopeptide antibiotics.

The implication of the cooperativity expressed in Figure 4 is that one functional group interaction between a ligand and a receptor can be improved electrostatically when additional interactions are introduced on the same template. Probably the best way to measure the exothermicity of an interaction is through the use of calorimetry, but cases where the binding of **X**, **Y** and **X-Y** can be measured with the necessary precision are not readily available. An alternative approach is to infer the strength of a given electrostatic interaction by the measurement of proton NMR chemical shifts (5,6). Thus when cell wall peptide analogues RCOO- bind to glycopeptide antibiotics such as ristocetin A and eremomycin, the chemical shift of an amide NH w_2 (Figure 5) is shifted appreciably down field (> 3 ppm for tightly binding ligands) due to a specific interaction with the carboxylate anion.

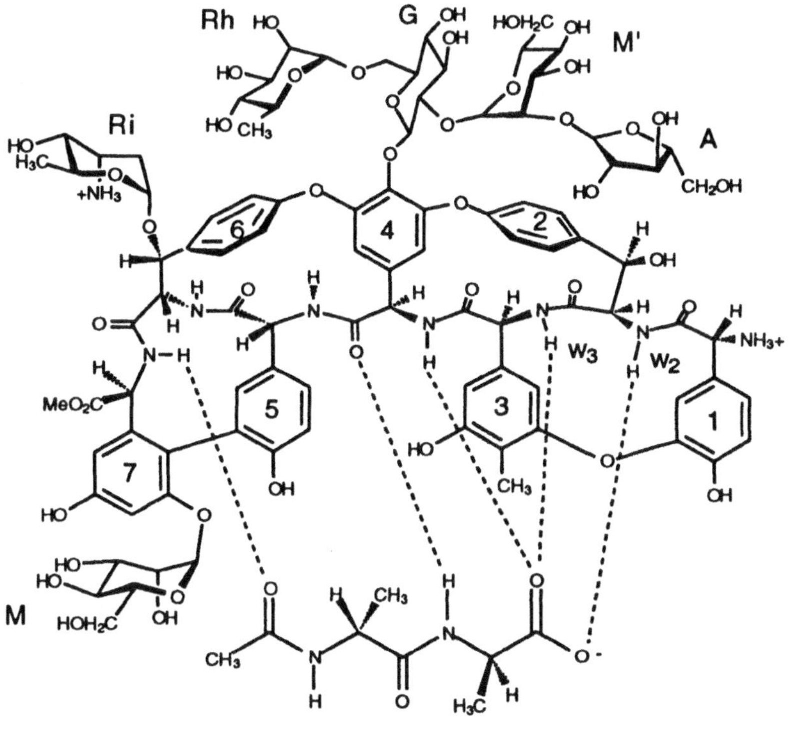

Figure 5. Exploded view of the ristocetin A complex with the dipeptide cell wall analogue N-Ac-D-Ala-D-Ala.

(A)

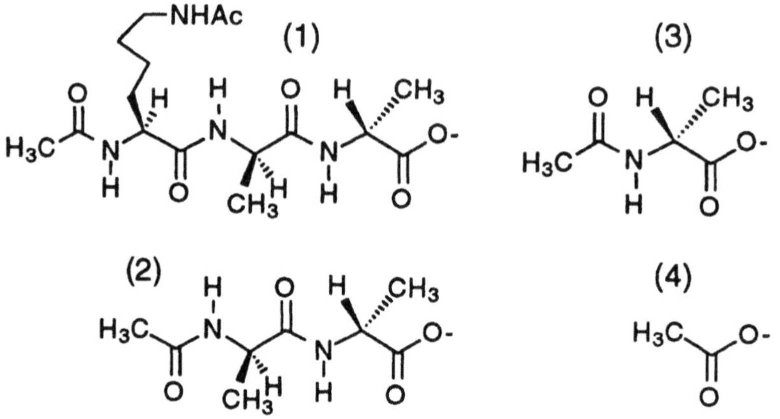

(B)

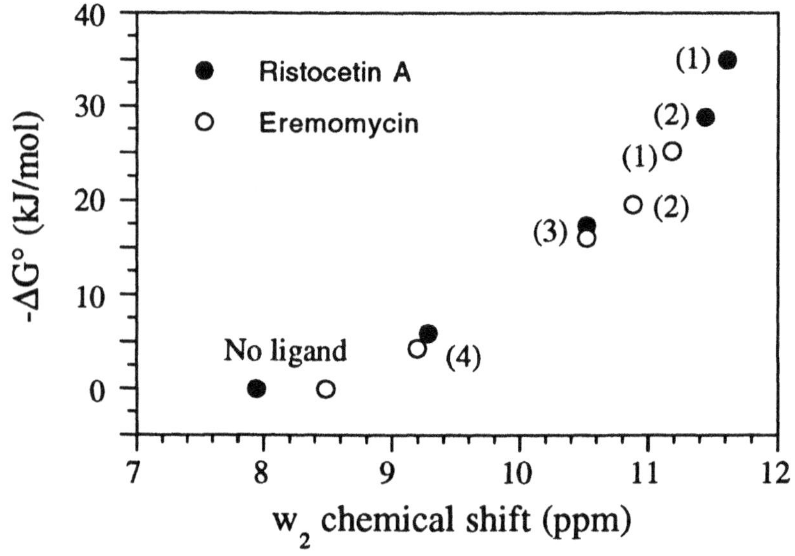

Figure 6. (A) Structure of cell wall analogues used in these studies. (B) Plot of ligand binding energy (ΔG° at 298 K) versus limiting chemical shift of antibiotic amide NH w_2 (pH 4.5) for studies with ristocetin A and eremomycin (4,5).

The magnitude of the downfield shift can be used as a measure of the strength of this electrostatic interaction. Larger binding shifts are taken to imply stronger electrostatic interactions in these complexes (as opposed to structural changes of different origin) since the carboxylate anion is a constant structural unit in a series of highly homologous ligands (Figure

6A) that closely mimic the natural cell wall peptidoglycan recognition sequence –L-Lys-D-Ala-D-Ala. A corresponding set of chemical shift changes are also observed for the amide NH w_3 in the same carboxylate binding pocket. In Figure 6B, $\Delta G°$ the overall ligand binding energy measured at 298 K is plotted against the chemical shift of w_2, as the ligand containing the carboxylate group is extended to contain increasing numbers of ancillary groups (Figure 6A) (5).

From these data it is apparent that the more extended ligands bind more strongly to the receptor antibiotic not only because the extending groups can add binding affinity, in the manner of a classical chelate effect, at a relatively small cost in entropy (since much of the exothermicity of binding comes from the interaction of the carboxylate group with three amide NHs), but also because these extending groups aid in holding the carboxylate anion more deeply in the electrostatic well which describes its interaction with the antibiotic. The data illustrate how multiple weak interactions on the same template act in a cooperative fashion to enhance the electrostatic binding energy of the carboxylate interaction.

Acknowledgements
We thank GLAXO, EPSRC and Zenova for financial support.

References.

1. Searle, M. S., Westwell, M. S., Williams, D. H. (1994) Application of a generalised enthalpy/entropy relationship to binding cooperativity and weak associations *J. Chem. Soc. Perkin Trans. 2* , in press.
2. Page, M. I., Jencks, W. P. (1971) Entropic contributions to rate accelerations in enzymic and intramolecular reactions and the chelate effect. *Proc. Natl. Acad. Sci. USA.* 68, 1678-1683.
3. Jencks, W. P. (1975) Binding energy, specificity, and enzymic catalysis: The circe effect. *Adv. Enzymol.* 43, 219-410.
4. Jencks, W. P. (1981) On the attribution and additivity of binding energies. *Proc. Natl. Acad. Sci. U.S.A.* 78, 4046-4050.
5. Groves, P., Searle, M. S., Westwell, M. S., Williams, D. H. (1994) Expression of electrostatic binding cooperativity in the recognition of cell wall peptide analogues by vancomycin group antibiotics. *J. Chem. Soc. Chem Comm.,* 1519-1520.
6. Sternberg, U., Brunner, E. (1994) The influence of short-range geometry on the chemical shifts of protons in hydrogen bonds. *J. Magn. Res. A*, 108, 142-150.

TOWARD COLUMNS AND CHANNELS FROM TORANDS AND MOLECULAR COILS

THOMAS W. BELL, AARON M. HEISS, HÉLÈNE JOUSSELIN AND RICHARD T. LUDWIG
Department of Chemistry, State University of New York
Stony Brook, New York 11794-3400, USA

1. Introduction

Self-assembled tubular structures capable of binding and conducting metal cations are of interest in the fabrication of permeation-selective membranes and the design of supramolecular devices, such as nanoscale transducers.[1-3] Artificial ion-transport assemblies can also serve as models for investigating structure and function of natural transmembrane pores and channels, such as Gramicidin aggregates.[4,5] Torands[6-14] are a recently explored class of relatively rigid, planar ligands capable of coaxial stacking; molecular coils[15,16] are analogous compounds capable of multiple-helical aggregation. Figure 1 shows the key compounds of interest in this chapter, which begins by reviewing the metal complexation and aggregation behavior of torand **1**. The formation of double-helical metal complexes by molecular coils **3** and **4** is then discussed. The chapter concludes with the synthesis of triaryltorands (**2**) in a planned approach to novel tubular mesophases.[1]

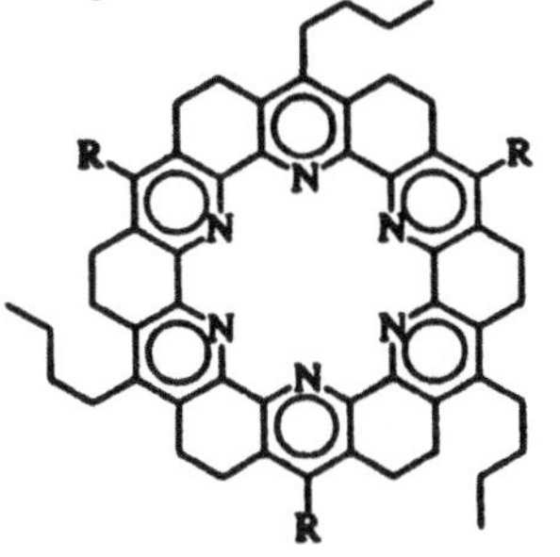

1 R = H
2 R = Aryl

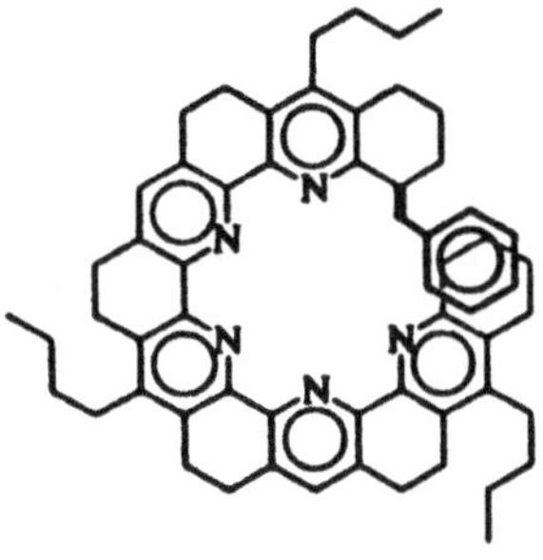

3

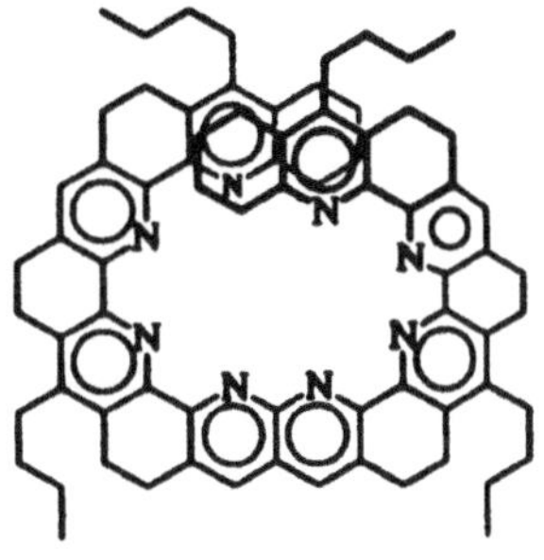

4

Figure 1. **Torands (1 and 2) and molecular coils (3 and 4).**

J. S. Siegel (ed.), Supramolecular Stereochemistry, 161–168.

2. Tri-n-butyl Torand (1)

Gram quantities of torand **1** are available *via* an 11-step synthesis from cyclohexanone and pentanal (3% overall).[6,11,12,17] This ligand binds alkali metals very tightly.[7,11,12] The log K_s values of 1:1 complexes with all alkali metal picrates exceed 11.0 in water-saturated chloroform, according to the results of extraction experiments. Cryptand competition experiments followed by ^1H NMR spectroscopy gave the following values for log K_s in this solvent: **1**•Li^+, 13.4; **1**•Na^+, 14.7; **1**•K^+, 14.3. While no evidence has been obtained for the existence of stable 2:1 complexes with alkali metal ions sandwiched between stacked torands, transient 2:1 complexes are suggested by rapid hopping of alkali metals between torands in solution.

The X-ray crystal structure of the lithium picrate complex of torand **1** (Fig. 2) also reveals the presence of stabilizing torand stacking interactions.[10] Whereas **1**•K^+(picrate)$^-$ and **1**•Rb^+(picrate)$^-$ crystallize as the 1:1 complexes with contact ion pairing,[9] **1**•Li^+(picrate)$^-$ crystallizes as the 2:2 complex, $1_2 \cdot Li^+{}_2(H_2O)_3$. A hydrated dilithium chain ($H_2O \cdot Li^+ \cdot H_2O \cdot Li^+ \cdot H_2O$) is threaded through coaxially stacked torand molecules. While **1** adopts a staggered conformation in the potassium and rubidium complexes, in the lithium complex the two torand molecules are complementarily folded, as shown in Figure 2. The supramolecular assembly is apparently stabilized by complementary metal-ligand, hydrogen bonding and van der Waals interactions, suggesting that longer chains or columns of stacked torands can be produced under suitable conditions.

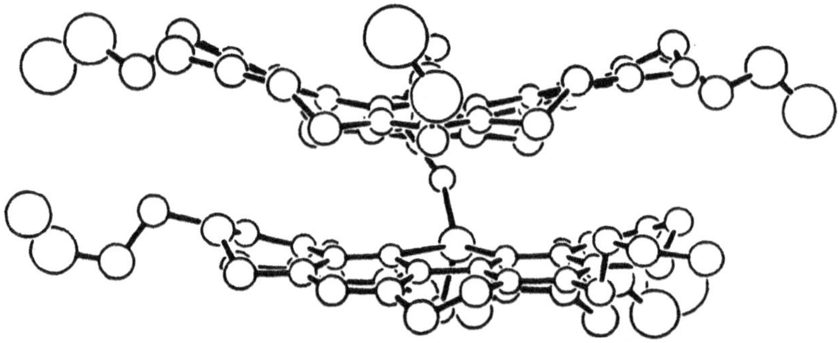

Figure 2. X-ray crystal conformation of $1_2 \cdot Li^+{}_2(H_2O)_3$[10].

Torand stacking interactions are also implied by the cation response of thin films of **1** at the air-water interface.[14] Even though torand **1** is substituted only with three short side-chains, it is sufficiently surface-active to form compressible thin films on water. Curves representing changes in surface pressure as a function of molecular area are dramatically affected by the presence and identities of metal cations in the aqueous subphase. Transitions at 100-120 Å^2/molecule suggest the formation of stacked-torand dimers in the presence of Li^+, Cs^+ and Pb^{2+}.

3. Molecular Coils (3 and 4)

The fused-ring molecular architecture of torands can be used to design open-ended or nonmacrocyclic helical ligands termed molecular coils or expanded heterohelicenes.[15,16] Molecular coil 3, consisting of five pyridine rings fused with six carbocyclic rings can be prepared in 12 steps and 2% overall yield. The benzylidene group of 3 serves as an effective ^1H NMR probe of complex structure; the binding of sodium by this ligand has been thoroughly studied by spectroscopic methods. A related molecular coil containing seven pyridine rings has similar complexation behavior, but it lacks the benzylidene NMR marker and its synthesis and isolation are more tedious.

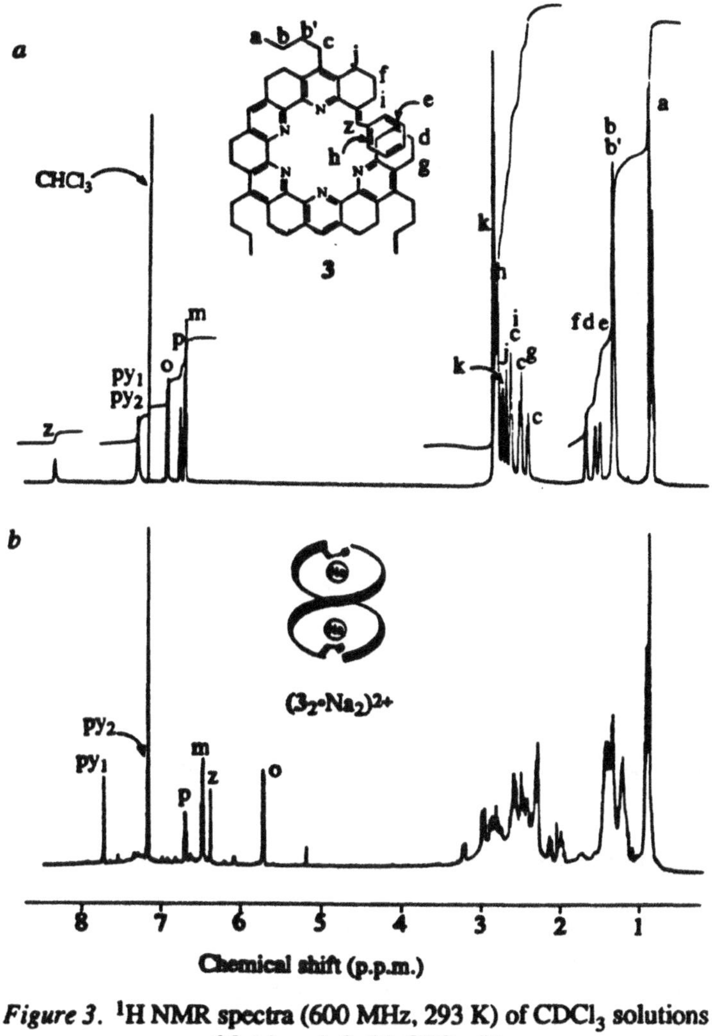

Figure 3. ^1H NMR spectra (600 MHz, 293 K) of $CDCl_3$ solutions of 3 (a) and $3_2(NaOTf)_2$ (b).

Molecular coil **3** extracts 0.5 equivalents of solid sodium triflate into $CDCl_3$, producing 3_2•NaOTf, while in methanol the complex $3_2(NaOTf)_2$ is formed. Microanalytical results for C, H, N and Na are consistent with these formulas and FAB mass spectra confirmed the molecular weights of these complexes. The ^{1}H NMR spectra of both complexes displayed pronounced upfield shifts of the aromatic and olefinic (benzylidene) peaks, as shown in Figure 3 for $3_2(NaOTf)_2$. The geminal protons of the CH_2 groups in **3** also become nonequivalent in these complexes, showing that helix inversion is slow on the NMR time scale. These features and the results of 2D-NMR studies are best explained by double-helical cations $(3_2•Na)^+$ and $(3_2•Na_2)^{2+}$, in which molecular coils intertwine to form roughly cylindrical metal-binding cavities.

Multiple-helical transition metal complexes of polypyridines are known,[18] but $(3_2•Na_2)^{2+}$ is the first example of double-helical complexation induced by the weaker and less directional interactions of an alkali metal ion with an organic ligand. The complexation behavior of compound **5** (Fig. 4) was then examined to determine whether helical preorganization of the ligand is a requirement for 2:1 complexation. As observed for molecular coil **3**, terdentate ligand **5** extracts 0.5 equivalents of sodium triflate into chloroform. Figure 4 shows that the formation of the 2:1 complex $(5_2•Na)^+$ is also accompanied by pronounced upfield shifts of the ^{1}H NMR resonances of the benzylidene group in **5**. While this 2:1 complex could not be isolated by crystallization, preorganization of just three pyridines in **5** suffices to stabilize a 2:1 complex of an alkali metal in solution. For ligand **5** a face-to-face structure is possible in the 2:1 complex with sodium, or the ligands can be partially intertwined, as proposed for $(3_2•Na)^+$ and $(3_2•Na_2)^{2+}$.

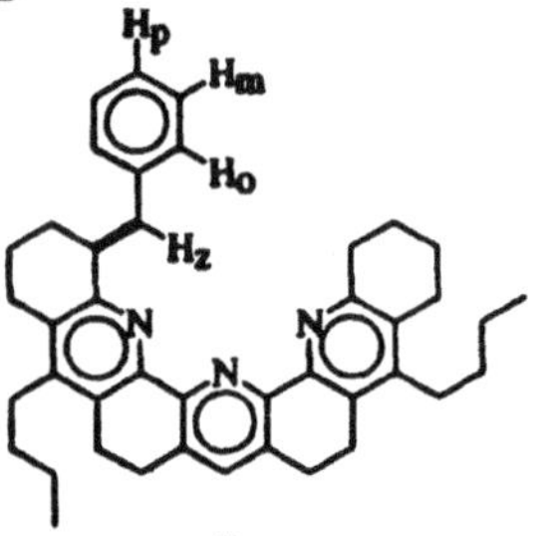

5

	δ (5)	δ $(5_2•Na)^+$
H_z	8.34	7.32
H_o	7.55	6.35
H_m	7.34	6.78
H_p	7.20	6.91

Figure 4. ^{1}H NMR shifts ($CDCl_3$) of benzylidene protons of compound 5 and its 2:1 complex with sodium triflate.

Molecular coil **4** contains 8 pyridine rings, two of which are fused together to form a 1,8-napthyridine unit in the center of the molecule. This compound was synthesized in 17 steps and 2% overall yield. Because the large cavity of **4** should bind large metal ions, **4** was treated with one equivalent of cesium picrate in a mixture of methanol and CH_2Cl_2. Concentration of the solution gave a complex of stoichiometry $4•Cs^+(picrate)^-$, according to elemental analysis. The FAB mass spectrum showed a peak at m/z 2426 consistent with the ion $(4_2•Cs_2)^+$; the ^{1}H NMR spectrum showed upfield shifts and broadening in the aliphatic region consistent with a double-helical 2:2 complex. Reaction of **4** with CsCl and NH_4PF_6 in methanol gave a complex of stoichiometry $\mathbf{4}_2(NH_4PF_6)_2CsPF_6$ with similar ^1H NMR features. Proposed double-helical structures for these complexes are shown graphically in Figure 5. The uneven curvature of **5** should cause an expanded central cavity to be formed upon intertwining of two molecular coils. This cavity can bind two cesium cations in a symmetrical 2:2 complex, or one cesium ion can be situated in the center with two ammonium ions bound at each of the tapered ends of the cavity.

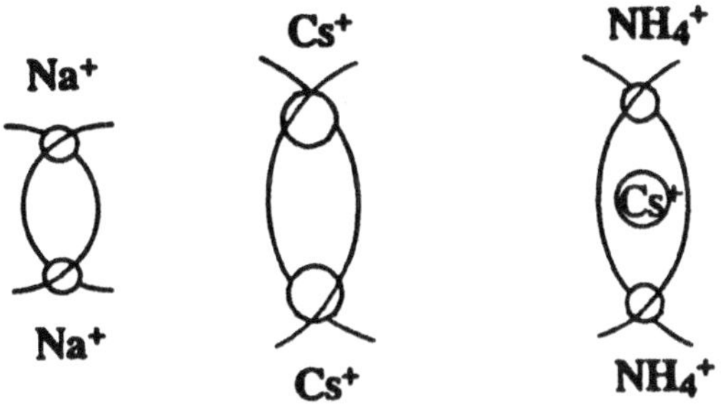

Figure 5. Diagrams comparing proposed structures of double-helical complexes of molecular coils **3** and **4**.

4. Triaryl Torands (2)

Molecular coils longer than **3** or **4** might intertwine to form channels that are stable even in the absence of metal cations, but this prospect is dimmed by the absence of efficient synthetic methods for homologation of fused pyridine rings. Torand stacking is an alternative, attractive strategy for self-assembled ion channels and tri-*n*-butyl torand **1** displays some propensity toward face-to-face association. Stacking interactions should be enhanced by increasing the molecular surface area, and attaching several long alkyl chains should give columnar discotic liquid crystals or tubular mesophases.[1] Synthetic efforts so far have failed to produce hexaalkyl torands, but a synthetic route to trialkyltriaryl torands (**2**) has recently been developed.[19] This method is outlined in Figure 6 for the synthesis of a tri-*n*-butyltris(*p*-nitrophenyl) torand (**2**, Ar = *p*-$NO_2C_6H_5$). As in the synthesis of tri-*n*-butyl torand **1**, the macrocycle is constructed in stages, building up two-thirds of the structure first and inserting the remaining third in the last step. The overall yield of this 11-step synthesis is 2 %, and efforts are in progress to develop a shorter, trimerization route to **2**. Reduction of the nitro groups

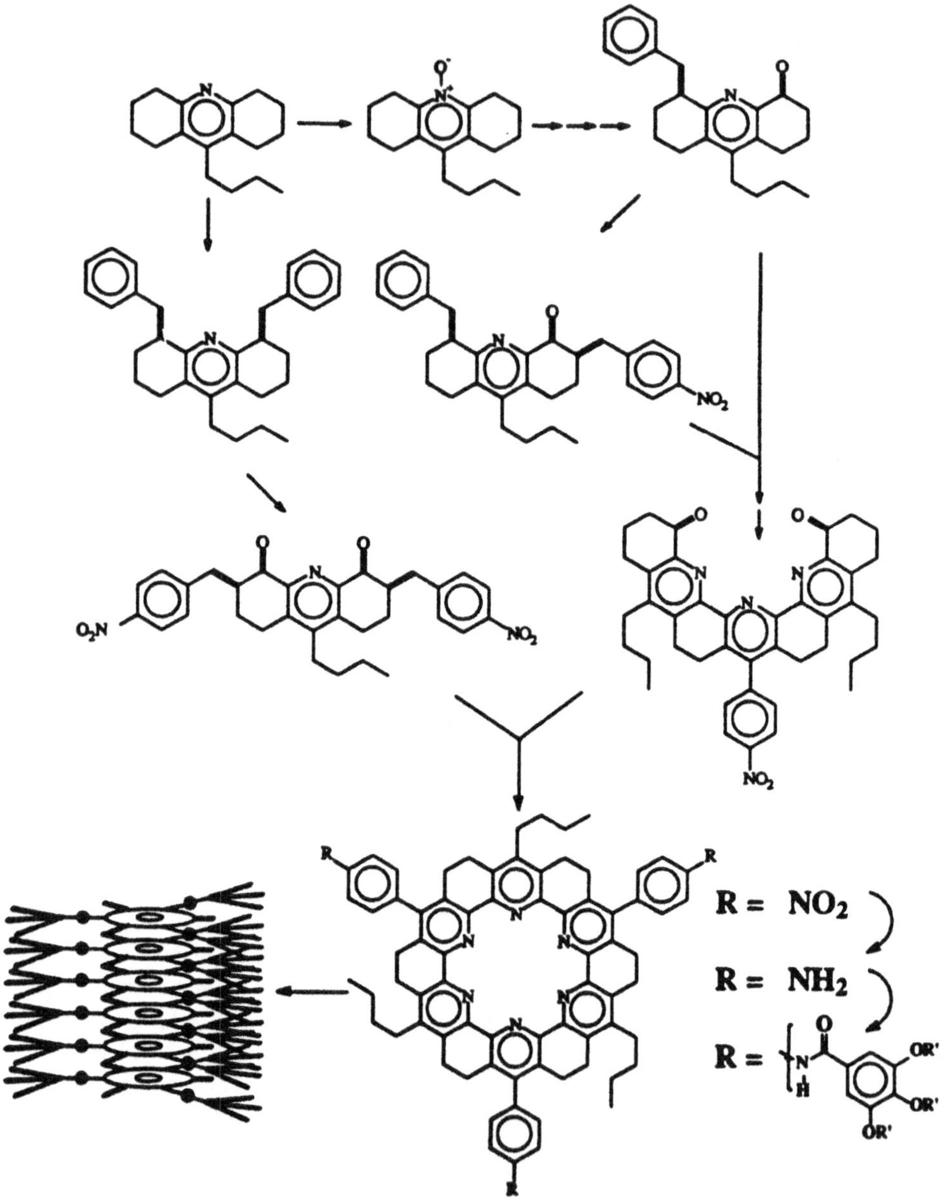

Figure 6. Approach to columnar mesophases via the synthesis of trialkyltriaryl torands (2).

and acylation of the resulting amino groups with 3,4,5-tris(*p*-n-dodecyloxybenzyl-oxy)benzoyl (DOBOB) or other mesogenic groups should afford tubular meso- phases,[1] as shown graphically in Figure 6.

5. Conclusion

Two general approaches to self-assembled, metal-binding tubular structures have been illustrated: stacking of torands and intertwining of molecular coils. Despite significant difficulties in the syntheses of these highly preorganized molecules, their tendency to form well-organized supramolecular assemblies with high affinities for cations makes them promising components for design of molecular ionic devices.

6. References

1. Lehn, J.-M. (1990) Perspectives in Supramolecuflar Chemistry–From Molecular Recognition Towards Molecular Information Processing and Self-Organization, *Angew. Chem. Int. Ed. Engl.* **29**, 1304-1319.
2. Whitesides, G. M., Mathias, J. P. and Seto, C. T. (1991) Molecular Self-Assembly and Nanochemistry: A Chemical Strategy for the Synthesis of Nanostructures, *Science (Washington, D.C.)* **254**, 1312-1319.
3. Conner, M., Janout, V. and Regen, S. L. (1993) Molecular Sieving by a Perforated Langmuir-Blodgett Film, *J. Am. Chem. Soc.* **115**, 1178-1180.
4. Salemme, F. R. (1988) Structural Polymorphism in Transmembrane Channels, *Science (Washington, D.C.)* **241**, 145, 230.
5. Lorenzi, G. P. (1991) An Alternative to Urry's Model for the Gramicidin Transmembrane Channels, *Chirality*, 315-317.
6. Bell, T. W. and Firestone, A. (1986) Torands: Rigid Toroidal Macrocycles. Calcium Sequestration by a Member of This New Ligand Class, *J. Am. Chem. Soc.* **108**, 8109-8111.
7. Bell, T. W., Firestone, A. and Ludwig, R. T. (1989) Exceptionally Stable Alkali Metal Complexes of a Torand, *J. Chem. Soc., Chem. Commun.* 1902-1904.
8. Bell, T. W. and Liu, J. (1990) Torand Synthesis by Trimerization–New Receptors for Guanidinium, *Angew. Chem. Int. Ed. Engl.* **29**, 923-925.
9. Bell, T. W., Cragg, P. J., Drew, M. G. B., Firestone, A. and Kwok, D.-I. A. (1992) Conformational Preference of the Torand Ligand in its Complexes with Potassium and Rubidium Picrate, *Angew. Chem. Int. Ed. Engl.* **31**, 345-347.
10. Bell, T. W., Cragg, P. J., Drew, M. G. B., Firestone, A. and Kwok, D.-I. A. (1992) A Supramolecular Assembly of Two Torands, Two Lithium Ions and Three Water Molecules, *Angew. Chem. Int. Ed. Eng.* **31**, 348-350.
11. Bell, T. W. (1992) From Crowns to Torands and Beyond, in S. R. Cooper (ed), *Crown Compounds: Toward Future Applications*, VCH Publishers, New York, pp. 305-318.

12. Bell, T. W., Cragg, P. J., Drew, M. G. B., Firestone, A., Kwok, D.-I. A., Liu, J., Ludwig, R. T. and Papoulis, A. T. (1993) Synthesis of New Torands and New Uses for Old Torands, *Pure and Applied Chem.* **65**, 361-366.
13. Bell, T. W., Beckles, D. L., Cragg, P. J., Liu, J., Maioriello, J., Papoulis, A. T. and Santora, V. J. (1993) Intrinsic Chromophores and Fluorophores in Synthetic Molecular Receptors, in A. Czarnik (ed), *ACS Symposium Series No. 538, Fluorescent Chemosensors of Ion and Molecule Recognition*, ACS Books: Washington (D.C.), pp. 85-103.
14. Bell, T. W., Boguslavsky, L. (1994) Cation Response of Torand Thin Films at the Air-Water Interface, *Langmuir*, **10**, 991-993.
15. Bell, T. W. and Jousselin, H. (1991) Expanded Heterohelicenes: Molecular Coils That Form Chiral Complexes, *J. Am. Chem. Soc.*, **113**, 6283-6284.
16. Bell, T. W. and Jousselin, H. (1994) Self-Assembly of a Double-Helical Complex of Sodium, *Nature*, **367**, 441-444.
17. Bell, T. W., Cho, Y.-M., Firestone, A., Healy, K., Liu, J., Ludwig, R. T., Rothenberger, S. D. (1993) in J. P. Freeman (ed), *Organic Syntheses, Collective Volume VIII*, Wiley, New York, pp. 87-93.
18. Constable, E. C. (1992) Oligopyridines as Helicating Ligands, *Tetrahedron*, **48**, 10013-10059.
19. Ludwig, R. T. (1992) The Synthesis of Functionalizable Torands, Ph.D. Thesis, State University of New York, Stony Brook, NY.

QMVIEW: AS A SUPRAMOLECULAR VISUALIZATION TOOL

Kim K. Baldridge*
San Diego Supercomputer Center
P.O. Box 85608 ,San Diego, California 92186-9784
Jerry P. Greenberg
San Diego Supercomputer Center
P.O. Box 85608, San Diego, California 92186-9784

Introduction

Utilization of computationally derived chemical and physical properties, in conjunction with experiments, has vastly enhanced the understanding of the properties of technologically important chemical structures. The application of computational techniques applied on the atomic scale such as a) the determination of reaction mechanisms, b) the study of the details of molecular forces and their role in structure determination, and c) the calculation of detailed potential energy surfaces and dynamics for reaction processes have led to advancements in areas such as materials chemistry, electronics, environmental chemistry, and medicinal chemistry. The ultimate goal of these calculations is the creation and understanding of 'designer' molecules that perform certain tasks within complex reaction chains and cycles. Because of the requirements that these types of molecular systems must display a special uniqueness of action or efficiency in response, the designer typically must meet strict criteria on specific structural tolerances. The resulting degree of complexity in these molecular blueprints increases at a rate only manageable by advanced high performance computing methods, such as massive parallelization, or ultrafast vectorization, as well as networked communications for data transfers.

Despite the extraordinary abilities of modern hardware technology and coding methods to manipulate the raw data that is calculated at an ever-increasing rate, the rate-limiting step in harnessing the intricacy and precision required to advance the desired chemical frontiers is ultimately the process of man-machine information transfer. Advancements in the development of versatile and facile three-dimensional networked visualization tools are essential to any successful human interface in this endeavor.

Quantum Mechanical View (QMVIEW) has been designed to provide the chemist with an expansive array of molecular perspectives. QMVIEW is an integrated visualization package which capitalizes on the increased capabilities of new graphics systems to profile three-dimensional molecules not only by their common ball-and-stick or space-filling models, but also by molecular orbitals, electron densities (differential and absolute), electrostatic potential gradients, vibrational normal modes, or regional hydro- or lipophillicity. Each profile can be adjusted, updated and presented three-dimensionally, fully colorized and in real time. All of these capabilities greatly

J. S. Siegel (ed.), Supramolecular Stereochemistry, 169–177.

facilitate the analysis of data produced from experimental and/or theoretical technique especially in the area of supramolecular chemistry where the molecules are large and awkward to visualize via human imagination.

Overview of program

QMVIEW is written primarily in C using the Silicon Graphics GL™ graphics library. Several of the quantum mechanical components (e.g., the orbital plotting procedure) are written in FORTRAN and interfaced with the main structure via file I/O. In addition, QMVIEW can act as a direct interface to the widely used quantum chemistry package GAMESS [1,2] where the latter is running on a supercomputer, via UNIX™ sockets written in C and called from within GAMESS.

User-interface communication is performed primarily through mouse activated icon boxes. In interactive mode, the user is initially presented with a screen as shown in Figure 1.

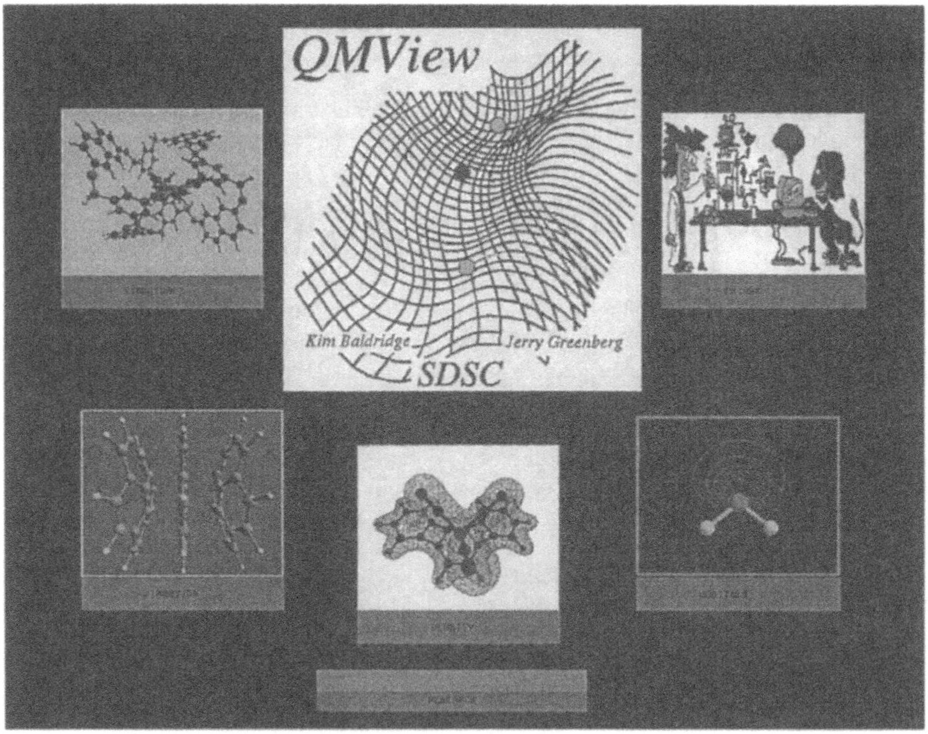

Figure 1. The first layer of the graphical user interface of QMVIEW. The icons represent QMVIEW functions (chosen by picking with the mouse):

Upper left:	Depiction of the molecular structure of kuratowskaphane.
Lower left:	Depiction of the bowl-to-bowl vibrational motion in corannulene.
Lower middle:	Depiction of the total electron density of lepidopterene.
Lower right:	Depiction of the molecular orbitals for water.
Upper right:	This icon allows one to view a) a depiction of a calculation as it is running on a supercomputer platform .
RESEARCH:	This icon allows visualization of user-generated input.

The five large icon boxes give basic demonstrations of QMVIEW's capabilities, which include 1) display and manipulation of molecular structure, 2) display of quantum mechanically generated vibrational motions, 3) display of quantum mechanically generated electron densities, molecular electrostatic potential and density difference maps, 4) display and manipulation of quantum mechanically generated molecular orbitals, both two-dimensional and three-dimensional, and 5) additional features explained in detail later. The icon on the very bottom of the QMVIEW interface, labeled 'research,' is intended for user-generated input for any of the above-listed capabilities. Activation of any of the interface icons will pop up a secondary interface intended for direct manipulation of the molecular system and associated properties. This interface is illustrated in Figure 2. Description of these capabilities and individual features are addressed below [3].

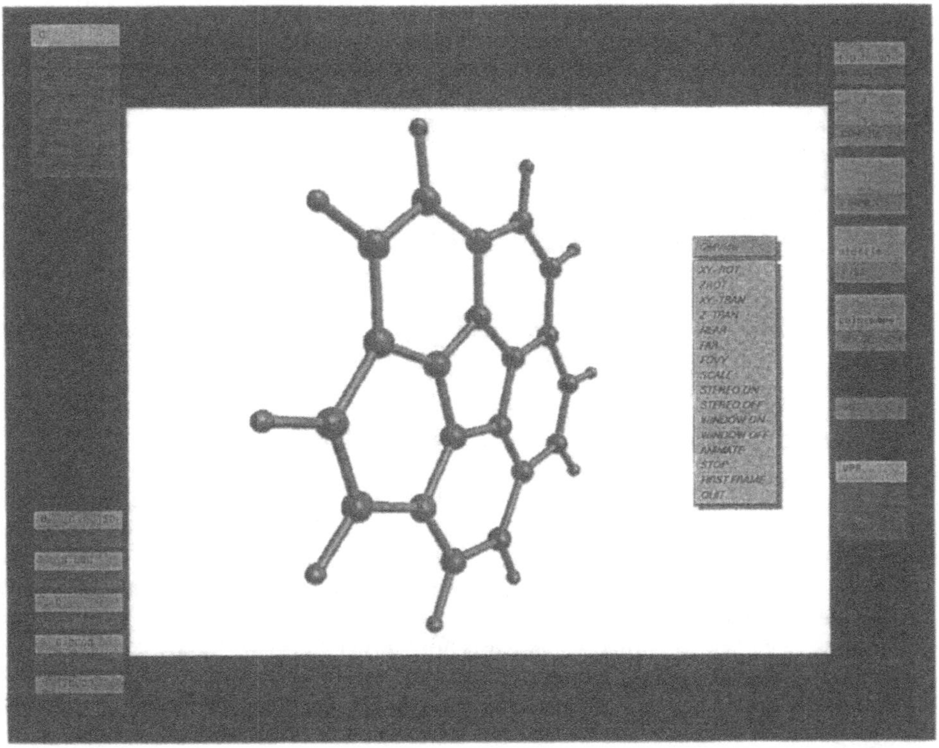

Figure 2. The second layer of the graphical user interface of QMVIEW. This layer may be accessed directly with command line arguments, thereby bypassing the first layer graphical user interface. Various functionality are depicted as icons along both left and right sides, as well as via a list that pops up from the right mouse button.

Display and manipulation of molecular structures

One way in which we can see into the microworld of molecules is via static visualization of basic structure. Experimental techniques, such as X-ray diffraction, give three-dimensional structure of atoms in terms of x,y,z coordinates data. X-ray crystallography, over the last eighty years or so, has developed into a very powerful experimental method for the study of three-dimensional chemical structure at atomic resolution. Applications of crystallographic databases has been greatly enhanced over more recent years with improved worldwide accessibility and gradually improving software systems. Indeed, with the databases themselves now well established, the development of sophisticated and user-friendly software interfaces is of major interest.

Using theoretical techniques, such as molecular mechanics or quantum mechanics, it is also very important to be able to display static structure given basic Cartesian coordinate data from any sort of minimization technique. Important applications include viewing of input for structural verification, visualization of intermediate structures from complicated minimization procedures, and viewing of minimized structures for comparative purposes. QMVIEW has the following types of static structure visualization capabilities:

i. Atoms as points, circles, or spheres with arbitrary size and color features.
ii. Bonds as simple lines, cylinders or cones of arbitrary size and color.
iii. Variable atomic surface characteristics: diffuseness, specularity, stereo, perspective, option for space filling models.
iv. Rotation, translation, and scaling of molecules
v. Stereo projection of images.
vi. Ability to query for structural information (bond lengths, angles, torsions).
vii. Ability to manipulate individual atoms, bonds, angles, or torsions.

Display and manipulation of vibrational frequencies

Static viewing of molecular structure can hold limited value for extracting any information about the dynamic motion of molecules. In general, molecules at room temperature, are not static, but vibrate and spin much faster than is observable. In addition, in the process of chemical transformation, also a dynamic process, atoms and molecules encounter one another, interact and go on to form new species. These processes take place on a time scale that is normally not observable by the human eye. Therefore, visualization techniques can be very useful for allowing one to study the dynamics of molecular motion, just as one would use spectroscopy as an experimental tool for understanding dynamic processes. This is particularly true for supramolecular systems, where it becomes more difficult to imagine what the various molecular vibrations will be.

In particular, QMVIEW can display, in real time, the vibrational frequencies as calculated from quantum chemical calculations. Given basic structural information as well as one or more calculated vibrational frequencies and corresponding modes as extracted from theoretical calculation, one can study visually the motion of individual vibrations in accord with their respective frequency. This kind of display can be especially enlightening in the study of transition state structures. In general, the calculation of transition state structures is very difficult, and even when one is found, it is not always the structure that one assumes belongs to the particular reaction of study. Visualization of the imaginary mode for the transition state can give some information in that regard.

A second very useful application is the display of quantum mechanically determined trajectories. There are two situations where this could be useful. One is the visualization of the series of configurations that are generated as one traverses over a potential energy surface in search of an energy minimum. A second, perhaps more useful application, is the visualization of a quantum mechanically determined minimum energy reaction path. Much information can be gleaned for viewing the change in structure and energetics from a transition state down to it reactants on the one hand, and products on the other hand. QMVIEW is capable of both of these tasks, given only the basic structural information at each of the steps of the trajectory.

Display of electron density and associated properties

Electron density, individual atomic charges, and molecular electrostatic potentials are properties of molecules which permeate nearly every facet of the study of reactivity of molecules. Such properties have been used as an index of reactivity of atomic sites in molecules, as an aid to understanding molecular moments, for understanding intermolecular forces, for interaction energy calculations in molecular complexes, for packing energy calculations in crystals, as well as a variety of other purposes.

Although total electron density plots can be visualized using QMVIEW, a more useful plot is a density difference plot. The latter map is the result of a subtraction of two pieces of electron density for a molecular system. For example, the density difference map might be a whole molecule's density minus the density of the individual atoms, or perhaps a whole molecule's density minus the density of molecular subunits larger than atoms. This kind of map can be very useful in the case of aromatic systems, where one may be interested in visualizing the extent of perturbation induced as one varies a substituent on the aromatic piece. In this case, the density difference map would be the whole molecule's density minus the density of the unsubstituted parent aromatic system. The resulting plot would show the buildup (depletion) of electron density in terms of negative (positive) contours.

Unlike net atomic charges, the molecular electrostatic potential (MEP) is a rigorously defined quantum mechanical property that can be visualized. These MEP maps show the energy of attraction or repulsion for a test positive charge, +q. The units of the contour lines are kcal/mole*e, where e is the electron charge. Therefore, negative contours show where electrophilic reagents will prefer to attack a molecule, while positive contours show the preferred sites of nucleophilic attack. These maps can be generated and viewed in three-dimensions by QMVIEW.

The display of any of these contour plots can be in terms of two-dimensional planes passed through the three-dimensional structure, or, in terms of three-dimensional contours of either net, solid [4] or transparent surfaces. In addition, one can create a transparent three-dimensional surface, with a superimposed two dimensional plane of particular contours. Contours can be color scaled with respect to extent of negative (positive) values. Figure 3 illustrates the total electron density of the anthracene photodimer, lepidopterene in terms of a net surface, as calculated using the Density

Functional Theory software package, DMol [5]. All transformation features as described for molecule manipulation work as well as for manipulation of a contoured structure.

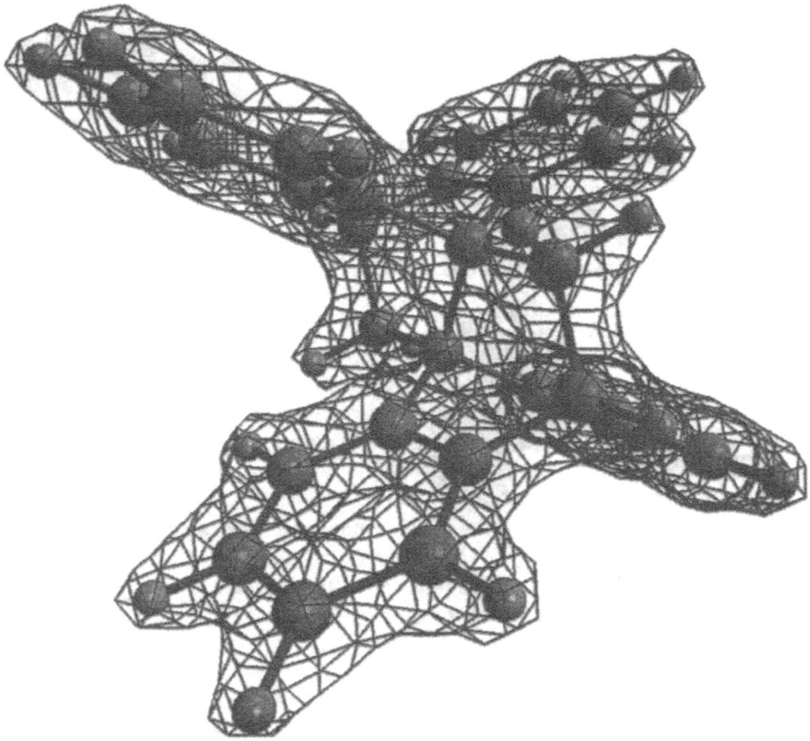

Figure 3. Depiction of the total electron density of lepidopterene as calculated using Local Density Functional Theory as implemented within DMol [5].

Display of molecular orbitals

Molecular orbitals are at the base of our interpretations of molecular structure, and we almost always express these orbitals in terms of, or in relation to, the orbitals of the constituent atoms. Study of the details of molecular orbitals can also be very important to the interpretation of many reaction mechanisms. In particular, for relatively large nonplanar supramolecular systems, subtle structural and electronic properties can not be understood without detailed inspection of the relationship among the molecular orbitals. Visualization of the 'interactive' set of molecular orbitals can clearly illustrate interactions between substructures, as well as the ordering of the molecular orbitals in terms of orbital energies.

QMVIEW contains a module that calculates contour plots of molecular orbitals expanded in a basis of s,p,d contracted Cartesian Gaussian [6] functions, or in a basis of s,p Slater [7] type functions, from the wavefunction of a previous quantum mechanical calculation. QMVIEW can then display the individual molecular orbitals either two-dimensionally, or three-dimensionally. Similar to the electronic map display, one can display in terms of two-dimensional planes passed through the three-dimensional structure, or, in terms of three-dimensional contours of either net, solid or transparent surfaces. The contours can be color scaled with respect to extent of negative (positive) values as well as whether the orbital is bonding or antibonding.

An added feature is the ability to display a sequence of molecular orbitals as a function of a reaction coordinate. This allows one to explicitly see the changes that can occur as a reaction progresses from on point to another on a potential energy surface.

Additional features

QMVIEW has been programmed to run in a distributed mode with other platforms. In distributed mode, a remote machine, such as a CRAY C90 or an Intel Paragon supercomputer computes the information that QMVIEW displays in real time. The connection between the SGI and the remote machine is via UNIX™ sockets written in C and called from within the quantum mechanics package. Once the appropriate information is provided, QMVIEW displays the first configuration that is produced by the quantum mechanics package, along with (optionally) two inset x-y graphs, one of which monitors the energetics as a function of the calculation iteration, and the other which monitors the root-mean-square of the minimization procedure with respect to geometry as a function of calculation iteration. This feature allows the user to monitor and check on long quantum chemical calculations to make sure it is proceeding correctly.

Output capabilities

QMVIEW uses the San Diego Supercomputer Center (SDSC) "vpr" program [8] to make hardcopy outputs of structures on color copiers and film. QMVIEW dumps an SGI image file to disk and "vpr" converts the raster image to the form required by the particular output device. The SDSC image tools may easily be incorporated in order to produce output on, for example, PostScript printers via the standard UNIX™ "lpr" utility.

QMVIEW also has several "camera control" options for making presentation videos (e.g., for the classroom or for seminar purposes). Rotations, translations, scaling and clipping may be programmed along with, for example, dynamical or vibrational progression. For example, a user may choose to display a vibrational mode while rotating the structure simultaneously (as seen in the video presentation at this poster session). They may then select the angle of rotation about one or more coordinate axes and also choose the number of vibrational iterations to skip over for each frame of film or tape. After each frame has been recorded, a "checkpoint" file is written that saves all the characteristics of the vibrational animation (transformation matrices, colors, bonds, etc.) so that the user may stop recording and continue at another time. When QMVIEW is restarted using the same input files, the checkpoint file is read automatically and the last configuration saved appears on the screen.

Using the "animate" feature, a user may predefine, within the input file, vector movements for all atomic positions where each vector set defines some type of motion (e.g. along an axis, diagonal, etc.). Applications of this feature include 1) the illustration of the superposition of conformers, and 2) the illustration of motion along a quantum mechanically calculated reaction path.

When all the motion options are selected, the user may select the "GO" menu item and the recording will proceed automatically. From our experience, we are able to record up to about one thousand NTSC resolution frames per hour to a videodisk depending on the size and type of data being displayed and on the particular SGI hardware being used. At SDSC, non-proprietary programs are available for controlling SONY laser disks directly from SGI workstations. The "record" command is called from within QMVIEW via the UNIX™ library routine "system" after a frame has been rendered and may be modified to suit a particular site.

Future Directions

Although the era of high-performance computing has allowed phenomenal capabilities for computing larger and more complex problems at ever-faster speeds, the scientific visualization of the complex data and the ability to visually teach fundamental chemical principles have captured the imaginations of scientists and educators in productive ways. QMVIEW will continue to be developed via collaborations in both research and education. The package is being distributed with the companion package, GAMESS, a widely distributed quantum mechanical package, on a nonprofit basis although, a stand-alone version plus a manual is also available by contacting the authors [9]. Beta test sites have already been successfully established at UCSD, ISU, and two industrial research sites. The currently available version is for the SGI platform, although future plans include a port to OpenGL™ so that other workstations can be used.

Acknowledgments

We thank the NSF (Grant No. ASC-8902827 and No. ASC-9212619) and the San Diego Supercomputer Center for support of this work.

References

1. GAMESS: General Atomic and Molecular Electronic Structure System, is an ab initio quantum chemistry package that allows study of chemical systems containing atoms through radon from the simplest closed shell case up to a general MCSCF case, permitting calculations at the necessary level of sophistication. For detailed information on this code and its parallel implementation, see references given in 2.
2. a) Schmidt, M.W., Baldridge, K.K., Boatz, J.A., Jensen, J.H., Koseki, S., Gordon, M.S., Nguyen, K.A., Windus, T.L., and Elbert, S.T. (1990) The General Atomic and Molecular Electronic Structure System, *QCPE Bull.* **10**, 52.
b) Schmidt, M.W., Baldridge, K.K., Boatz, J.A., Elbert, S.T., Gordon, M.S., Jensen, J.H., Koseki, S., Matsunaga, N., Nguyen, K.A., Su, S., and Windus, T.L. (1993) The General Atomic and Molecular Electronic Structure System, *J. Comp. Chem.*,**14**, 1347-1363.
3. Baldridge, K.K., Greenberg, J.P. (1995) QMVIEW: A Computational Visualization Tool at the Interface Between Molecules and Man, *J. Mol. Graphics*, in press.
4. a) Lorensen, W.E. (1987) Marching Cubes: A High Resolution 3D Surface Construction Algorithm, *Computer Graphics* **21**, 163-165.
b) We would like to acknowledge Jonathon Shade, SDSC, and Michael Bailey, SDSC, for their help in the coding and implementation of these methods into QMVIEW.
5. a) VernonClark, R., Battersby, T., Gantzel, P., Chadha, R., Baldridge, K.K., and Siegel, J.S. (1995) Pi-sigma-pi Through-Bond Coupling and 'Long' C-C Single Bonds, *J. Am. Chem. Soc.*, *in preparation.*
b) DMol is a program developed at Northwestern for quantum mechanical calculations using Density Functional Theory: Delley, B. (1990) Density Functional Theory of Molecules, *J. Chem. Phys.* **92**, 508-514.
6. a) Boys, S.F. (1950) Electronic Wavefunctions I. A General Method of Calculation for the Stationary States of Any Molecular System, *Proc. Roy. Soc. (London)* **A200**, 542-554.
b) Shavitt, I. (1962) Methods in Computational Physics, Wiley, New York.
7. a) Slater, J.C. (1929) The Theory of Complex Spectra, *Phys. Rev.* **34**, 1293-1323.
b) Slater, J.C. (1930) Cohesion in Univeriant Metals, *Phys. Rev.* **35**, 509-529.
8. Nadeau, D.R., Elvins, T.T., and Bailey, M.J. (1991) Image Handling in a Multi-vendor Environment, *IEEE Computer Society Press Reprint*, 276-279.
9. Dr. Kim Baldridge/Dr. Jerry Greenberg
P.O. Box 85608
San Diego, CA 92186-9784
kimb@sdsc.edu
jpg@sdsc.edu

Cis-trans-trans-1,2,3,4-Tetracyclohexylcyclohexane: a Monocyclic Unconstrained Cyclohexane Existing in a Twist-Boat Conformation

Ishay Columbus, Shmuel Cohen and Silvio E. Biali*

Department of Organic Chemistry

The Hebrew University of Jerusalem, Jerusalem 91904, Israel.

Introduction. The conformational analysis of cyclohexyl rings remains one of the central topics in organic stereochemistry. The conformational preferences of alkylcyclohexanes can be summarized by a series of rules: (i) Monocyclic nonconstrained alkyl-substituted cyclohexanes prefer the chair (**1**) over the twist-boat (TB, **2**) conformation. (ii) An equatorial disposition of a substituents R (**3**) is favored over an axial one (**4**).[1,2] In the last years we have been engaged in a research project aimed to test the boundaries of these rules. Our work concentrated on monocyclic polyalkyl-substituted cyclohexanes. Since in these systems no special structural feature (such as an intramolecular hydrogen bond) is capable of stabilizing the high energy conformation (with the possible exception of attractive van der Waals interactions), the reversal of the usual conformational behavior can only be attained by the selective relative destabilization of the "normal" conformation.

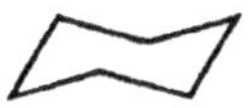

1

2

3

4

J. S. Siegel (ed.), Supramolecular Stereochemistry, 179–183.

Axial/equatorial stability reversal in polyalkylcyclohexanes. We have shown previously that in contrast with *all-trans*-1,2,3,4,5,6-hexaethylcyclohexane (**5**) which exists in a "normal" conformation in which all groups are disposed equatorially,[3] in *all-trans*-1,2,3,4,5,6-hexaisopropyl-cyclohexane (**6**) all the alkyl groups are located at axial positions with their methine protons *anti* to the cyclohexyl methine hydrogens.[4] The equatorial conformation of **6** is destabilized by unavoidable eclipsing interactions which result when several isopropyl groups are oriented in a homodirectional fashion. According to MM2 calculations, at least four neighboring isopropyl groups are needed for the relative stabilization of the axial form in *all-trans* systems.[4] In monoalkyl cyclohexane systems, the substituent is located in an equatorial position, but as molecular mechanics calculations indicate, if the substituent is U-shaped the axial position may be preferred.[5]

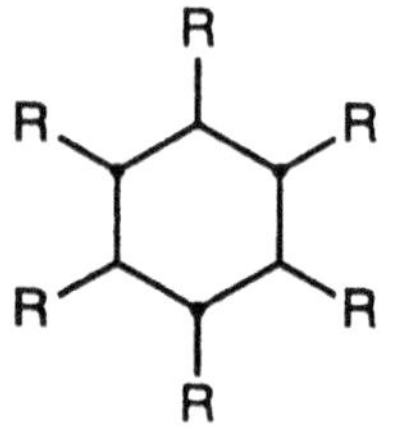

5 R = Et
6 R = i-Pr

Chair/TB stability reversal. A classical example of a conformational flexible system adopting a nonchair conformation is the dihydroxycyclohexane **7** in which the TB form is stabilized by an intramolecular hydrogen bond.[6] In some hydrocarbons (e.g., twistane) the cyclohexane ring is constrained to exist in a TB conformation by the 1,4 carbon bridges. Only two types of unconstrained monocyclic hydrocarbon systems are known were the TB form is energetically preferred over the chair conformation. The first type comprises tert-butyl substituted cyclohexanes like *cis*-1,4-di-tert-butylcyclohexane in which the chair conformation is destabilized by the presence of an axial tert-butyl group.[7] The second type comprises the per(cycloalkylated)cyclohexanes studied by Fitjer's group.[8] In the latter systems, the groups directly attached to the spiro atoms are located in the chair form far from the "natural" axial and equatorial positions due to the geometrical requirements of the spirocycloalkyl groups and this special geometry affects the chair/TB energy gap. This is shown by dodecamethylcyclohexane, which although sterically similar to the per(cycloalkylated) systems, exists in a chair conformation.[9] Not surprisingly, the accepted view is that "extreme substitution" is needed for the relative stabilization of the TB form.[1b]

For a monocyclic polyalkycyclohexane to exist in a TB conformation, the steric hindrance of the chair form must be larger than the destabilization of the TB form. Our recent molecular mechanics calculations indicate that the relatively unhindered 1,1,3,3,5,5-hexaisopropyl cyclohexane (**8**) and *cis-trans-trans*-1,2,3,4-tetraisopropylcyclohexane (**9**) prefer the TB over the lowest energy chair form by 0.9 and 2.2 kcal mol^{-1}, respectively.[10] In the case of **9** the *cis-trans-trans* disposition of the substituents is unique in relatively destabilizing the two possible chair forms (with one and three axial substituents) and rendering the TB the lowest energy conformation.

***cis-trans-trans*-1,2,3,4-tetracyclohexylcyclohexane.** System **9** could be prepared in principle by catalytic hydrogenation of the unknown 1,2,3,4-tetraisopropylbenzene. Since our efforts to prepare the compound by methylation of a 1,2,3,4-tetramethylbenzene derivative proved unsuccessful, and an isopropyl and a cyclohexyl group are approximately isosteric in the vicinity of their methine groups,[11] the target was changed to *cis-trans-trans*-1,2,3,4-tetracyclohexylcyclohexane (**10**).

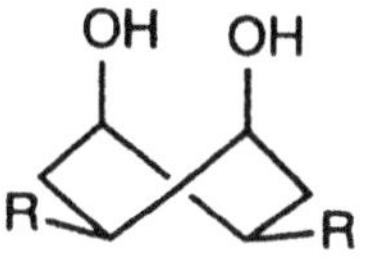

7

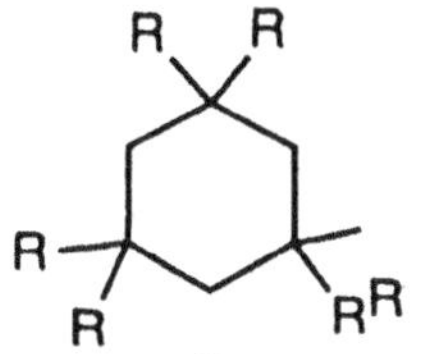

8

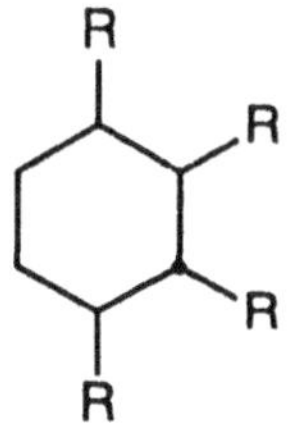

R = i-Pr

9 R = i-Pr
10 R = c-C_6H_{11}

Hydrogenation of 1,2,3,4-tetraphenylbenzene with Ru/C or Rh/C (1:1 w/w, c-C_6H_{12}, 600 psi H_2, 250°C, 24 hrs) afforded a main product which displayed in the ^{13}C NMR spectrum both aromatic and aliphatic signals, indicating only partial reduction of the aryl rings of the starting material, and was characterized by X-ray crystallography as 2,3,6-tricyclohexylbiphenyl. Exhaustive hydrogenation of 1,2,3,4-tetraphenylbenzene was accomplished by using Pd/C catalyst (1:1 w/w, c-C_6H_{12}, 600 psi H_2, 170°C, 24 hrs) yielding a mixture of the *cis-trans-cis* and *cis-trans-trans* stereoisomers of 1,2,3,4-tetracyclohexylcyclohexane. X-ray diffraction analysis of a single crystal of the *cis-trans-cis* isomer indicates that all rings adopt the usual chair conformation. However, as shown by the X-ray data, in the *cis-trans-trans* isomer the central ring exists in a twist-boat conformation.[12] This represents the first X-ray structure of an unconstrained monocyclic polyalkylcyclohexane which exists in a TB conformation. The dihedral angles defined by the carbons of the central ring of one of the two enantiomeric forms present in the crystal are 37.2°, 16.4°, -48.8°, 24.0°, 28.6° and -60.1°. The external cyclohexyl rings occupy the pseudoequatorial, pseudoaxial, isoclinal and pseudoequatorial positions of the TB forms, and have HCCH dihedral angles of -173.4°, -75.4°, -72.4°, -73.8°, respectively.

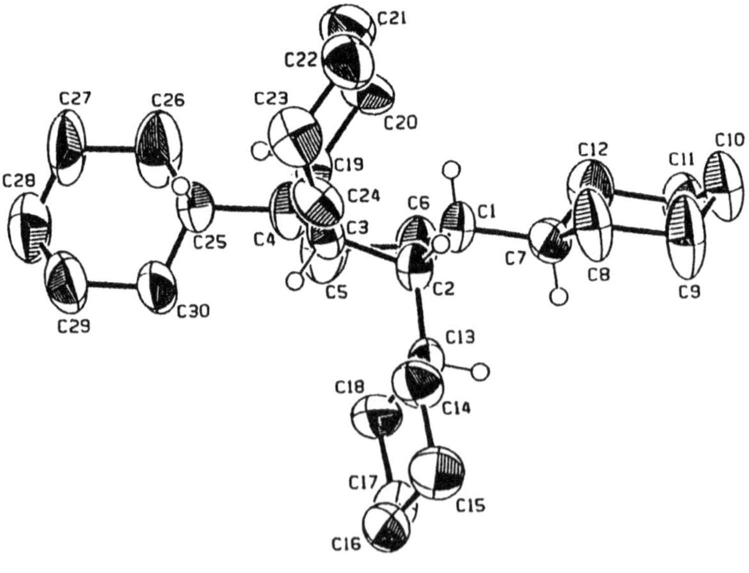

Crystal structure (side view) of *cis-trans-trans*-1,2,3,4-tetracyclohexylcyclohexane

Conclusions. There is no need for "extreme substitution" for achieving the relative stabilization of the TB conformation. As shown both computationally and experimentally, moderately hindered systems such as **9** and **10** adopt a TB conformation.

References and Notes.

1. (a) Mann, G. Z.(1990), *Z. Chem.* **30**, 1. (b) Anderson, J. E. (1992), in *The Chemistry of Alkanes and Cycloalkanes*, Patai, S., Rappoport, Z., Eds., Wiley, Chichester, ch. 3.

2. For a review on nonchair conformations of cyclohexane see: Kellie, G.M. and Riddell, F.G. (1974), *Top. Stereochem.* **8**, 225.

3. Immirzi, A. and Torti, E. (1968), *Atti Accad. Naz. Lincei, Cl. Sci. Fis. Nat. Rend.* **44**, 98.

4. Golan, O., Goren, Z., and Biali, S.E. (1990), *J. Am. Chem. Soc.* **112**, 9300.

5. Biali, S.E. (1992), *J. Org. Chem.* **57**, 2979.

6. Stollow, R.D., McDonagh, P.M., and Bonaventura, M.M. (1964) *J. Am. Chem. Soc.* **86**, 2165; Stollow, R.D. (1964) *J. Am. Chem. Soc.* **86**, 2170.

7. See for example: Unwalla, R.J., Profeta, S., Jr., and Cartledge, F.K., (1988) *J. Org. Chem.* **53**, 5658.

8. Fitjer, L., Scheuermann, H. J., Klages, U., Wehle, D., Stephenson, D.S., and Binsch, G. (1986) *Chem. Ber.* **119**, 1144; Traetteberg, M., Bakken, P., Seip, R., Fitjer, L., and Scheuermann, H.J. (1987) *J. Mol. Struct.*, **159**, 325; Fitjer, L., Klages, U, Kühn, W., Stephenson, D.S., Binsch, G., Noltemeyer, M., Egert, E., and Sheldrick, G.M. (1984) *Tetrahedron* **21**, 4337.

9. Wehle, D., Scheuermann, H.-J., and Fitjer, L. (1986) *Chem. Ber.* **119**, 3127.

10. Wieser, J., Fitjer, L., and Biali, S.E. submitted for publication.

11. A recent comparative study of tetraisopropylethene and tetracyclohexylethene (Columbus, I. and Biali, S.E. (1994) *J. Org. Chem.* **59**, 3402) shows that both compounds adopt very similar conformations.

12. Columbus, I, Cohen, S., and Biali, S.E. *J. Am. Chem. Soc.* in press. For other studies on polycyclohexyl systems see: Columbus, I. and Biali, S.E. (1993) *J. Org. Chem.* **58**, 7029.

The Conformation of Coumarin-Crown Ethers with Spin-Spin and Site Exchange Simulations

Çakıl ERK
Technical University of Istanbul, Faculty of Art & Sciences, Department of Chemistry,Maslak,80626,Istanbul,TURKEY.

Abstract. The conformation of the fluorogenic host molecules of coumarin-15-crown-5 ethers were examined by ^{1}H and 2D NMR. The gem. and vic. coupling constants were estimated to confirm the *a,±g,a* unit conformations clearly. The proton site exchange rate simulations proved the rate of ± *gauche* interconvertions are not more then 800-1000 Hz which governed the role of molecular recognition.
Key Words: Macrocyclic ethers,coumarin-crowns,NMR,coupling constants,pseudorotation rates.

1. Introduction

The conformationally rigid cavitands are usually slow in forming their complexes whereas the flexible coronands are relatively fast. However, the rapid cation-ligand exchange of complex formation causes the association constants of coronands in the range of K_a ; 10^3-10^5 depending on the solvent [1]. The macrocyclic ether chemistry was oriented to synthesis of some coronands with specific side-groups. The lipophilicity and functionality of the terminal groups of various molecular structures initiated an interesting field [1,2]. The enhanced cation transport properties could be furnished with the different type of terminals likewise the chromophores of ion responsive systems [3].

Recently we initiated the work of design and characterization of the macrocyclic ethers with chromogenic coumarin moieties, Scheme.1. [4]. Our studies aimed the practical applications of the such molecules so that the complexation behavior of this molecules, with alkali and alkaline earth cations could also be applied for analytical procedures of ion selective membrane transport, UV spectroscopy and particularly spectrofluorometric cation determinations. We revealed the coumarin-crown/cation interactions with the fluorescence spectroscopy in CH_3CN. The remarkable changes on fluorescence spectral intensity upon cation addition is quite apparent and surprisingly selective for the cations. The success of this studies is not only the role of the fluorophores but also the hidrophilicity of such systems for recognition [4]. On the other hand NMR spectra of such molecules enabled us to study on their conformations in solutions.

J. S. Siegel (ed.), Supramolecular Stereochemistry, 185–190.

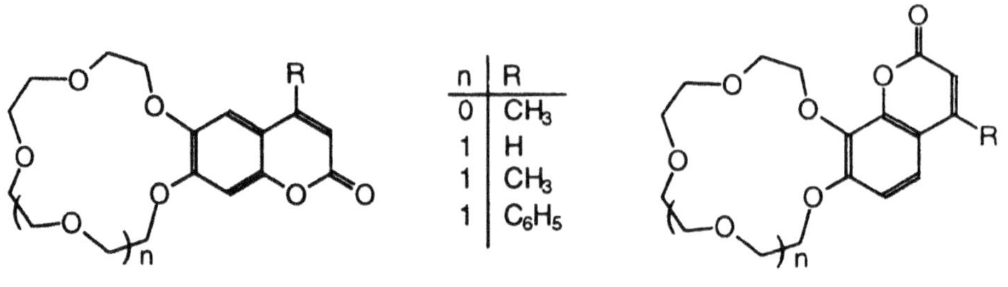

Scheme.1

2. Experimental

Macrocyclic systems were available from the earlier works [4]. Proton and 2D NMR (COSY, INV. CORR.) studies were done on a 500 MHz BRUKER spectrometer. Spin-spin coupling simulations were tried with LAOCOON II. Chemical exchange program was from University of Kaiserslautern as a PC form of program of Prof. Binsch from University of Munich.

3. Results and Discussion

The present work deals with the characterization of conformation of five oxygen member coumarin crowns of cyclic polyoxyethylene bridges via NMR. We studied briefly on the high field ^{1}H NMR spectra of -CH_2-CH_2-O groups attached to aromatic ring. The chemical shifts were well resolved and existed in 3.8-4.5 ppm range for 7,8-derivatives and in 3,7-4,3 ppm for 6,7-derivatives which are not only due to ring current and hinderance effects but also, sterically induced charge polarizations which causes deshielded protons bound to shielded carbons. This is briefly proved from the ^{1}H-^{13}C inverse correlation spectra that the ^{13}C of 1C atom is observed at 69.8 ppm while the 2C is observed at 74.0 ppm in $CDCl_3$ for the compound of n=1, R=Me (see Scheme.1 and figs.1,2). The anizotropy of the benzo-α-pyrone ring would accompany to overall pseudorotations of cyclic polyoxaethylene backbone. Accordingly, geminal methylene protons were chemically equivalent but magnetically non-equivalent exhibiting an AA'BB' (or AA'XX') type of spectra. Therefore, the proton spectra mostly relays on the geminal coupling constants which are not observed but predictable quantum mechanically. The vicinal and geminal coupling constants were estimated by simulations using the observed spectra to conclude the conformation ambiguously in solution, Table.1. The average coupling constants of two conformers found are agreeable with the chemical exchange rate calculations proving the preferred *anti-gauche-anti* $(a, \pm g, a)$ unit conformations. Due to enantiomeric and diastereotopic relationships between the pseudorotamers depicted below the simulated spectra were tried, Scheme.2.

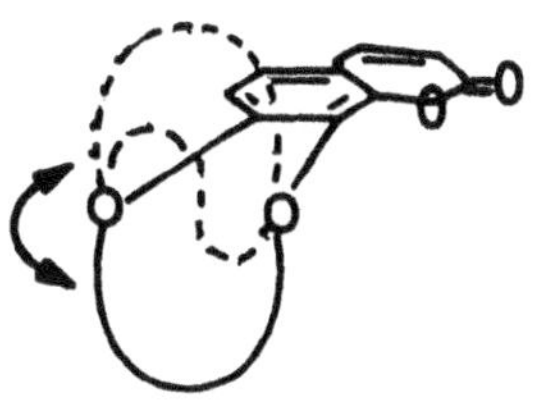

Scheme.2

Table. 1. The partial ^1H NMR of coumarin-15-crown-5 (n=1,R=Me) at 500 MHz (#)

Proton	J_{gem}	$J_{a'b}$	J_{ab}	$J_{ab'}$	J	J'	υH	K(*)
$1H_a$	2.0	9.0	5.0	3.0	6.0	5.0	2210	>.1
$1H_b$	2.0	9.0	5.0	3.0	6.0	5.0	2029	>.1
$2H_a$	6.0	9.0	3.0	3.0	6.0	3.0	2190	≈1
$2H_b$	6.0	9.0	3.0	3.0	6.0	3.0	2036	≈1
$3H_a$	5.0	5.0	6.0	1.0	3.0	6.0	1917	>10
$3H_b$	7.0	5.0	6.0	1.0	3.0	6.0	1904	>10

(*) K and data in Hz, N=J' + J, L=J - J', $J=1/2(J_{ab'} + J_{ba'})$, $J'=J_{ab}=J_{a'b'}$
(#) The results are similar to the dioxo-cyclopolyoxa crowns which have no coalescence down to -100 °C in CD_2Cl_2. see refs. 7-9.

However, the rapid rotamer interconvertion of the *gauche* conformers of the O-C-C linkage was proved with a site exchange simulation program for a non-first-order-band shape calculations [5]. This may in fact do not originate symmetry pairs of oxaethylene conformational units undergoing the interconvertion displaying the AA'BB' spectra. This is exemplified in Table.1. Therefore, the interconvertion (pseudorotation) rates are already depended on the location of macrocyclic back-bone and aromatic terminals causing a torsional network which play important roles. Accordingly, coalescence rates, K, are varied within 250 - 1000 Hz limiting values depending on segmental location which are not more then 4.0-5.0 kcal mol^{-1} of barrier energy at a relevant temperature [6]. The similar results were obtained with the other analogous of the of n=0,1 and R=H, Me, Ph series, Scheme.1. [4b].

This is a good measure of the evidence of cation binding effect for the molecular dynamics of macrocyclic hosts. Accordingly, such type of coumarin-crowns possessed the cation-ligand association constants with Na^+ ion around the

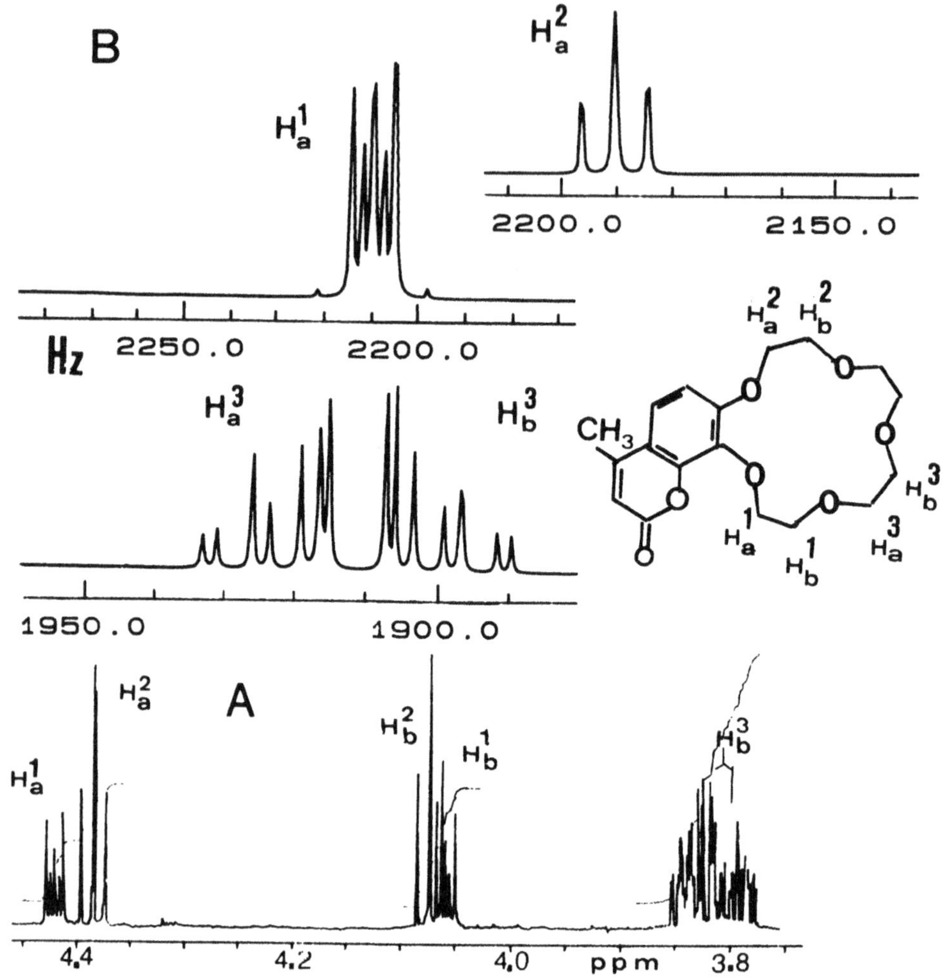

Fig.1. A. 500 MHz proton spectra of coumarin-crown (n=1,R=Me) in $CDCl_3$. B. The simulated spectra of terminal ether protons of the same compound.

logk$_a$; 4.55 - 5.08 for a 1:1 ratio of Complexing in water solutions obtained with a Na^+ ion selective electrode. That means that they are as strong as their benzocrown analogues [4b,c].

However, the fluorescence responses of coumarin-crowns found upon complexation completely structure depended which ruled us to investigate the solution conformations primarily. We therefore explained why the 7,8-derivatives of hindered terminals, OCH_2- in molecules mostly gave the complex formation-enhanced fluorescence spectra whilst the 6,7-derivatives gave the complex formation-enhanced quenching spectra by this way, Fig.3, [2,4d].

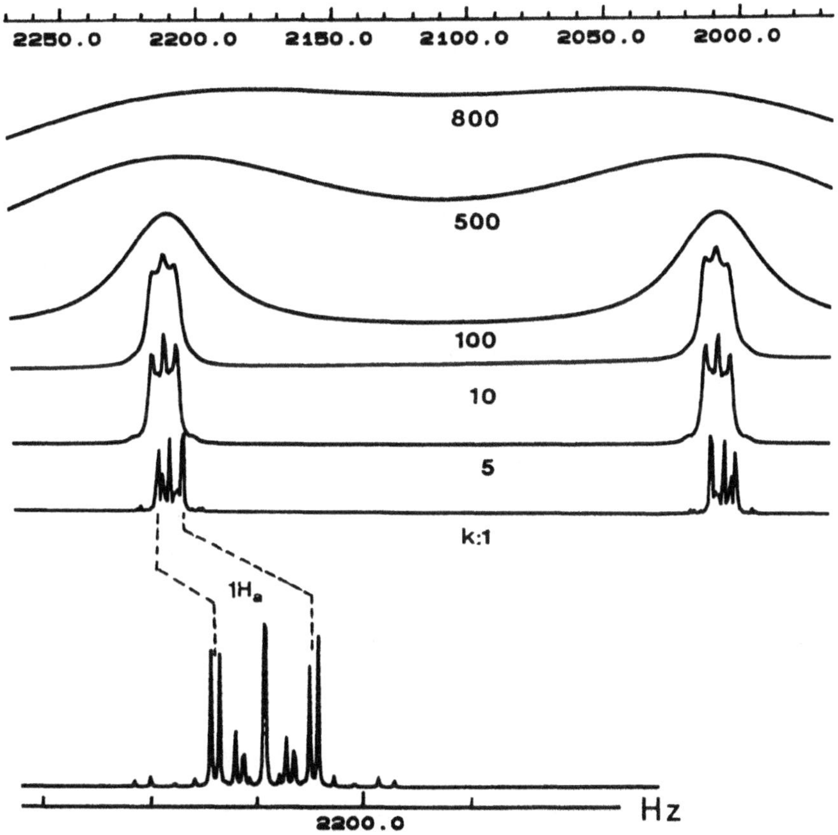

Figure 2. The simulated exchange spectra of compound n=1, R=Me with rates in Hz, T_2 = 5.0 sec

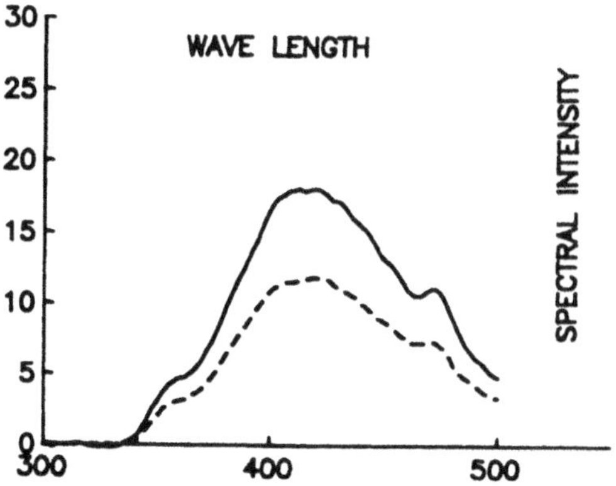

Figure 3. Fluorescence emission spectra of compound n=1, R=Me (.....) and with NaN_3 (—) in CH_3CN.

Acknowledgement,. Research Project supported by Technical University of Istanbul and TUBİTAK / NATO grant for the NATO Advanced Workshop at Hveragerdi, Iceland,1994 are acknowledged.

References

1. Inoue, Y., and Gokel, G. W. (1990) *Cation Binding by Macrocycles.* Marcel Dekker, Basel.
2. a. Vögtle, F.(1991) *Supramolecular Chemistry,* J. Wiley & Sons, Sussex. b. Schneider, H.J. and Durr, H. (1991) *Frontiers in Supramolecular Organic Chemistry and Photochemistry,* VCH, Weinheim.
3. a. Erk, Ç. (1990) The Ion-Dipole Interaction Mechanism of Macrocyclic Ethers with ^{13}C Dipole Relaxation Time Measurements, Part.6. J. Phys.Chem.**94**, 8617-8621. b. Erk, Ç. (1991) An Improved Synthesis of Cyclic Polyether-Esters by Cyclization. *Liebigs.Ann.Chem.*,1083-1084.
4. a. Bulut, M.and Erk, Ç. (1992) The Synthesis of Novel Crown Ethers.Part.I. *Synth. Commun.* **22**,259-263. b. Göçmen, A., Bulut, M. and Erk, Ç. (1993) Synthesis and Charaterization of Coumarin Crown Ethers, *Pure Appl Chem.***65**,447-450. c. Göçmen A., and Erk, Ç. (1993) The potentiometric determination of stability constants of macrocyclic ethers using linear regression techniques. *Fresenius, J. Anal. Chem.*,**347**,471-474. d. Göçmen, A., Bulut, M., and Erk, Ç. (1994) Fluorescence Spectra of Coumarin-Crown Ethers and their Salt Solutions. Spectrosc. Lett., submitted.
5. a. Binsch, G. (1968) *Topics in Stereochem.* **3**,97-192. b. Stevenson, D.S. and Binsch. G., (1970) Automated Analysis of High Resolution NMR Spectra.Part.II. J.*Mag. Res.*,**37**,409-430.
6. Kaplan, J.I. and Fraenkel, G. (1980) *NMR of Chemically Exchanging Systems*. Acad.Press, New York.
7. Steigel, A. (1982) NMR 15, Dynamic NMR Spectroscopy, in Diel, P., Fluck, E. and Kosfeld, R. (eds), p 1, Springer, Berlin.
8. Anet, F.A.L.(1978) *Dynamic Nuclear Magnetic Resonance Spectroscopy,* in Jackman,L.M. and Cotton, F.A. (eds), p 543, Acad Press, New York.
9. a. Erk, Ç. (1986) The 13C Dipole-Dipole Relaxastion and Pseudorotation of Some Macrocyclic Ethers and a Polyoxalactone. *Appl.Spectrosc.***41**,100-103. b. Erk, Ç. (1987) Conformation of Dioxo-polyoxa cycloalkanes by ^{1}H NMR. *Spectrosc.Lett.*,**20**,167-176. c. Erk, Ç. (1989) ^{13}C Dipole-Dipole Relaxation Time and Internal Motions of Oligocyclic Ethers. *J. Mol.Liqs.*,**40**,1-15.

CLATHRATION PROPERTIES OF PROPELLER-SHAPED MOLECULES

PAOLO FINOCCHIARO
Istituto Chimico, Facoltà di Ingegneria
Università di Catania, Viale A. Doria, 6
95125 Catania (Italy).

A common feature to many important host molecules such as triphenylmethanes, *trans, anti, trans, anti, trans*-perhydrotriphenylene, tri-*o*-thymotide, cyclotriveratrylene, and hexa-host, is their ability to assume trigonal symmetry in the solid state [1]. It is thus conceivable that other classes of molecules able to assume a propeller-like conformation will display inclusion ability.
However, among the great variety of aryl-substituted triphenylmethanes studied by us and which are known to adopt a propeller conformation with trigonal symmetry in the solid state as well in solution, only the parent triphenyl-methane itself (TPM), triphenyl-chloromethane, triphenyl-carbinol and TPM derivatives which possesses two hydroxyl groups in the para position, form inclusion compounds.
The host/guest molar ratio of the TPM clathrates is always 1:1 and competitive inclusion experiments indicate that aniline seems to be the guest preferred by TPM [2]. On the other hand, the stoichiometry of the clathrates formed by the para-hydroxy derivatives is dependent upon the solvents and the hosts used [3]. No doubt that this different behaviour is due to the different architecture of the hosts. In the C_6H_6-TPM clathrates the host maintains a propeller-shaped conformation and the guest molecules are located on the C3 axis of the complex [4], whereas the para hydroxy substituted triphenyl methanes form a layer-like structure originated by the connection through hydrogen bonding of each hydroxyl group to the hydroxyl groups of two other molecules, thus forming an interlocking assembly [5].
1,1,2,2-tetrakis-arylethanes (TAE) also adopt a propeller-shaped conformation in solution, as well in the solid state.
TAE **1-5**, however behave very differently: **1** forms clathrates with benzene, toluene, dioxane in a host/guest molar ratio of 2:3, 3:1 and 4:1, respectively, but does not include xylenes and 1,2-dichloroethane. TAE **2** does not form inclusion compounds at all, whereas TAE **3** forms clathrates in a molar ratio 1:1 with almost any kind of solvents (benzene, toluene, xylenes, mesitylene, cyclohexyl-amine, 2-butanone, cyclododecene, ethylacetate, (+)-α-pinene, etc.). TAE **4** and **5**, although very close in chemical structures, are also different in their clathration properties; compound **4** does not form at all inclusion compounds whereas TAE **5** is a good host for benzene and ethylacetate which are enclathrated in a molar ratio host/guest of 2:1 [2].
Why such compounds, which all certainly adopt a propeller-shaped conformation, behave such differently in the formation of clathrates? A tentative hypothesis which can be laid on the table is connected with symmetry and with the number of possible stereoisomers present in such compounds.
In fact, TAE **2** and **4** which both exist as one *dl* pair, behave in the same manner: *i.e.* they do not form clathrates. On the contrary, TAE **3** and **5**, which both are expected to be present as 20 stereoisomers, form strong inclusion complexes.

J. S. Siegel (ed.), Supramolecular Stereochemistry, 191–194.

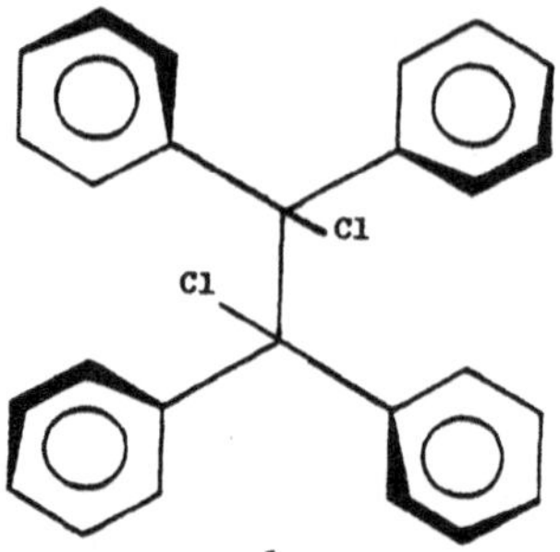

1

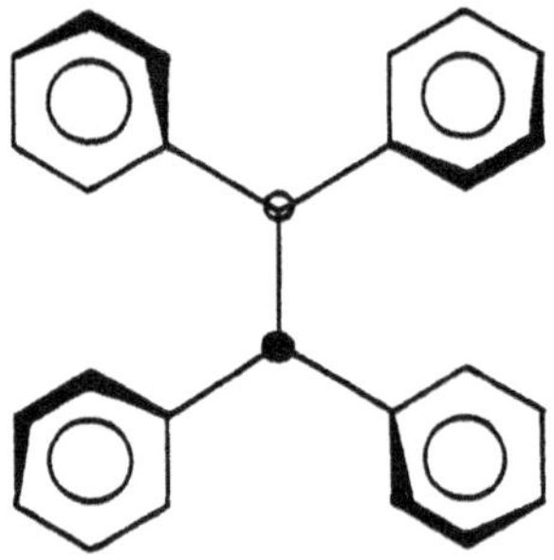

2

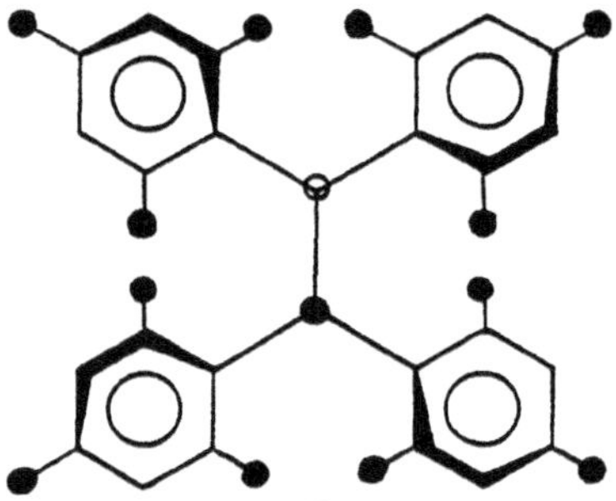

3

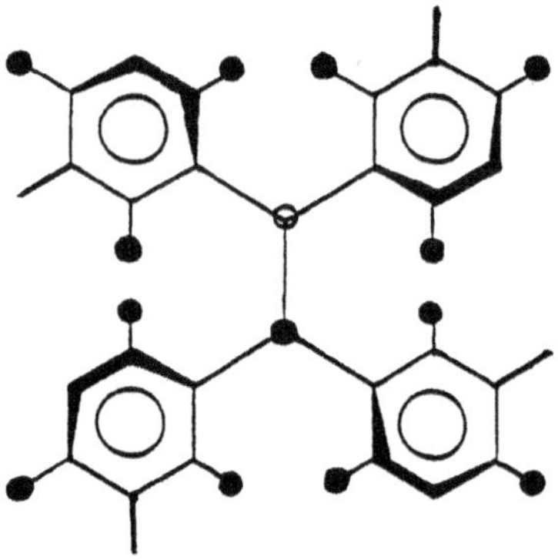

4

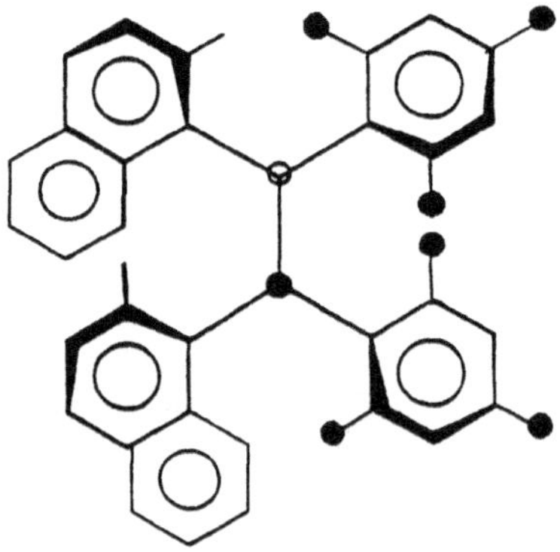

5

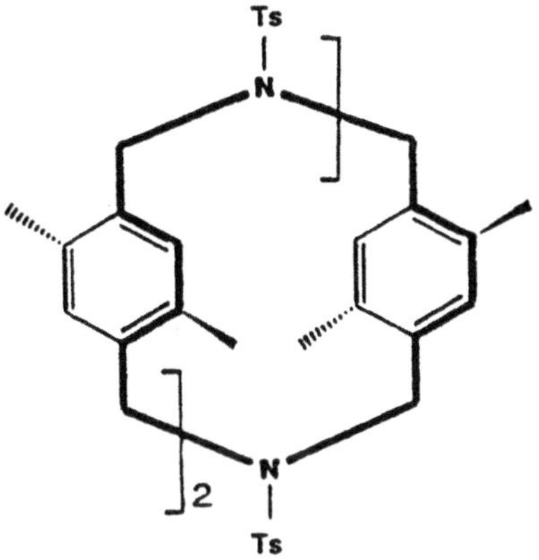

6

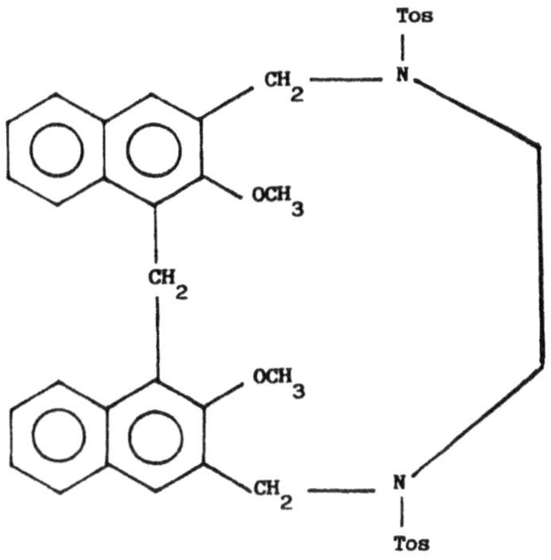

7

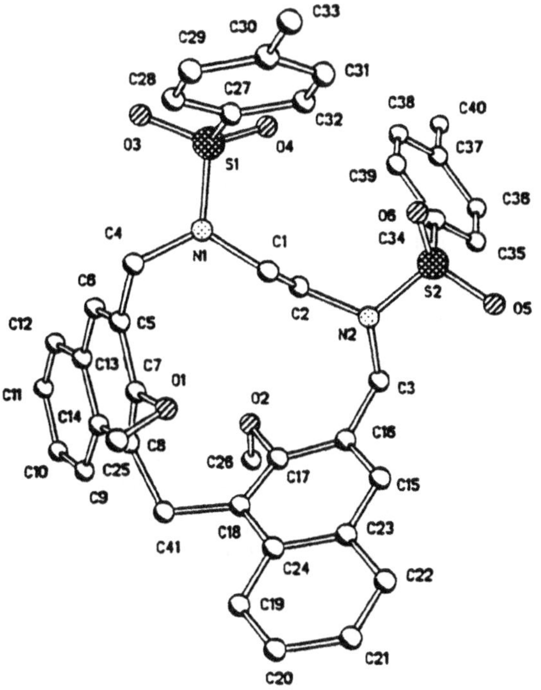

Fig.1. Perspective view of macrocycle 7

N,N',N"-Tritosyl-5,8,14,17,23,26-hexamethyl-2,11,20-triaza [3.3.3] paracyclophane (6) possesses a trigonal symmetry and thus it is not surprising that the three-fold symmetric azamacrocycle **6** was isolated as a crystalline adduct with CH_2Cl_2 [6].
The X-ray structural analysis of **6** crystallized with one mole of CH_2Cl_2 was undertaken and the geometry of the host was thus elucidated. Host **6**, in the solid state possesses a trigonal symmetry with a central cavity in which guests, in principle, can be included. However, the diameter of the internal cavity was found to be too small even for accomodating guests of a small dimension. On the other hand, the molecular packing of single units of **6** leaves in the crystal lattice intermolecular voids of suitable size for clathrating the molecule of CH_2Cl_2 [6].
With the aim of enlarging the central cavity of N-tosyl azamacrocycles, and in order to introduce some rigid "walls", we synthesized the macrocycle **7** which was found to form 1:1 inclusion compounds with $CHCl_3$, CH_3CN and toluene [7].
The solid state structure of **7** solvated with $CHCl_3$ was solved by X-ray diffraction techniques and the molecular conformation of host **7** is reported in Figure 1.
The two naphthyl rings are twisted in the same direction with respected to the reference N(1), N(2) and C(41) plane and by virtue of the presence of the two tosyl groups open voids are formed; such void spaces are defined by the tosyl and naphthyl rings within the same molecular unit, but they are not large enough to incorporate guest molecules. Thus the guests are present in the crystal lattice and not inside the cavity of the host [7].

References

1. MacNicol, D.D. (1984) Structure and Design of Inclusion Compounds: the Hexa-host and symmetry considerations, in J.L. Atwood, J.E.D. Davies and D.D. MacNicol (eds), *Inclusion Compounds*, Academic Press, London, vol. 2, pp. 123-168.
2. Recca, A. Bottino, F.A., Libertini, E. and Finocchiaro, P., (1979) Guest selectivity properties of some aromatic inclusion compounds, *Gazz. Chim. It.* **109**, 213-215.
3. Driver, J.E., and Mok, S.F. (1955) The Condensation of Phenols with Nitrobenzaldehydes, *J. Chem. Soc.* 3914- 3918; Driver, J.E. and Lai, T.F. (1958) Solvated Triphenylmethane derivatives containing bromine, *J. Chem. Soc.* 3219-3220.
4. Allemand, A. and Gerdil, R. (1975) The Structure of the Molecular Complexes of Triphenylmethane with Benzene, Thiophene, Pyrrole and Aniline, *Acta Crystallogr.* **A31**, S130.
5. Barlow, G.B. and Clamp, A.C. (1961) Molecular Complexes based on Derivatives of 4,4'-Dihydroxy-triphenylmethane, *J. Chem. Soc.* 393-395.
6. Bottino, F., Finocchiaro, P, Lipkowski, J. Mamo, A., Pappalardo, S. and Suwinska, K. (1991) Crystal and molecular structure of N,N',N"-Tritosyl-5,8,14,17,23,26-hexamethyl-2,11,20-triaza [3.3.3]-paracyclophane 1:1 Dichloromethane Clathrate, *J. Inclus. Phenom.* **11**, 41-48.
7. Failla, S., Finocchiaro, P., Belsky, V.K., Zavodnik, V.E. and Sobolev, A.N. (1993), Synthesis of N,N'-Ditosyl-8,19-dimethoxy-2,5-diaza-[6.1]-naphthyl-cyclophane: Crystal and Molecular structure of its 1:1 chloroform solvate, *J. Inclus. Phenom.* **15**, 247-258.

CONTROL OF SELF-ASSEMBLY BY ACID-BASE CHEMISTRY:

Reversible Encapsulation of Xenon by Two Identical Half-Shells

R.M. GROTZFELD, N.R. BRANDA, C. VALDES, J. REBEK, JR.
Department of Chemistry, Massachusetts Institute of Technology
77 Massachusetts Avenue, Cambridge, MA 02139

Recent progress in the field of molecular recognition has led to the introduction of self-assembling systems of varying complexity. A simple system in which two identical subunits are held together by hydrogen bonds has previously been developed.[1] Since then, two similar capsule-forming compounds have been synthesized (figure 1).[2] These are based on p-N,N-dimethylaminophenyl- or carbethoxysubstituted glycolurils which are readily available from the corresponding benzil and dihydroxytartaric acid (disodium salt) respectively by condensation with urea. By using other alcohols for the esterification such as isoamyl alcohol or cyclohexanol, soluble glycolurils (e.g. in pyridine or dioxane) are obtained which are promising candidates for the synthesis of more complex self-assembling systems. Alkylation of the glycolurils **1a** and **1b** with 1,2,4,5-tetrakisbromomethylbenzene gave the new self-complementary compounds **2a** and **2b**, both showing enhanced solubility in organic solvents. As expected, both compounds can encapsulate methane, ethane, and ethylene as evidenced by ^{1}H-NMR.

J. S. Siegel (ed.), Supramolecular Stereochemistry, 195–197.

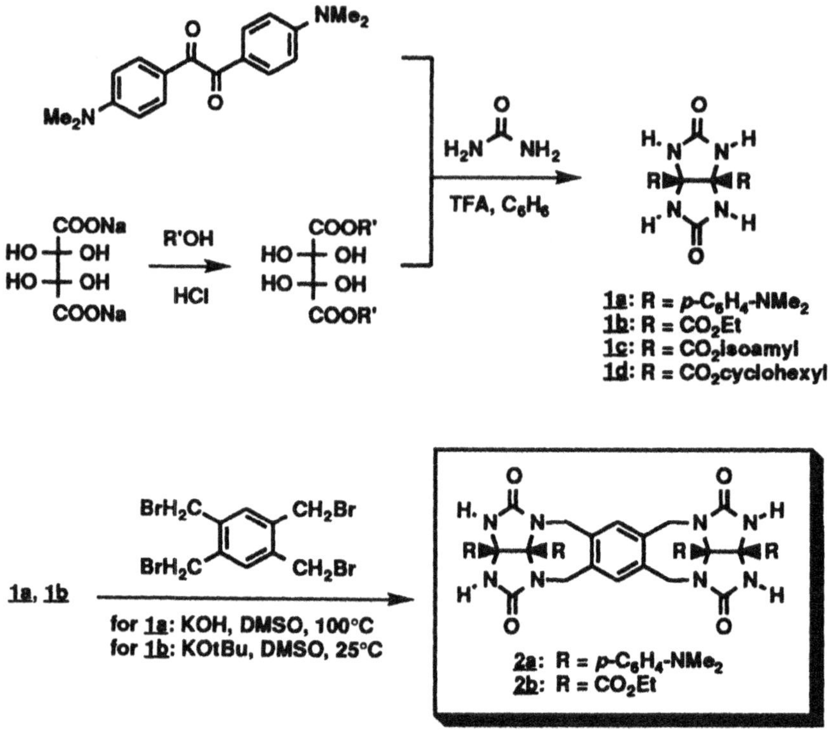

Figure 1: Synthesis of glycolurils and the new self-complementary molecules.

In DMF, compound **2a** does not form a dimer, but dimerization can be induced by addition of small guest species such as methane, ethylene, or xenon. Downfield shift of the N-H proton signal in the hydrogen bound dimer allows monitoring of capsule formation by ^{1}H NMR. Direct observation of the bound guest is possible in the case of methane in which the signal for encapsulated methane appears at -0.92 ppm. For the more soluble compound **2b**, bound xenon can be observed directly by ^{129}Xe NMR in $CDCl_3$.

Compound **2a** was designed so that the basicity of peripheral dimethylamino groups on the glycoluril moieties would allow control of self-assembly by changes in acidity of the surrounding medium. Upon addition of acid (e.g. *p*-toluenesulfonic acid) the NMR-signals characteristic for the xenon binding dimer begin to disappear. A large excess of acid completely shifts the equilibrium to the presumably multi-protonated monomer. Addition of base (Na_2CO_3) regenerates the spectrum of the xenon bound dimer, thus demonstrating the reversibility of the dissociation process (figure 2).

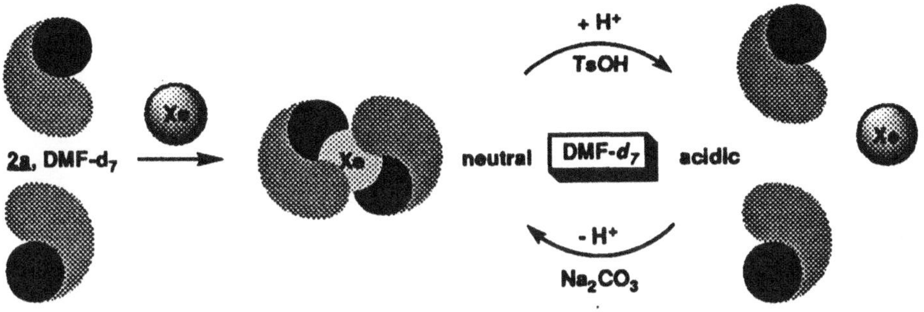

Figure 2: Induced dimerization of **2a** and reversible encapsulation of xenon

References:

1. (a) Wyler, R., de Mendoza, J. and Rebek, J., Jr. (1993) A Synthetic Cavity Assembles Through Self-Complementary Hydrogen Bonds, *Angew. Chem. Int. Ed. Engl.* **32**, 1699-1701. (b) Branda, N., Wyler, R. and Rebek, J., Jr. (1994) Encapsulation of Methane and Other Small Molecules in a Self-Assembling Superstructure, *Science* **263**, 1267-1268.
2. Branda, N., Grotzfeld, R.M. and Rebek, J., Jr. (1994) Control of Self-Assembly and Reversible Encapsulation of Xenon in a Self-Assembling Dimer by Acid-Base Chemistry, *J. Am. Chem. Soc.*, in press.

STRUCTURAL DIVERSITY IN A FAMILY OF SYNTHETIC OLIGOAMIDES

Adrian P. Bisson,[a] Fiona J. Carver,[a] Christopher A. Hunter,* [a] Graham S. Lynch,[a] N. J. Osborn,[a] Duncan H. Purvis,[b] Richard J. Shannon,[a] Kathryn Tse [b]
[a] Department of Chemistry, University of Sheffield, Sheffield, UK.
[b] Department of Chemistry, University of Otago, Dunedin, New Zealand.

1. Introduction

Biological systems use a relatively small number of molecular building blocks to generate a vast array of structurally and functionally diverse molecular architectures. The keys to the diversity found in nature are the oligomeric nature of the covalent framework and the range of different non-covalent structural motifs which are possible within this family of molecules [1,2].

We have been investigating the properties of a family of synthetic oligomeric molecules based on two simple building blocks, iso-phthalic acid and 1,1'-bis-(3,5-dimethyl-4-aminobenzene)cyclohexane. The molecules exhibit a surprising variety of different non-covalent structural motifs which confers a rich structural and functional chemistry on this system. The principle interactions which determine the properties of the molecules are H-bonds and π-π interactions which are also two of the most important forces which control three-dimensional structure, recognition, reactivity and self-assembly in biological systems.

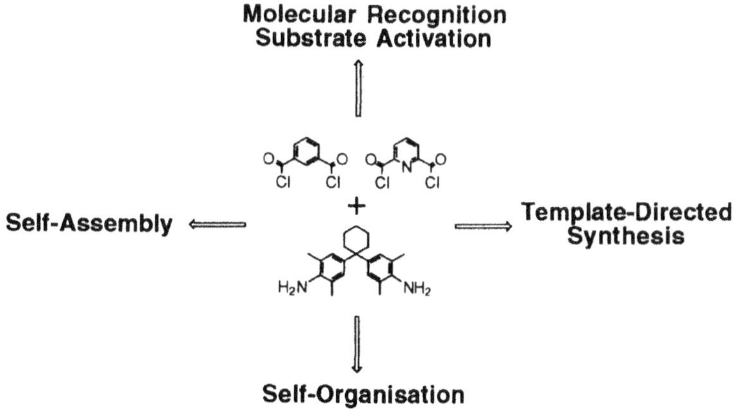

Figure 1.

J. S. Siegel (ed.), Supramolecular Stereochemistry, 199–202.

This contribution illustrates some of the structural and functional properties which we have characterised in this system to date (Figure 1). These observations are particular to specific members of the family of molecules, and the results begin to map out some interesting structure-function relationships which can be rationalised on the basis of the properties of the individual building blocks.

2. Molecular Recognition

The macrocyclic dimers have large open cavities which bind complementary substrates with high selectivity (Figure 2). *p*-Benzoquinone is recognised via H-bonds and edge-to-face π-π interactions [3]. The intermolecular H-bonds activate the bound quinone towards electrochemical reduction by the polarising the quinone carbonyl groups [4].

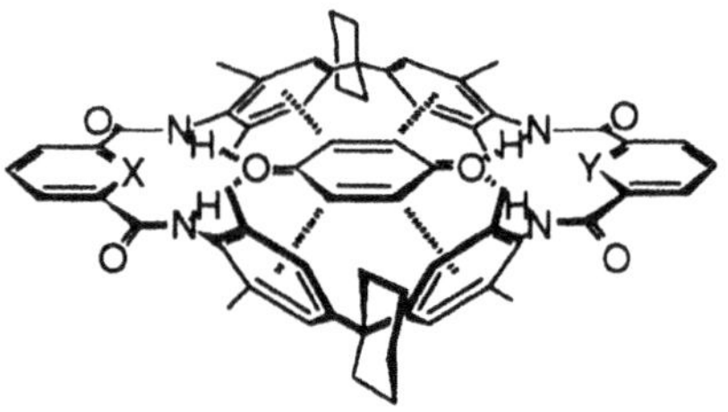

X = CH, Y = CH $K = 1200\ M^{-1}$ in $CDCl_3$
X = N, Y = CH $K = 1800\ M^{-1}$ in $CDCl_3$
X = N, Y = N $K = 230\ M^{-1}$ in $CDCl_3$

Figure 2.

3. Template-Directed Synthesis

Intermolecular H-bonding interactions template the formation of [2]-catenanes (two interlocked cyclic dimers, Figure 3) in the high dilution macrocyclisation reactions [5-7]. Catenane formation is sensitive to substituents: low yields of catenane are obtained with 2,6-pyridyl groups (which promote formation of the macrocyclic dimer) [8] and with 2,6-diethylaniline groups (which are too bulky to thread through the macrocyclic cavity).

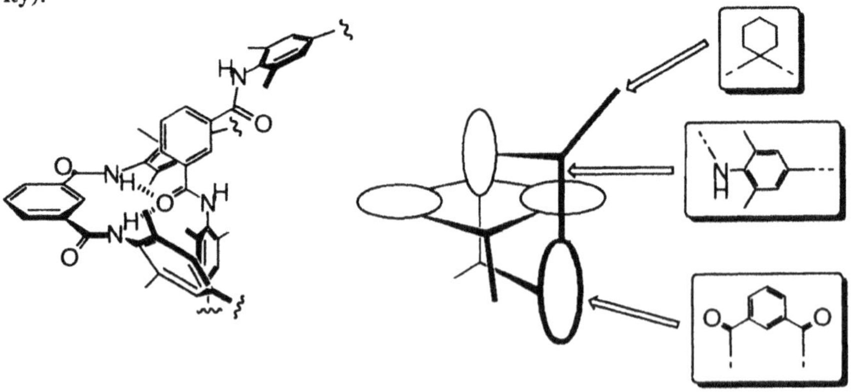

Figure 3.

4. Self-Assembly

Acyclic linear iso-phthaloyl oligomers dimerise in solution to form 'zipper' complexes. Oligomers which are perfectly complementary in length and in H-bonding and π-π interaction recognition sites form more stable 1:1 'zipper' complexes as shown in Figure 4 [9]. The intermolecular interactions along these complementary oligomers are cooperative which leads to very stable complexes whose structure parallels that of double-stranded nucleic acids.

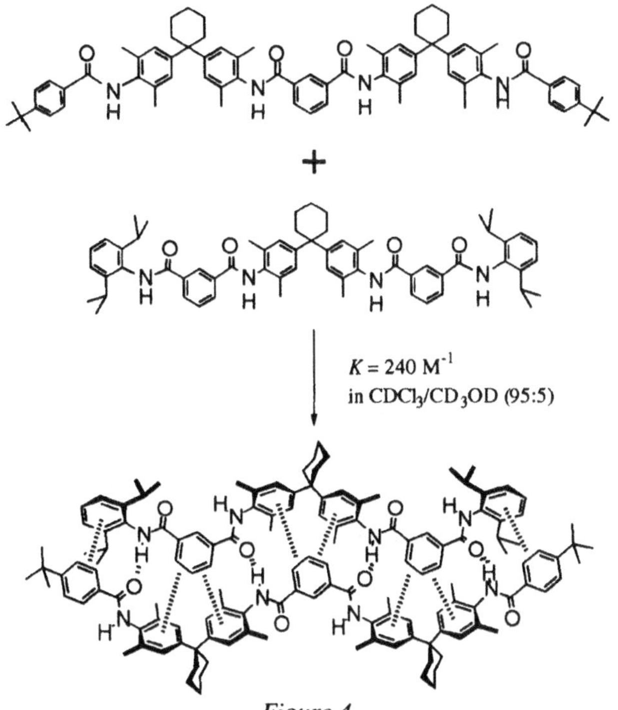

Figure 4.

5. Self-Organisation

Introduction of 2,6-pyridyl residues into the linear oligomers abolishes intermolecular 'zipper' complexation. Instead, intramolecular H-bonds cause the molecules to fold up to generate cavities for substrate binding (Figure 5). Similar properties have also been observed in the cyclic oligomers: potentially flexible molecules adopt stable organised structures in solution, and the preferred structure depends on the covalent structure (or sequence) of the oligomer [10].

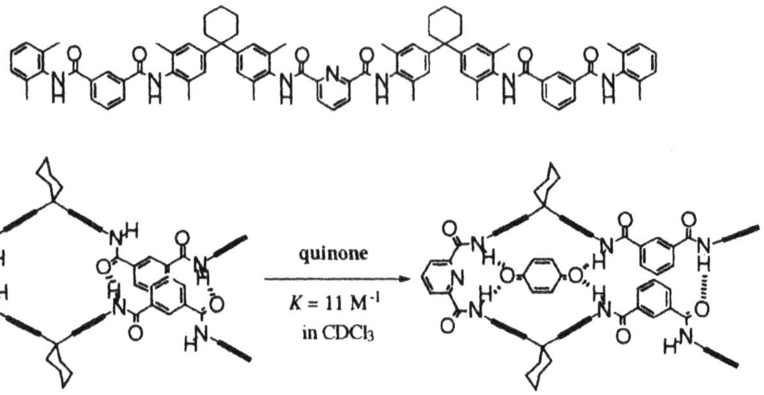

Figure 5.

6. References

1. Saenger, W. (1984) *Principles of Nucleic Acid Structure*, Springer-Verlag, New York.
2. Fersht, A. R. (1985) *Enzyme Structure and Mechanism*, Freeman, New York.
3. Hunter, C. A. (1991) Molecular recognition of *p*-benzoquinone by a macrocyclic host, *J. Chem. Soc., Chem. Commun.*, 749-751.
4. Brooksby, P. A., Hunter, C. A., McQuillan, A. J., Rowan, A. E., Purvis, D. H. and Shannon, R. J. (in press) Supramolecular activation of *p*-benzoquinone, *Angew. Chem. Int. Ed. Engl.*.
5. Hunter, C. A. (1992) Synthesis and structure elucidation of new [2]-catenane, *J. Am. Chem. Soc.* **114**, 5303-5311.
6. Hoss, R. and Vogtle, F. (1994) Template Syntheses, *Angew. Chem. Int. Ed. Engl.* **33**, 375-384.
7. Leigh, D. A. unpublished results.
8. Carver, F. J., Hunter, C. A. and Shannon, R. J. (1994) *J. Chem. Soc., Chem. Commun.*, 1277-1280.
9. Bisson, A. P., Carver, F. J., Hunter, C. A. and Waltho, J. P. (in press) Molecular zippers, *J. Am. Chem. Soc.*.
10. Hunter, C. A. and Purvis, D. H. (1992) A Binary Quinone Receptor, *Angew. Chem. Int. Ed. Engl.* **31**, 792-795.

DIASTEREOSPECIFIC RECOGNITION OF DYES BY SALT CRYSTALS: A CASE OF PLUNDERED STEREOCHEMISTRY IN POSTWAR EUROPE*

BART KAHR, MICHAEL P. KELLEY
Department of Chemistry, Purdue University
West Lafayette, IN 47907-1393
USA

Abstract. As early as March 1944, the British were making plans for the occupation of Germany.[1] A chief concern of these planners was the capture of technologies that would assist the Allied nations in the defeat of Japan and in the postwar economy. The assets of IG Farben, the chemical conglomerate that enabled Germany to fight in the First and Second World Wars, were among the most sought after targets. In this paper we discuss how incomplete crystal structures of triarylmethane dyes, captured from IG Farben by British Intelligence, led to an incorrect analysis of the stereospecific recognition of these dyes by simple ionic salts. We review our work on the determination of these stereospecific association mechanisms within the context of earlier errors and remind the reader of consequential political and economic events that serve as the requisite background for a complete understanding of an otherwise insignificant stereochemical puzzle.

---

*in appreciation of Professor Kurt Mislow

J. S. Siegel (ed.), Supramolecular Stereochemistry, 203–221.

1. Dye Inclusions in Simple Ionic Crystals

During the last several years we have been studying mixed crystals of simple ionic salts that contain regular inclusions of synthetic organic dyes.[2] We recorded the optical spectra[3] of these crystals and probed their internal structure with synchrotron X-ray radiation.[4] These studies enabled the design of dye inclusions in K_2SO_4 using chromophores with large fluorescence quantum efficiencies.[5] We have recently developed these crystals into a family of solid state dye lasers,[6] thereby demonstrating that ideas for some technologically important materials may lie undiscovered in the literature.

One material to which we have paid particular attention is a host/guest system in which K_2SO_4 crystals include acid fuchsin (**1a**) and its congeners (**1b** and **1c**). Equipped with modern semi-empirical quantum mechanical methods for molecular geometry optimization[7] and the interrogation of excited states,[8] as well as polarized absorption spectra from thin crystal sections, we developed a model for type **1** compounds included in the {110} growth sectors of K_2SO_4. Our previously published conclusions are summarized below (see Figure 1):[3]

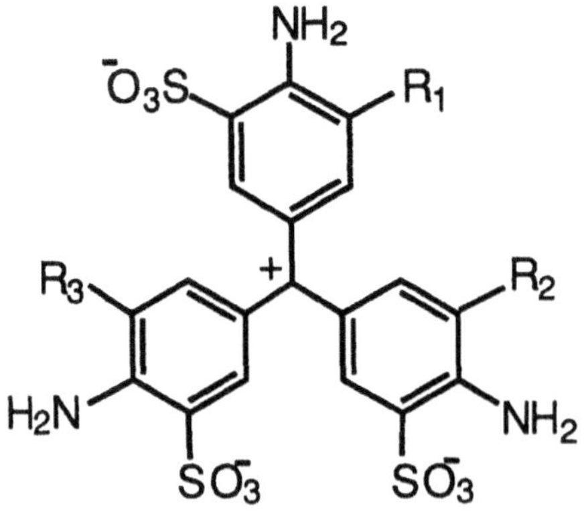

1a: R_1=CH_3; R_2, R_3 = H
1b: R_1, R_2, R_3 = H
1c: R_1, R_2, R_3 = CH_3

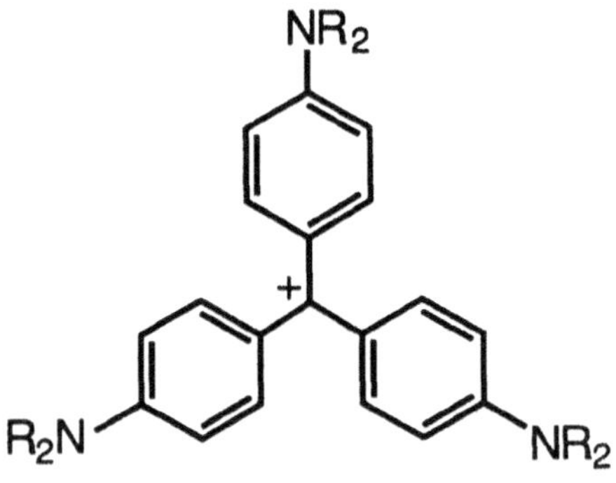

2: R = H
3: R = CH_3

(1) The acid fuchsin molecular skeleton (neglecting the methyl group as in **1b**) presents itself in two nearly equi-energetic conformations, having C_1 and C_3 symmetries.[9]

(2) Absorption spectra of the mixed crystals show two visible transitions, an observation consistent with the C_1 isomer, rather than the C_3 isomer which has a first excited E state.

(3) Polarized absorption spectra from oriented thin crystals enabled the determination of the experimental transition moments. These directions were in good agreement with a model in which the sulfonate groups of a C_1 isomer were least-squares fit to a triad of K_2SO_4 sulfate groups related to one another by a and c lattice translations.

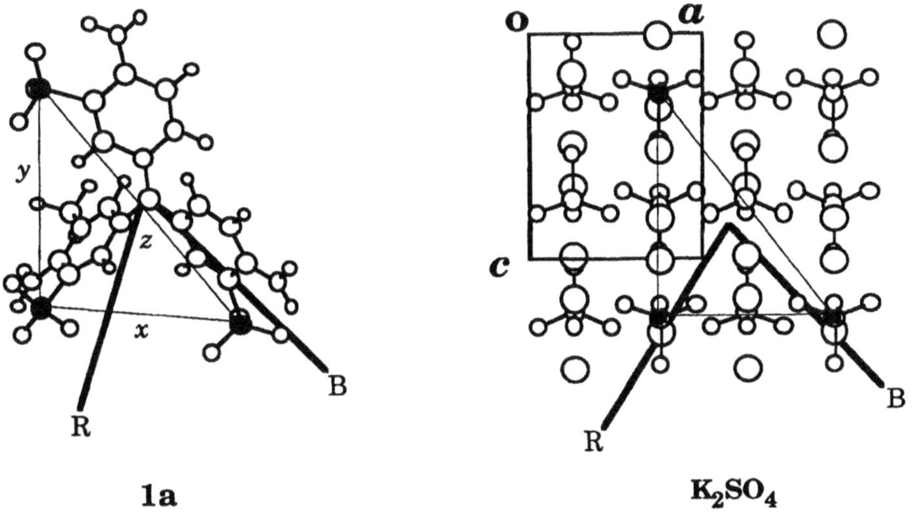

Figure 1. Left: Calculated structure of the C_1 diastereomer of **1a**. Right: K_2SO_4 lattice showing the dye's binding site. Sulfonate and substituted sulfate S atoms are black circles. Calculated low energy (R) and high energy (B) transition moments are shown superimposed on **1a** and corresponding experimental transition moments are shown superimposed on the K_2SO_4 lattice.

2. The Labor of Harold Buckley

During the 1930s, Buckley, a crystallographer who worked in Manchester,[10] wrote a series of papers describing the habit modification of simple salts when grown in the presence of commercial dyes.[11] During the next ten years he recorded the results of 16,000 crystallizations of salts with dyes.[12] It was during the course of these habit modification studies[13] that Buckley discovered some of the rare dye inclusions such as acid fuchsin in K_2SO_4[14] described in the previous section.

In the 1930's, the conformational analysis and the electronic structure of dyes were poorly understood; optical spectroscopies were exotic rather than routine.[15] Buckley, therefore, was not able to provide a molecular view of dyes regularly included in the host crystals eventhough he clearly realized that his observations undoubtedly contained structural information about the stereoselective interactions of organic dyes with growing crystal surfaces. He conceded, however, that the meaning of his work "will remain hidden to all but the most careful scrutiny"[11b] while describing himself as "an indifferent organic chemist."[12]

3. John Whetstone's Puzzle

One other scientist, John Whetstone, a researcher at the Nobel Explosives Division of the Imperial Chemistry Industries (ICI), had taken a serious interest in Buckley's work. He was initially motivated to control the habits of NH_4NO_3 crystals for use in explosives and fertilizers.[16] Following Buckley, Whetstone studied 120 dyes as habit modifying agents. Acid fuchsin and its congeners (**1a-c**) were especially effective.[17]

Whetstone, too, in the course of his habit modification work, discovered several dye inclusion systems based on NH_4NO_3 and KNO_3. Not an "indifferent organic chemist", Whetstone developed structural models for various including dyes. He superimposed these "scaled" molecules, drawn on celluloid, over projections of the salt lattices. Possible structures were suggested when the positively and negatively charged functionalities of the dyes were well-superimposed upon salt cations and anions, respectively.

Of Whetstone's structural propositions, not all of which we have yet reconsidered,[18] one captured our interest; NH_4NO_3 containing **1b**. At the outset, we should state that, to date, we have been unable to reproduce this inclusion in our laboratory. In a 1954 paper Whetstone showed a drawing of **1b** superimposed on the (100) plane of the room temperature NH_4NO_3 crystal structure.[19] A slightly modified figure was published in 1956 and it is reproduced below.[20] Remarkably, Whetstone had made a stereochemical choice in clearly representing the asymmetric C_1 diastereomer. He did no quantum mechanical calculations and no quantitative spectroscopy (he only "eyeballed" the dichroism) and yet was intuitively able to make a stereochemical decision that in our labs necessitated considerable forethought.

How did Whetstone do it? There is very little text associated with these figures, all of it unrevealing. Our only clue comes from a strange citation to triarylmethyl cation structures found in another Whetstone paper.[21] That citation reads:

"*B.I.O.S.* microfilm, *R*20, P.B. 73,654, *FDX* 226".

As in any mystery, some background must be supplied before we can reveal the meaning of the clue. A full understanding of the relationship of the work of Buckley and Whetstone, and of the meaning of *B.I.O.S. microfilm*, necessitates that we take a look at the European synthetic dyestuffs industry before and after the Second World War.

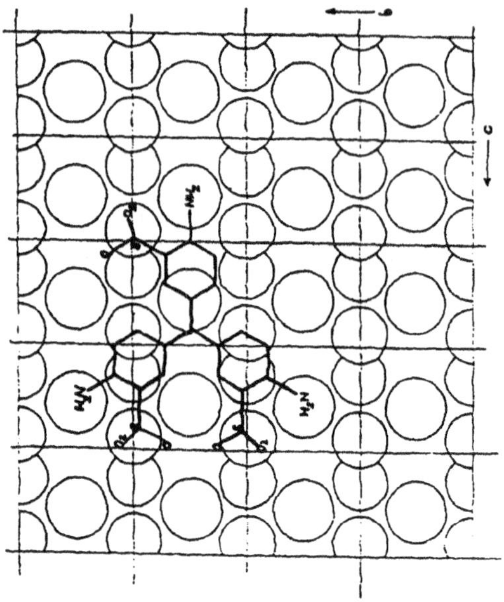

Figure 2. Figure from reference 17 showing asymmetric C_1 diastereomer of **1b** superimposed on a (100) face of NH_4NO_3 such that $-SO_3^-$ groups fall upon NO_3^- positions. The image was rotated by 90° in order to maintain a presentation consistent with the other figures.

4. The Rise of the German Dye Industry[22,23]

Buckley's habit modification studies were possible only because of his access to thousands of commercial dyes. Some were obtained from the neighboring ICI corporation in Manchester, but most of the dyes were purchased from German firms. At this time Germany led the world in the production of synthetic dyes, and Buckley's work underscores the success and commercial importance of the German synthetic dye industry. Here, we give a brief account of the rise of this industry focusing on the early British-German synergy.

In the first half of the Nineteenth Century, all coloring matter was derived from animal, vegetable, and mineral sources. In fact, the first reported dye inclusion of a simple salt involved a red compound, hematoxylin, found in logwood bark.[24] It is widely conceded that the birth of the synthetic dyestuffs industry began in 1856 in London when William Perkin (1838-1907) accidentally synthesized a mixture that stained silk *mauve*. Perkin was trying to make synthetic quinine by oxidizing aniline. The latter was first made available from coal tar by his teacher Auguste von Hofmann (1818-1892). Perkin's aniline was impure and contained toluidenes. When the mixture was oxidized he produced, by condensation, some triarylmethane dyes, among other things. Perkin temporarily left academic research to make his fortune as inventor of the synthetic purple.

Despite this early advance in Britain, the Germans quickly dominated the synthetic dyestuffs industry, in large measure because German chemists based their preparations on the emerging science of structural organic chemistry. Auguste von Hofmann took the lead as a consultant for a dye firm known as AGFA (Aktiengesellschaft für Anilinfabrikaten, Berlin).

By the turn of the century, international competition in the production and distribution of dyes was effectively quashed by AGFA and five other German firms: BASF, Bayer, Cassella, Höchst, and Kalle. In order to obviate competition amongst one another, the German firms formed a trust that was inspired by Rockefeller's Standard Oil. Each company kept its identity; however, dyestuffs and dyestuffs alone, were subject to collective regulations. The trust, formed in stages between 1904 and 1925,[25] was known as IG Farben (Interessen Gemeinschaft Farbenindustrie Aktiengesellshaft).[26]

5. IG Farben in World War I, Especially at Oppau[27]

IG Farben gained a mythical status during the First World War under the leadership of engineer Carl Bosch (1874-1940). Bosch's reputation was sealed when he surmounted considerable technical problems in the industrial scale-up of Haber's otherwise academic synthesis of ammonia from the elements. A plant was constructed at Oppau, near the BASF Ludwigshafen headquarters, that started producing synthetic ammonia in 1913. In principle, IG Farben's Ammoniaxlaboratorium, Oppau, was now in a position to provide an inexhaustible supply of $NaNO_3$ for gunpowder by oxidizing ammonia. Germany was free to make war without the tenuous connection to the Chilean saltpeter reserves which at any time might be severed by the British fleet. In the last year of the war the Oppau plant synthesized 90,000 tons of nitrates.[28]

Bosch was a capable negotiator as well as an engineer. He served as the IG Farben representative at Versailles and went to dramatic lengths to protect IG assets from Allied destruction. When foreign investigators demanded to see the Oppau plant in operation, Bosch refused, maintaining that the factory was a commercial installation, and that the fertilizers produced would be required to stave-off famine in Germany. In secret meetings with French officials, Bosch succeeded in protecting his prized Oppau plant by conceding valued technologies.

In 1921, an explosion of unknown origin destroyed the Haber-Bosch plant at Oppau. More than 600 people were killed. BASF spared no expense in its reconstruction and achieved full production within 3 months.

6. IG Farben in World War II[29]

It has been said that IG began preparations for the Second World War before the ink was dry on the Treaty of Versailles. Bosch had freed Germany from nitrate importation. He realized, however, that genuine military self-sufficiency required the production of synthetic rubber and synthetic fuel.

Once again, IG realized its technological vision. It succeeded in the hydrogenation of coal to make lubricating oils and gasoline. Synthetic rubbers based on butadiene and styrene polymers, called buna rubbers, were also discovered between the wars. According to Sasuly, "When armored vehicles rolled on buna-S tires...powered by synthetic gasoline, the German General staff knew that rearmament was in its last stages. Now the war could start whenever Hitler gave the signal."[30]

Between 1939 and 1941 IG subsumed the chemical industries of Austria, Poland, Czechoslovakia, France, and the Baltic States. It gained influence as an international economic force while playing a key role in stabilizing the greater-Germany. IG profited from Hitler, and Hitler profited from IG. This relationship matured at Auschwitz where IG built a huge fuel and rubber factory for which it contracted nearby slave labor from the SS.[31]

7. British Intelligence Objectives Subcommittee[1,32]

IG Farben was dissolved after WWII and its chief officers were convicted of war crimes at the Nuremberg trials.[33] Allied occupation forces were determined, this time, not to agree to any deal that would protect German technologies and patents. Plans for the acquisition of German "know-how" were thorough and swift.

The British and Americans created special scientific intelligence units whose function was to acquire technical information of military value. Eisenhower's SHAEF (Supreme Headquarters, Allied Expeditionary Force) established the T-forces soon after the Normandy invasion, whose mission was to secure and guard targets. The British and American Combined Chiefs of Staff created CIOS (Combined Intelligence Objectives Subcommittee) whose duties were to coordinate the exploitation of the targets. CIOS spawned sister agencies in Washington (Technical Industrial Intelligence Committee, TIIC) and in London to organize the booty.

BIOS (British Intelligence Objectives Subcommittee), the English TIIC counterpart, collected information in so-called BIOS Reports. The Americans compiled FIAT (Field Information Agency, Technical) Reports. Together, technical data was collected on an astonishing range of subjects both military and non-military. BIOS and FIAT reports are available on microfilm at the Library of Congress. The Special Libraries Association has published an index[34] which correlates BIOS and FIAT report numbers with so-called "PB numbers."[35]

Fifty British and American investigators reached the IG complex at Ludwigshafen-Oppau on 25 March 1945. Two-thirds of the plant was already destroyed by the time they arrived. Nevertheless, local Germans were organized to dig through the rubble for documents. A valuable treatise on acetylene chemistry prepared by Walter Reppe (1892-1969) was unearthed during this search.[36] Many unimportant documents were collected and organized as well.

8. *B.I.O.S microfilm, R20,* **P.B. 73,654,** *FDX* **226**

B.I.O.S. microfilm, *R20*, P.B. 73,654, *FDX* 226, contains several thousand pages of technical information, mainly on fertilizers and paraffin oxidation chemistry. Buried within this mass is a 12 page report prepared by Dr. Heinz Krebs,[37] dated 15 June 1943, on preliminary X-ray crystallographic studies of parafuchsin (**2**) and crystal violet (**3**) salts. This is the report to which Whetstone referred (see section 3).

Krebs, a wartime employee at the Oppau explosive factory, may have been urged to work triarylmethane dyes for reasons similar to those of Whetstone; perhaps the action of dyes as anti-caking agents for nitrate crystals was being considered. In his report, Krebs provides an academic reason for the structure determinations. He was interested in distinguishing which resonance structure (charge localized on the methyl carbon or charge localized at the amino nitrogens), was a better representation of the solid state structure.[38]

For the chloride of parafuchsin (**2**·**Cl**$^-$) Krebs reported the following data: hexagonal system, a = 14.13 Å, c = 24.20 Å, space groups $P6_3mc$ (no. 186) or $P6_3/mmc$ (no. 194), Z = 6. The first choice requires a site symmetry of C_s (m). The second choice requires a site symmetry C_{2v} ($mm2$) or C_{2h} ($2/m$). In either case, these symmetries are incompatible with propeller-shaped ground states of triarylmethyl cations.

Krebs considered the structure of crystal violet chloride (**3**·**Cl**$^-$) in greater detail and reported the following data: a = 16.02 Å, c = 7.86 Å, space groups $P31c$ (no. 159) or P-$31c$ (no. 163), Z = 2. In $P31c$ the site symmetry would be C_3 (3), compatible with what we understand today about the symmetry of triarylmethyl cations. The choices in the centrosymmetric supergroup are S_6 (-3) and D_3 (32). While the propellers may have D_3 symmetry, the impossibility of an inversion center precludes S_6. In either space group, Krebs requires layers of cationic crystal violet molecules separated by 3.9 Å. His sketch of this structure is shown as Figure 3. He compares the 3.9 Å separation favorably with the 3.7 Å molecular separation in hexamethylbenzene and the 3.4 Å separation between layers in graphite.

Krebs apologized for his incomplete report that was hastily prepared because he was summoned to the Wehrmacht (German armed forces).[39]

Oppau fell within the French occupation zone. We do not know whether access to Krebs' data prompted Cécile Stora, a crystallographer at the Sorbonne,[40] to begin preliminary X-ray studies of **2** and **3** in 1949.[41] Her data on crystal violet compares favorably with Krebs'. She reports a trigonal structure *P*31*c* or *P*-31*c* and concludes that the "molécule en grande partie plane". We believe that she meant that the ~3.9 separating distance precluded pyramidal dye ions and was not a statement about the aryl ring dihedral angles. Stora's work was for many years the only reference to the X-ray structure of **3** in the open literature. Subsequent crystal violet spectroscopists interpreted her results in two ways: some assumed that the molecule was not pyramidal in its coordination about the central carbon,[42] while others inferred that the aryl rings occupied a common plane.[43] Whetstone did not cite Stora and was apparently unaware of her work.

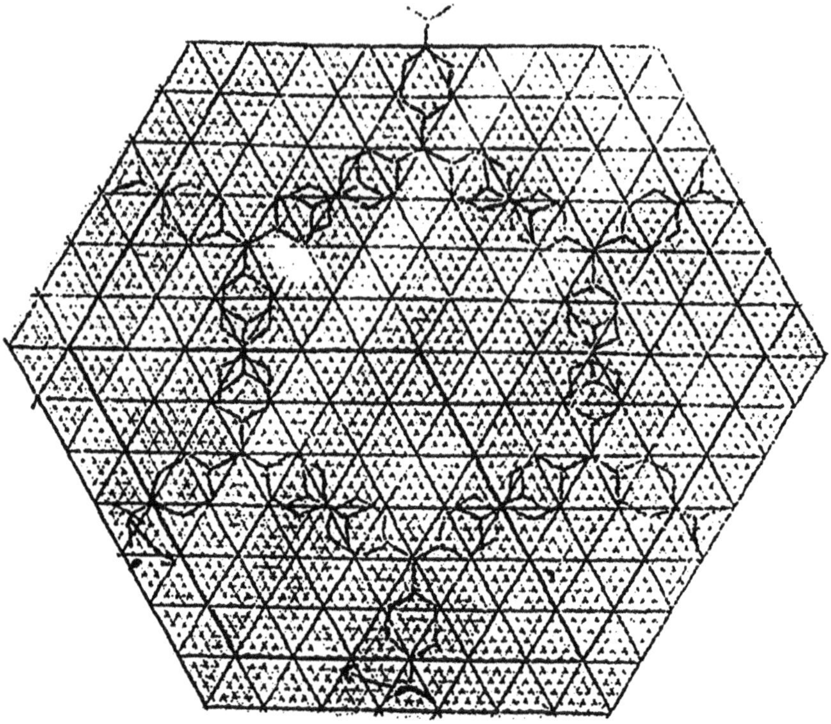

Figure 3. Kreb's sketch of the crystal violet chloride (**3**·Cl^-) structure from *BIOS microfilm.*

9. Whetstone's Solution

We constructed a planar model of the acid fuchsin (**1a**) molecule by driving the aryl ring torsion angles in our AM1 optimized structure to 0°. In this structure all of the triarylmethyl carbon atoms essentially occupy a common plane. We least-squares fit the sulfonate S atoms to the nitrate N atoms (x,y,z ; x,y,z+1 and x,y+2,z) in the NH_4NO_3 lattice. In this way we get an excellent fit (ave. d_{S-N} = 0.36 Å). Our model is virtually identical with Whetstone's model. Shown in Figure 4 is slightly different model in which sulfonate S atoms and amino N atoms of the dye were fit to nitrate N atoms and ammonium N atoms of the salt, respectively. Compare Figures 2 and 4.

These analyses indicate that Whetstone, upon examining the Krebs/BIOS report, concluded that *triarylmethyl cation dyes are planar* in that all the phenyl ring carbon atoms lie in a common plane, the trigonal coordination plane of the central carbon. This conclusion is not required by Kreb's sketchy data. Whetstone, as had others, apparently over-interpreted the meaning of the interlayer separating distance and chose the asymmetric diastereomer for the wrong reason.

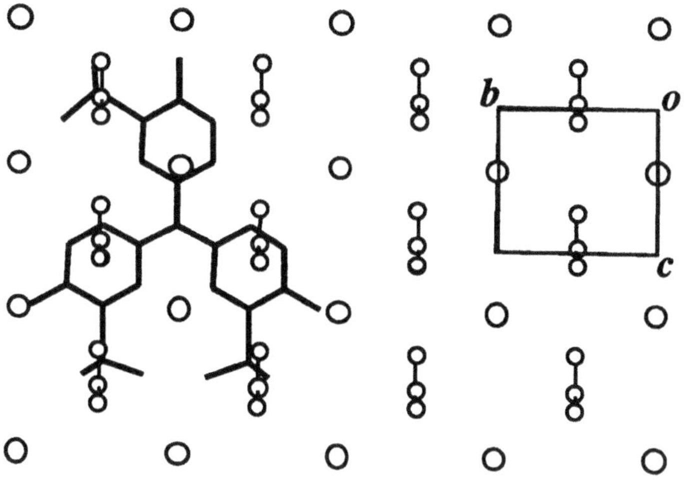

Figure 4. Flat model of **1b** fit onto the (100) face of NH_4NO_3. In this case, sulfonate S atoms and amino N atoms of the dyes were fit to nitrate N atoms and ammonium N atoms, respectively. Hydrogen atoms have been removed for clarity.

Of course, there is now overwhelming evidence that triarylmethyl cations cannot be flat due to the mutual steric repulsion of the *ortho*-hydrogen atoms; however, it took some time for this view to gain universal acceptance. In 1942, G.N. Lewis first suggested that triarylmethyl cation rings cannot be mutually planar due to steric hindrance.[44] Whetstone cites and rejects Lewis' predilection for a propeller structure. He prefers a flat carbon skeleton in which the *ortho*-hydrogen repulsions were accommodated by out-of-plane C-H bond bending.[45]

Seel independently proposed a triphenylmethyl propeller geometry with twist angles of 20°;[46] Szwarc came to a similar conclusion in 1947.[47] However, experimental evidence for the mutual non-planarity of the aryl rings of triarylmethyl cations first came from IR studies in 1957.[48] X-ray crystallographic analysis later quantified this departure from D_{3h} symmetry.[49] Twist angles in triphenylmethyl perchlorate[50] and tri-*p*-methoxyphenylmethyl chloride,[51] were 32° and 30°, respectively. A refined X-ray structure for **2** with average twist angles of 32° was first reported in 1971.[52]

10. Elusive X-ray Structure of Crystal Violet Chloride

Crystal violet has probably stained more hands than any other dye in history as the active ingredient in carbon paper.[53] Despite Krebs' and Stora's early work, the first refined crystal structure of this important ion was reported as the tetraphenylborate salt, as late as 1989.[54] To date, a refined structure of crystal violet chloride (**3·Cl⁻**) has not been published.

We have tried to reproduce the Krebs-Stora trigonal phase. Commerical (Aldrich) **3·Cl⁻** was recrystallized from acetone solutions by diethyl ether vapor diffusion. We obtained prisms with a green, metallic sheen characteristic of dye single crystals. However, these did not prove to be the elusive trigonal structure. We measured a monoclinic lattice for a monohydrate with a = 10.316(3) Å, b = 21.639(5) Å, and c = 21.353(9) Å, β = 94.04(3)°, Z = 8, space group $P2_1/c$. The average aryl ring twist angles for the two independent cations were 33.7° and 34.2°. At the time this manuscript was submitted we were still searching for the trigonal, anhydrous polymorph that inspired Whetstone's stereochemical decision.[55]

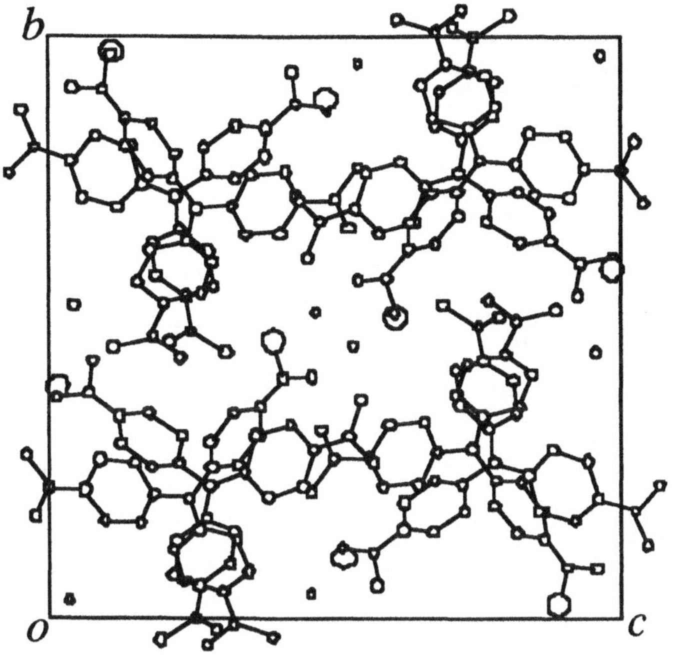

Figure 5. Unit cell of monoclinic structure of crystal violet chloride monohydrate. Large isolated circles represent Cl atoms. Small isolated circles represent oxygen atoms in water molecules. All hydrogen atoms have been deleted for clarity.

Acknowledgment

We received research support from the following sources during the preparation of this manuscript: NSF (CHE-9457374), ACS-PRF (27665-AC3), Exxon Education Foundation, and the Purdue Research Foundation in the form of a fellowship to M. P. K. We are most grateful to Professor Roald Hoffmann for calling to our attention F. Seel's independent proposal of the triphenylmethyl propeller geometry.

References

1. Stokes, R. G. (1988) *Divide and Prosper: The Heirs of I. G. Farben under Allied Authority, 1945-1951*, University of California Press, Berkeley.
2. Kahr, B.; Chow, J. K.; Peterson, M. L. (1994) Organic Hourglass Inclusions, *J. Chem. Ed.* **71**, 584-586.
3. Kelley, M. P.; Janssens, B.; Kahr, B.; Vetter, W. (1994) Recognition of Dyes by K_2SO_4 Surfaces: Choosing Organic Guests for Simple Salts, *J. Am. Chem. Soc.* **116**, 5519-5520.
4. Dudley, M.; Vetter, W.; Kahr, B. (1993) X-ray Topographic Studies of Hourglass Inclusions, *National Synchrotron Light Source Report*, (eds. Hurlburt, S. L.; Lazarz, M. L.) Brookhaven National Laboratory, Upton, New York.
5. Kelley, M. P.; Kahr, B. to be submitted for publication.
6. Elliott, D.; Rifani, M.; Jay, M. J.; Kelley, M. P.; Kahr, B. to be submitted for publication.
7. Dewar, M. J. S.; Zoebisch, E. G.; Healy, E. F.; Stewart, J. J. P. (1985) AM1: A New General Purpose Quantum Mechanical Molecular Model, *J. Am. Chem. Soc.* **107**, 3902-3909; Stewart, J. J. P. Quantum Chemistry Program Exchange, Bloomington, Indiana, No. 455.
8. Ridley, J.; Zerner, M. C. (1973) An Intermediate Neglect of Differential Overlap Technique for Spectroscopy: Pyrrole and the Azines, *Theor. Chim. Acta*, **32**, 111-134; (1976) Triplet States via Intermediate Neglect of Differential Overlap: Benzene, Pyridine, and the Diazines, **42**, 223-236.
9. For reviews of the stereochemistry of molecular propellers, see: Mislow, K.; Gust, D.; Finocchiaro, P.; Boettcher, R. J. (1974) *Fortschr. Chem. Forsch.* **47**, 1-28; Mislow, K. (1976) *Acc. Chem. Res.* **9**, 26-33.
10. For biographical information and an almost complete bibliography see: Deer, W. A. (1961) Memorial of Harold Eugene Buckley, *Am. Min.* **46**, 481.
11. (a) Buckley, H. E. (1933) Systematic Habit-Variation in $KClO_3$ Crystals Produced by Dyes, *Zeit Krist.* **85**, 58-73; (b) (1934) Molecular Configuration and its Relation to Modification of Crystal-Growth, **88**, 381-411; (c) (1935) Some New Features in Habit-Modification Shown by $KClO_4$ Crystals, **91**, 375-401.
12. Buckley, H. E. (1951) Studies in the Growth and Habit Modification of Crystals, *Memoirs and Proceedings of the Manchester Literary and Philosophical Society*, **92**, 77-123.
13. For our review of Buckley's habit modification studies, see: Kelley, M.; Chow, J. K.; Kahr, B. (1994) Critical Evaluation of Dyes as Crystal Growth Inhibitors, *Mol. Cryst. Liq. Cryst.* **242**, 201-214.
14. Buckley, H. E. (1934) The Oriented Inclusion of Impurities in Crystals, *Zeit. Krist.* **88**, 248-255.
15. During the end of his career, Buckley began to use optical spectroscopies to identify dye inclusions in salts. He published one spectrum and declared that "A new branch of Chemistry seems nearby" (ref 11).
16. Butchart, A.; Whetstone, J. (1949) The Effect of Dyes on the Crystal Habits of Some Oxy-Salts, *Discussions of the Faraday*

Society, 243-254; Whetstone, J. (1951) Modification of Crystal Habit of Inorganic Salts with Dyes, *Nature*, **168**, 663-664.
17. Whetstone, J. (1952) Solution to the Caking Problem of Ammonium Nitrate and Ammonium Nitrate Explosives, *Ind. Eng. Chem.* **44**, 2663-2667; British Patent (1953) 665,478.
18. Davey, R. J.; Polywka, L. A.; Maginn, S. M. (1991) *Adv. Ind. Cryst.* (eds Garside et al.) Butterworth-Heinemann, Oxford, pp 150-165.
19. Whetstone, J. (1954) The Adsorption of Dyes by Crystals, *Discussions of the Faraday Society*, 132-140.
20. Whetstone, J. (1956) Crystal Symmetry, and the Adsorption of Dyes by Growing Crystals. Part I. Ammonium Nitrate IV. *J. Chem. Soc.* 4841-4847.
21. Whetstone, J. (1955) The Crystal Habit Modification of Inorganic Salts with Dyes: Part 2. - The Relationship between the Structure of Crystal and Habit Modifying Dyes, *Trans. Faraday Soc.* **51**, 1142-1153.
22. Travis, A. S. (1993) *The Rainbow Makers: The Origins of the Synthetic Dyestuffs Industry in Western Europe*, Lehigh University Press, Bethlehem; Beer, J. J. (1959) *The Emergence of the German Dye Industry*, Urbana, University of Illinois Press; Haber, L. F. (1958) *The Chemical Industry During the Nineteenth Century*, Clarendon, Oxford.
23. There is an excellent and extensive literature on the synthetic dyestuffs industry in general and IG Farben in particular. We have tried to select representative citations for each of the following 5 subsections. There is a great deal of redundancy in the information contained within the acknowledged monographs. Much of the information can be found in several places.
24. de Sénarmont, H. (1854) Versuche über die Künstliche Erzeugung von polychromismus in krystallisirten Substanzen, *Ann. Phys.* (Leipzig) **167**, 491-504.
25. Haber, L. F. (1971) *The Chemical Industry, 1900-1930*, Clarendon, Oxford.
26. Two other companies, Griesheim-Elecktron and Farbwerk Mühlheim, were later added to IG.
27. Borkin, J. (1978) *The Crime and Punishment of I G Farben*, The Free Press, NY.
28. Ibid. p 29.
29. (a) Sasuly, R. (1947) *I G Farben*, Boni and Gaer, New York; (b) Hayes, P. (1987) *Industry and Ideology. I G Farben in the Nazi Era*, Cambridge University Press, Cambridge.
30. ref 29a, p 88.
31. Among IG's most notorious crimes was the production, through one of its subsidiaries, of the insecticide Zyklon B. The prussic acid in Zyklon B was ordinarily accompanied by an odoriferous compound to alert humans to its presence. The SS ordered large amounts *without the indicator*. This modified poison product was better suited to the mass murder of those not forcibly engaged in the war-time chemical industry, among others.[27]
32. Gimbel, J. (1990) *Science, Technology, and Reparations. Exploitation and Plunder in Postwar Germany*, Stanford University Press, Stanford.

33. DuBois, J. E. Jr. (1952) *The Devil's Chemists*, The Beacon Press, Boston.
34. *Numerical Index to the Bibliography of Scientific and Industrial Reports, Vols. 1-10, 1946-1948*, (1949) Ann Arbor, Edwards Bros.
35. Maynard, C. W. (1955) *The Chemistry of Synthetic Dyes and Pigments*, (ed. Lubs, H. A.) ACS, New York.
36. ref 31, pp 11-12.
37. Krebs was a leading crystallographer who started an X-ray diffraction school in Bonn and later wrote a popular book: Krebs, H. (1968) *Fundamentals of Inorganic Crystal Chemistry* (trans. Walter, P. H. L.) McGraw-Hill, London.
38. Krebs sites a discussion between two contemporaries: Eistert, B. (1938), *Tautomerie and Mesomerie*, Ferdinand Enke, Stuttgart; and Wizinger, R. (1933) *Organische Farbstoffe*, Ferdinand Dümmlers, Berlin.
39. "Infolge Einberufung zur Wehrmacht konnte sie nicht beendigt werden."
40. Ewald, P. P. (1962) *Fifty Years of X-ray Diffraction*, International Union of Crystallography, International Union of Crystallography, Utrecht.
41. Stora, C.; Eller, G. (1949) Caractères radiocrystallographiques de quelques dérivés du triphénylméthane, *Comp. Rend.* **229**, 766-768; Stora, C. (1958) Sur la stéoréchimie aux rayons X de la pararosaniline et du violet hexaméthylé. *Comp. Rend.* **246**, 1693-1695.
42. Lueck, H. B.; McHale, J. L.; Edwards, W. D. (1992) Symmetry-Breaking Solvent Effects on the Electronic Structure and Spectra of a Series of Triphenylmethane Dyes, *J. Am. Chem. Soc.* **114**, 2342-2348.
43. Angeloni, L.; Smulevich, G.; Marzocchi, M. P. (1979) Resonance Raman Spectrum of Crystal Violet, *J. Raman Spec.* **24**, 305-310.
44. Lewis, G. N.; Magel, T. T.; Lipkin, D. (1942) Isomers of Crystal Violet Ion. Their Absorption and Re-emission of Light, *J. Am. Chem. Soc.* **42**, 1774-1782.
45. Brunings, K. J.; Corwin. A. H. (1944) Effect of Substituents on the Structure of Dipyrrylmethenes. Some Relationships between the Dipyrryl- and the Triphenylmethane Dyes, *J. Am. Chem. Soc.* **66**, 337-342.
46. Seel, F. (1943) Zur quantentheoretischen Begründung der Stabilität von Kohlenstoffradikalen, *Naturwissenschaften*, **31**, 505. His contributions are also summarized by Hoffmann et al. See: Hoffmann, R.; Bissell, Farnum, D. G. (1969) The Balance of Steric and Conjugative Effects in Phenyl-Substituted Cations, Radicals, and Anions, *J. Phys. Chem.* **73**, 1789-1800.
47. Szwarc, M. (1947) The Resonance Energy of the Benzyl Radical, *Discussions of the Faraday Society*, **2**, 39-46.
48. Sharp, D. W. A.; Sheppard, N. (1957) Complex Fluorides. Part VIII. The Preparation and Properties of Salts of the Triphenylmethyl Cation: The Infrared Spectrum and Configuration of the Ion, *J. Chem. Soc.* 674-679.
49. For an overview of triarylmethyl radical crystal structures see: Kahr, B.; Van Engen, D.; Gopalan, P. (1993) Crystal Structure and Magnetic Susceptibility of a Hydrocarbon Free Radical: Tris(3,5-di-*tert*-butylphenyl)methyl, *Chem. Mat.* **5**, 729-732.

50. Gomes de Mesquita, A. H.; MacGillavry, C. H.; Eriks, K. (1965) The Structure of Triphenylmethyl Perchlorate at 85°C, *Acta Cryst.* **18**, 437-443.
51. Andersen, F.; Klewe, B. (1965) Structure Investigation of Tri-*p*-methoxyphenylmethyl Carbonium Ion, *Acta Chem. Scand.* **19**, 791-796.
52. Koh, L.L.; Eriks, K. (1971) The Crystal Structure of a Stable Carbonium Ion, Tri(p-aminophenyl)carbonium Perchlorate, $(H_2NC_6H_4)_3C^+ \cdot ClO_4^-$, *Acta Cryst. Sect. B.* **B27**, 1405-1413.
53. Wingate, P. J. (1982) *The Colorful DuPont Company*, Serendipity Press, Wilmington, Delaware.
54. Spangler, B. D.; Vanysek, P.; Hernandez, I. C.; Rogers, R. D. J. (1989) Structure of Crystal Violet Tetraphenylborate, *Cryst. Spectr. Res.* **19**, 589-596.
55. If and when we find the trigonal phase, we will report the details of the monoclinic structure at that time. Atomic coordinates and structure factors for the monohydrate may be obtained by writing to the authors directly.

SELECTIVITY ENHANCEMENT BY CONCAVE REAGENTS

U. LÜNING* and W. SCHYJA
Institut für Organische Chemie der
Christian-Albrechts-Universität
Olshausenstr. 40, D-24098 Kiel (Germany)
Fax: +49-431-880-1558

Concave Reagents can be made by combination of standard reagents of organic chemistry with macrocyclic structures as concave shielding [1, 2, 3]. Their geometry resembles that of a lamp with the light bulb as the reactive center and the lamp shade being the concave shielding. Concave Acids and Bases have been tested in model reactions [4, 5].

The addition of an alcohol to a ketene can be catalyzed by (concave) bases. The alcohol is activated by the formation of a hydrogen bond to the basic centre (e. g. to a pyridine nitrogen atom). Under standardized reaction conditions, rates of catalysis have been measured as k_{obs} [5].

$$\text{rate} = k_{obs}\,[\text{ketene}]$$

with $k_{obs} = [\text{ROH}]^n\,[\text{base}]$, concentrations chosen: [ROH] = 50 mM, [base] = 50 mM

The observed rate constants for the addition of ethanol depend on the basicity *and* the geometry of the bases. A Brønsted plot (see Fig. 1) shows the basicity dependence for two classes of Concave Pyridines **1** and **2** in comparison to pyridine itself [5].

When the addition of other alcohols to diphenylketene is catalyzed by (concave) pyridines a size dependence of the reactivity is found (see Fig. 2). But the rates do not correlate with the sterical parameters E_s of the alcohol residues R. If indeed the rate determining step is the activation of the alcohol by formation of a hydrogen bond, the site for this interaction is not in α-position to the residue R of the alcohol ROH (at the oxygen atom) but at the hydrogen atom of the hydroxyl group. Unfortunately E_s parameters do not exist for alkoxy residues OR. Therefore the logarithms of the rates of catalysis for the addition of alcohols ROH to diphenylketene were plotted against the E_s parameters for residues CH_2R.

J. S. Siegel (ed.), Supramolecular Stereochemistry, 223–226.

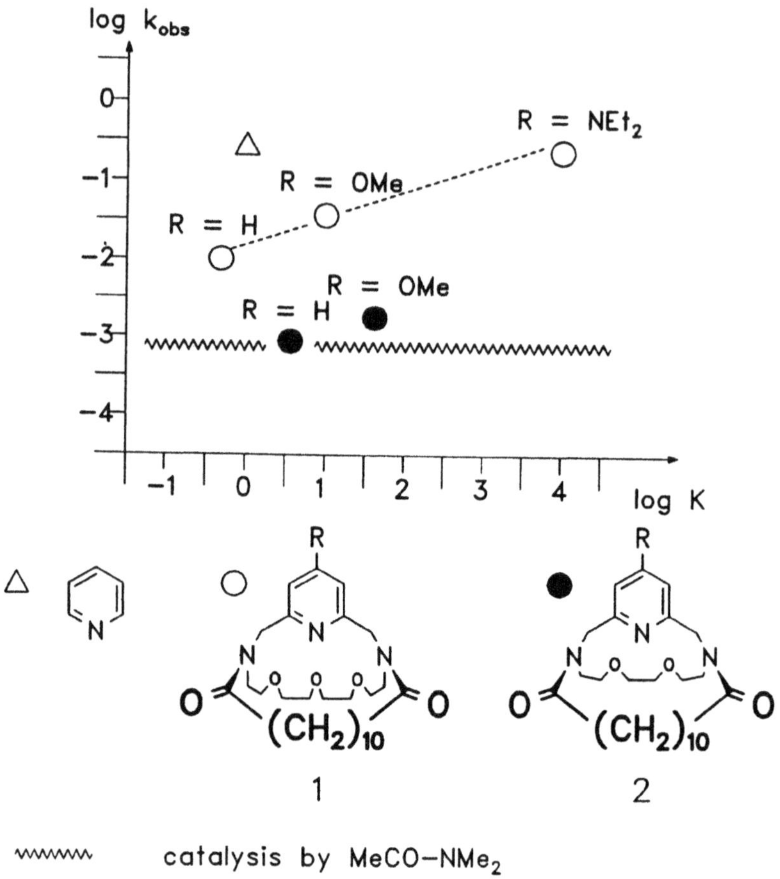

Figure 1. Basicity dependence (log K, relative pK_a scale) of the logarithms of the rates of catalysis log k_{obs} for the base catalyzed addition of ethanol to diphenylketene (Brønsted plot).

While for pyridine and the larger Concave Pyridines **1** the rates for the addition of isopropanol and *tert*-butanol are considerably smaller than the rates for the addition of primary alcohols, the smaller Concave Pyridines **2** exhibit hardly any selectivity. If **2** is compared to other catalysts and to the catalysis by an amide, it becomes obvious that not only the selectivity of **2** for the addition of various alcohols is small but also the absolute rate of catalysis is slow and only comparable to that of an amide (see Fig. 1). It therefore can be assumed that the smaller Concave Pyridines **2** catalyze via their amide bridgeheads.

From the data of Fig. 2 selectivities can be calculated. Table 1 compares these calculated values with data obtained by direct competitions.

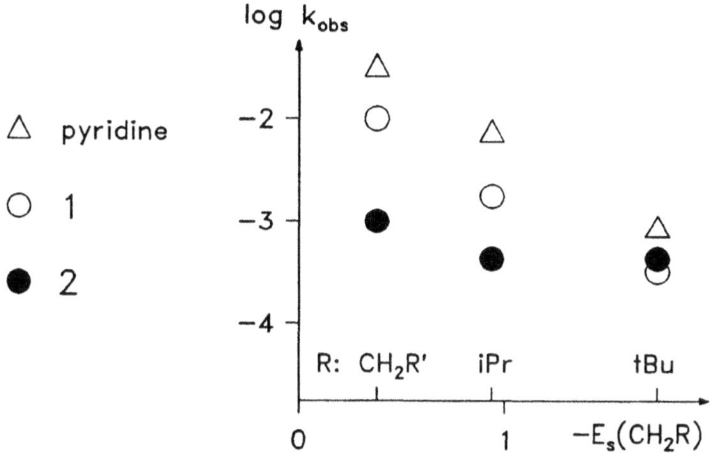

Figure 2. Correlation between the logarithm of the rate of catalysis log k_{obs} and sterical parameters $E_s(CH_2R)$ for the base catalyzed addition of alcohols ROH to diphenylketene.

Table 1. Comparison of selectivities for the addition of ethanol or isopropanol to diphenylketene obtained by calculation from the k_{obs} values ($1°/2°_{calc}$) or by direct competition ($1°/2°_{comp}$). It must be remarked that the experiments cannot be carried out identically because in the competition experiment either the concentration of each alcohol [ROH] can be chosen identically or the overall alcohol concentration $\Sigma_i[R_iOH]$ can be kept constant.

base	pyridine	**1**	**2**
$1°/2°_{calc}$	3.8	5.3 - 5.9	1.8 - 1.9
$1°/2°_{comp}$	4.2	8.3 - 11.5	3.5 - 4.6

Because the selectivities of the direct primary/secondary competitions $1°/2°_{comp}$ were even larger than the calculated selectivities $1°/2°_{calc}$ intramolecular 1°/2° competitions have been investigated, too. Indeed, the 1°/2° selectivity for the diphenylacetylation of 1,2-propanediol was in favor of the acylation of the primary OH group. As for the intermolecular competition the larger Concave Pyridines **2** again were most selective (see Table 2).

Table 2. Intramolecular competition $1°/2°_{intra}$ for the base catalyzed addition of diphenylketene to the primary and secondary OH group of propanediol.

base	pyridine	**1**	**2**
$1°/2°_{intra}$	7	8 - 15	3

The most remarkable selectivities have been found for the diphenylacetylation of the hydroxyl groups of certain sugar derivatives. Tables 3 and 4 list the results for the selective functionalization of the glucose derivative **3** and the chinovose derivative **4**. In both sugars all non-protected hydroxyl groups are secondary. In addition they are all equatorial! With good yields only one OH group is functionalized when a Concave Pyridine is used as catalyst.

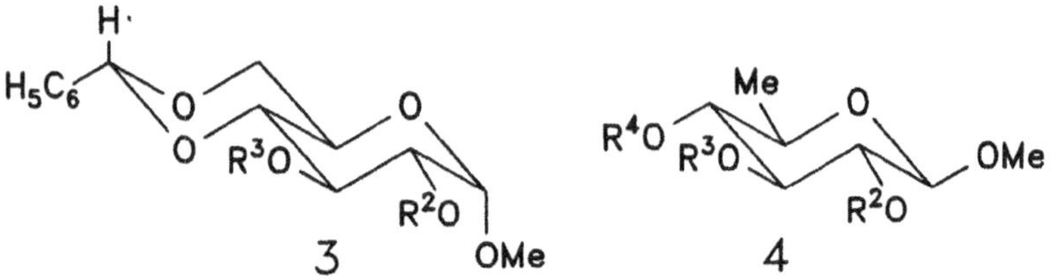

3 4

Table 3. Yields (%) of mono- and bis(diphenylacetates) of the glucose derivative **3**.

	uncatalyzed	pyridine	**1** (R = NEt_2)
2-acylated	42	50	67
3-acylated	36	12.5	7
2,3-bis(acylated)	3	12.5	-
ratio (2-acyl./3-acyl.)	1.2	4	9.6

Table 4. Fraction (%) of (presumably) 3-diphenylacetylated **4**. One tri-, three di- and three monoacylated products can be formed.

pyridine	**1** (R = OMe)	**1** (R = NEt_2)
29	60	48

In the future we will extend the application of our Concave Catalysts to other polyhydroxy compounds (other sugars, disaccharides, cellulose etc.).

[1] Lüning, U. (1987) Concave Reagents: Syntheses of Macrobicyclic Pyridines, *Liebigs Ann. Chem.* 949 - 955.

[2] Lüning, U., Baumstark, R. and Schyja, W. (1993) Concave Reagents 14. Concave Pyridines with a 2,6-Diarylpyridine Core, *Tetrahedron Lett.* 34, 5063 - 5066; and references cited.

[3] Lüning, U. (in press), Concave Acids and Bases, *Top. Curr. Chem.*

[4] Lüning, U. and Müller, M. (1992) Concave Wrapped Protons: Reagents for Contra-Thermodynamic Protonations (Concave Reagents 10), *Angew. Chem.* 104, 99 - 102; *Int. Ed. Engl.* 31, 80 - 82; and references cited.

[5] Lüning, U., Baumstark, R. and Schyja, W. (1991) Concave Reagents 9: Alcoholysis of Diphenylketene catalyzed by Concave Pyridines and Open-Chain Analogs, *Liebigs Ann. Chem.* 999 - 1002.

FROM *p-tert*-BUTYL-CALIX[6]ARENE to *p-tert*-BUTYL-[1^6]META-CYCLOPHANE

Isolation and Crystal Structure of a Mono Bridged Penta-Phosphoester

J. -B. REGNOUF de VAINS[a*], R. LAMARTINE[a], S. PELLET-ROSTAING[a], M. PERRIN[b], A. THOZET[b], S. LECOCQ[b],

a) Laboratoire de Chimie Industrielle; b) Laboratoire de Cristallographie; Université Claude Bernard, 43, Bd. du 11 Novembre 1915, 69622 Villeurbanne, France.

Calixarenes are cyclic phenolic oligomers mixing complexation abilities for neutral or ionic guests and a good functionalization flexibility with a high thermal stability. Since the pionnering work of Gutsche [1], they became a promising chemical platform for the building of a new and large family of synthetic receptors [2].

A part of our investigations in this field is dedicated to the replacement of the -OH groups by other reactive functionalities, such as NO_2 and NH_2 or halogens, in view to enlarge our ability to build new supramolecular devices.
In this sense, we thought that such transformations could be efficiently carried out starting from a [1^n] metacyclophane.
We found in fact that Biali et al. [3] prepared in good yields from the *p-tert*-butyl-calix[4] and [8]arenes the corresponding [1^4] and [1^8] metacyclophanes by a procedure involving the activation of the phenolic OH group into its di(ethyl) aryl phosphoester, followed by a reductive cleavage step by means of potassium in liquid ammonia. Surprisingly, we

J. S. Siegel (ed.), Supramolecular Stereochemistry, 227–232.

found that the *p-tert*-butylcalix[6]arene was not subjected to this study, although Reinhoudt and coworkers [4] employed this procedure without preparing of its fully OH-depleted analog **1**.

We tried unsuccessfully to obtain **1** by different methods involving reduction of the activated Caryl-O bond, notably the hexa-tosylate with hydrazine/Pd/C or the hexaphosphoester **2** with titanium metal.

By an adaptation of the Biali's method (Figure 1), the *p-tert*-butyl-calix[6]arene **4** [5] was treated by a large excess of $ClP(O)(OEt)_2$ and aqueous NaOH in PTC conditions, affording the hexaester **2** and a compound **3** we analyzed as a hexa-esterified species bearing one bridging phosphoester unit, in 20% and 17% yields respectively. Replacing aqueous NaOH by solid NaOH [4] changed those values to 35% and 12%. Reacting **3** in the same PTC conditions did not produce **2**, allowing us to certify that **3** and **2** are co-products in this reaction.

The conversion of **2** into **1** was performed by reducing the Caryl-O-bond with potassium metal in liquid ammonia at -70°C, following the previously, but slightly modified, described procedure[3a]. **1** was isolated with a yield of 33%. The same procedure applied on **3** did not afford **1** but a mixture of polar compounds which are actually under investigations.

^{1}H-NMR was not sufficient enough, at room or low temperature, to give clearly the structure of **2**. At 225 K, the structure begins to rigidify, but the resonance signals are still too much overlaped to be analyzed; at 298 K, **2** can be described with broad signals at the related aromatic and alkyl positions.

In the case of 3, the room temperature ^{1}H-NMR spectrum is perfectly well defined in the aromatic region, presuming a rigidified conformation: ten resonance signals (proton ratio: 1, 1, 1, 2, 1, 2, 1, 1, 1, 1) are located as doublets (J = 2.2 Hz) at 6.22, 6.27, 6.76, 7.15, 7.21, 7.24, 7.37, 7.60, 7.63 and 7.69 ppm respectively. Attempts to attribute directly those peaks were unsuccessful. Three bridging positions can be considered (**3 a**, **b** and **c**), in each of which many conformations can be proposed. The X-ray crystal structure analysis of **3** is in fact in agreement with the hypothesis

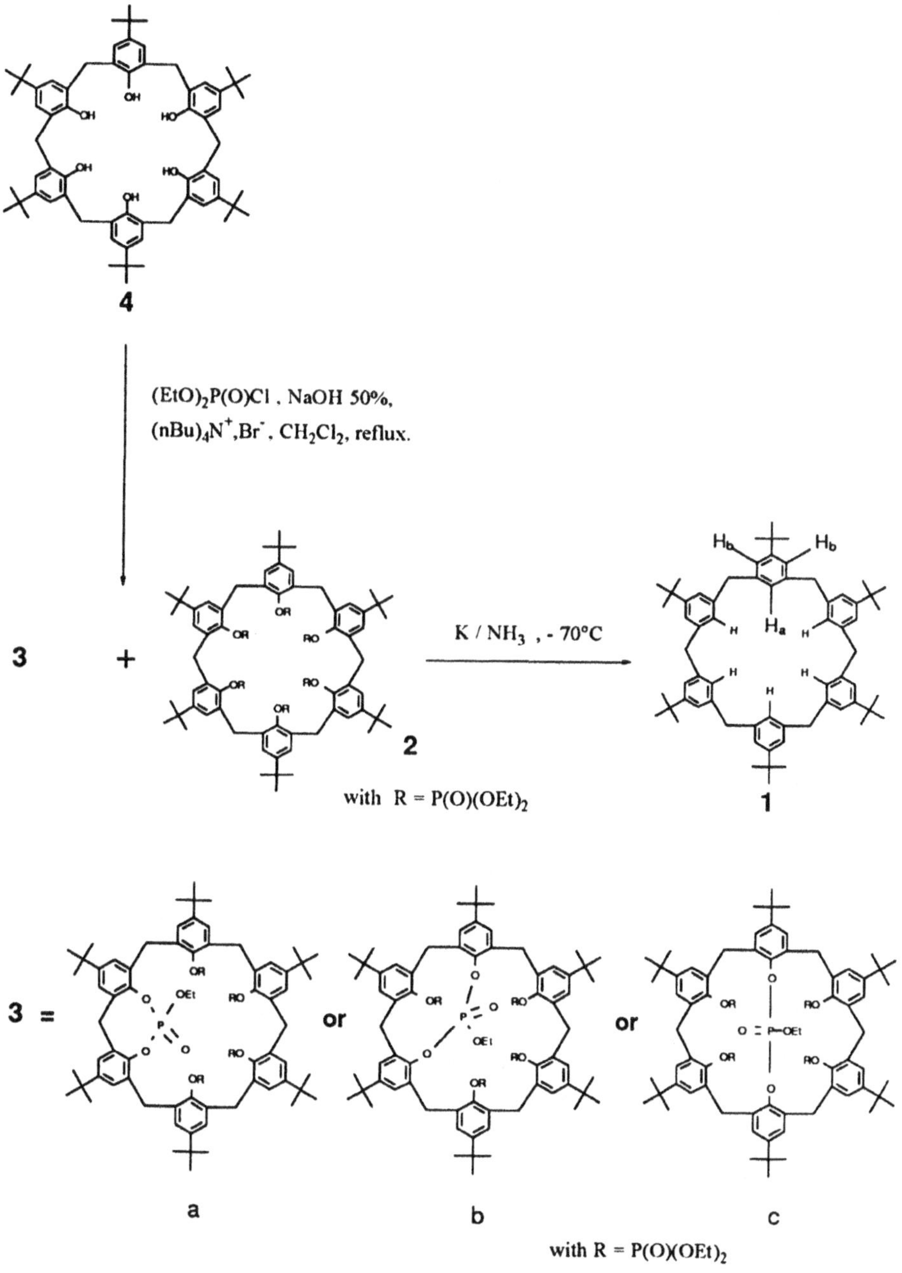

Figure 1. Synthetic pathway from
p-tert-butylcalix[6]arene to *p-tert*-butyl[1^6] metacyclophane.

3a where the bridging phosphoester group is spanning two adjacent phenol subunits. The molecule crystallizes in the triclinic system, space group $P\bar{1}$, showing two independant molecules slightly differenciated by the positions of their pendant ethoxy groups (Figure 2).

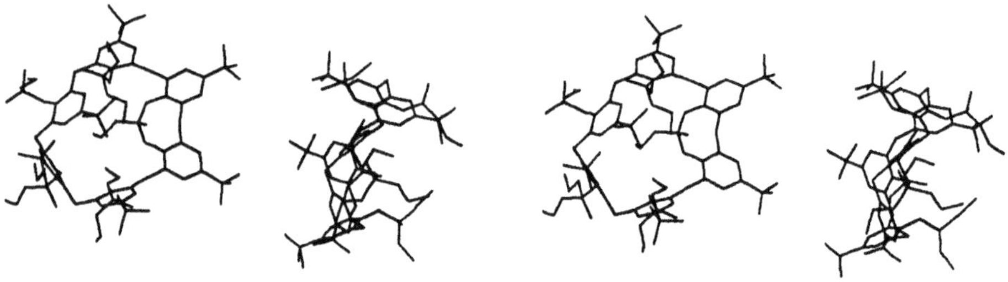

Figure 2. Stereoscopic view of the two independant molecules in the crystal.

Both molecules display the residual OEt group of the bridge positionned inside the cavity. The two substituted phenols adjacent to the bridge are symetrically positionned and oriented, presuming the presence of a mirror. This symetry is in fact broken by the fifth and the sixth substituted phenol which are conversely oriented with regard to the plane of the crown.

^{1}H-NMR spectroscopy of **1** shows the high equivalence in the Ar-CH_2-Ar and aromatic protons families, characterized by three singlets at 3.81, 6.65 (Ha) and 6.96 (Hb) ppm respectively, in agreement with the fact that the removal of the hydroxyl hydrogen bonds may result in an increase of the conformational flexibility of this new molecular system. ^{13}C-NMR and DEPT experiments confirmed the structure of **1**.

Other analytical techniques (UV, IR, MS, Elem.Anal.) gave satisfactory results for **1**, **2** and **3**.

Studies of conformational and chemical properties of **1**, **2** and **3** are presently under investigation, as well as the search of new synthetic pathways to change **4** into **1** and in a general manner, any calixarene into their corresponding [1^n] metacyclophanes.

We actually pursue the structure refinement of **3** and try to correlate these results with high resolution NMR conformation studies in solution.

REFERENCES

1. Gutsche, C. D. (1989) Calixarene, in J. F. Stoddart (ed), *Monographs in Supramolecular Chemistry, Series*, Royal Society of Chemistry, Cambridge.
2. a) Shinkai, S. (1993) Calixarene as the Third Supramolecular Host, in Gokel, G.W. (ed), *Advances in Supramolecular Chemistry*, J. Press Inc., Vol 3, pp 97-130.
3. a) Grynszpan, F., Goren, Z.and Biali, E.(1991) Partially OH depleted calix[4]arene *J. Org. Chem.* **56**, 532-536; b) Goren, Z. and Biali, S. E. (1990) Multiple reductive cleavage of calixarene diethyl -phosphoesters: a route to [1^n]metacyclophanes, *J. Chem. Soc., Perkin Trans. I*, 1484-1487.
4. Janssen R. G., Verboom W., Harkema S., van Hummel G. J., Reinhoudt D. N., Pochini A., Ungaro R., Prados P.and de Mendoza J. (1993) Selective phosphorylation of *p-tert*-Butylcalix[6]arene, *J. Chem. Soc., Chem. Commun.* 506-508.
5. Gutsche, C.D., Dhawan, B., Konig, M and Stewart, D. (1989) p-tert-Butylcalix[6]arene, *Org. Synth.* **68**, 238-242.
6. Markovski, L. N. Kal'chenko,V. I. and Parkhomenko, N. A. (1990) O-Phosphorylated *p-tert*-Butylcalixarene, *Zhurnal Obshchei Khimii* **60**, 2813; Byrne, L.T., Harrowfield, J. M., Hockless, D. C. R., Peackey, B.J., Skelton, B.W.and White, A. M. (1993) Calixarene Phosphorylation: Cation Control of Reaction Pathways, *Aust. J. Chem.* **46**, 1673-1683; Moran, J. K. and Roundhill D. M. (1992) Introduction of Bridging Ethyl Phosphate Substituent onto the Lower Rim of Calix[6]arenes, *Phosphorus, Sulfur and Silicon*, , **71**, 7-12; Grynszpan, F., Aleksiuk, O. and Biali, S. E. (1993) Phosphorus Polybridged Calixarenes, *J.Chem. Soc., Chem. Commun.* , 13-16.

Two Shells About A Nut:

Carceplexes, Hemicarceplexes and Highly Stable Ternary Complexes

J.C. Sherman
Department of Chemistry, 2036 Main Mall
University of British Columbia
Vancouver, British Columbia, Canada V6T 1Z1

1. Introduction

Carceplexes are closed surface molecules that permanently entrap guest molecules within their shells. We have shown [1] that the reaction to form carceplex **2**•guest (*Figure 1*) is subject to a template effect such that pyrazine is a one million times better template that *N*-methyl-2-pyrrolidinone (NMP). This and 22 other *template ratios* were determined by running the carceplex reaction in the presence of pairs of template molecules and measuring the product ratios by integration of the host and guest signals in the ^{1}H NMR spectra of the mixtures. The template ratios reflect the relative rates of the guest determining step (GDS, the step beyond which no guest exchange can occur) for each template molecule. We suggested that one of the driving forces for carceplex formation is favorable van der Waals interactions of the template molecule with the walls of the forming cavity in the transition state of the GDS.

2

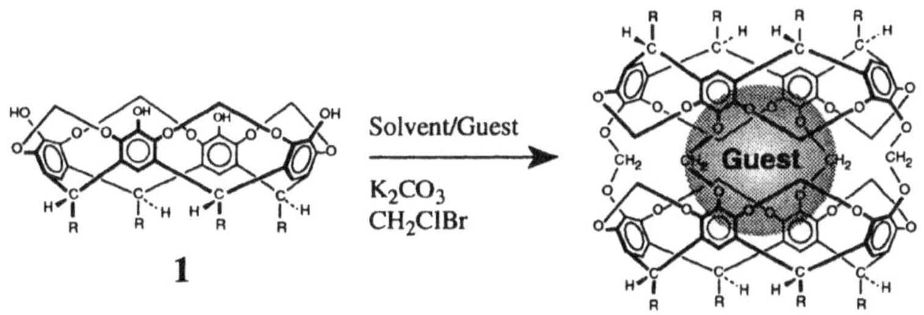

1

2•Guest

Figure 1. Synthesis of carceplex **2**•guest.

J. S. Siegel (ed.), Supramolecular Stereochemistry, 233–238.

2. Pyrazine Orientation

To further probe the powerful template ability of pyrazine, we have investigated its orientation within the shell. We synthesized an asymmetric carceplex, represented schematically in *Figure 2*, and found that there were two signals for entrapped pyrazine in the ^{1}H NMR spectrum, each consisting of a doublet with 1.2 Hz coupling. We concluded that pyrazine must be oriented as shown below (**A**) since only this orientation affords two sets of non-equivalent, *meta*-coupled signals. The activation barrier to rotation of the pyrazine about the C_2 axes of the shell is 19 kcal/mol and represents the energy needed for the narrow equator of the shell to accommodate the 7.1 Å versus 5.6 Å cross section of pyrazine.

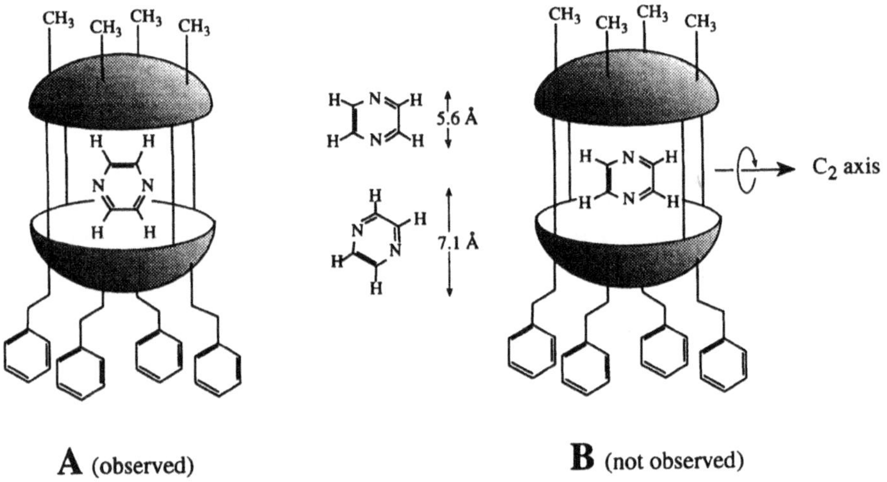

Figure 2. Two potential orientations of pyrazine in asymmetric carceplex **2**•pyrazine.

3. Hemicarceplex

We next investigated the template effect on the formation of the less symmetric hemicarceplex **4**•guest from triol **3** [2] as represented in *Figure 3*. This reaction is statistically more demanding than the carceplex reaction since, for example, there are nine ways the first bridge can be formed, only three of which can lead to product. We performed the same competition experiments using triol **3** as we did using tetrol **1** to generate template ratios. The ratios are tabulated in Table 1 along with the carceplex template ratios and reaction yields. The yields obtained exceed the statistical yields for both carceplex **2**•guest and hemicarceplex **4**•guest. A plot of the log of the two sets of template ratios yields a line with a correlation of 0.97. This means that there is a correlation between the forces at play in the transition states of the GDS's for the carceplex and hemicarceplex reactions.

2

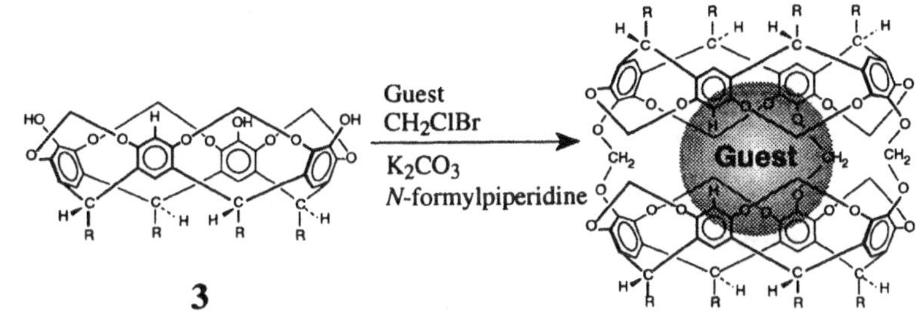

3

R = CH_2CH_2PH

4•Guest

Figure 3. Synthesis of hemicarceplex **4**•guest.

TABLE 1. Template ratios for carceplex **2**•guest and hemicarceplex **4**•guest

Guest	% Yield of Hemicarceplex **4**•Guest	% Yield of Carceplex **2**•Guest	Template Ratio for Hemicarceplex **4**•Guest	Template Ratio for Carceplex **2**•Guest
pyrazine	56	87	170 000	1 000 000
1,4-dioxane	57	68	52 000	290 000
DMSO	51	63	6 200	70 000
THF	39	50	1100	12 000
acetone	40	51	620	6 700
pyrrole	48	73	360	1 000
1,3,5-trioxane	16	24	10	100
DMA	14	15	2	20
NMP	11	5	1	1

4. Reaction Intermediates

To further probe the forces involved in the transition state of the GDS for the carceplex reaction, we are trying to isolate the reaction intermediates that are represented schematically in *Figure 4*. Premature workup of the carceplex reaction yields mostly starting tetrol **1**, product carceplex and the tris-bridged intermediate. We found that there was no guest exchange upon going from the tris-bridged intermediate to product. Thus, the GDS occurs prior to the formation of the final bridge.

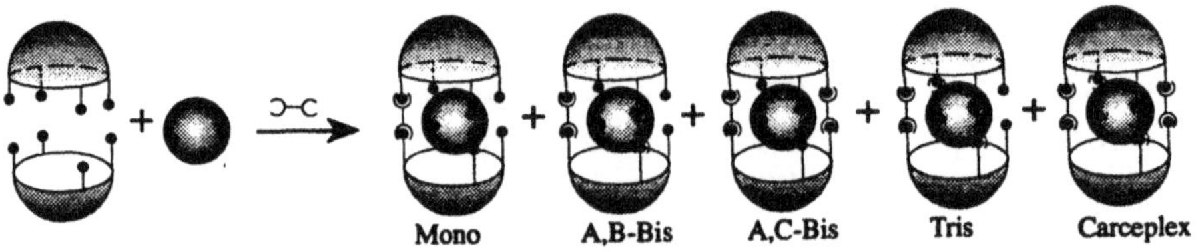

Figure 4. Carceplex reaction intermediates.

5. Ternary Complex

With the isolation of mono and bis-bridged intermediates proving to be somewhat elusive, we decided to investigate the potential interaction of the starting tetrol with the template molecules, speculating that there might be a complex formed prior to bridging. We found that, in the presence of DBU as base, a complex is formed between tetrol **1** and our template molecules. A ^{1}H NMR spectrum of the pyrazine complex is shown in *Figure 5* along with spectra of the "free" tetrol + DBU species and carceplex **2**•pyrazine, for comparison. The complex has the stoichiometry 2:1 tetrol:guest and is in slow exchange on the ^{1}H NMR timescale.

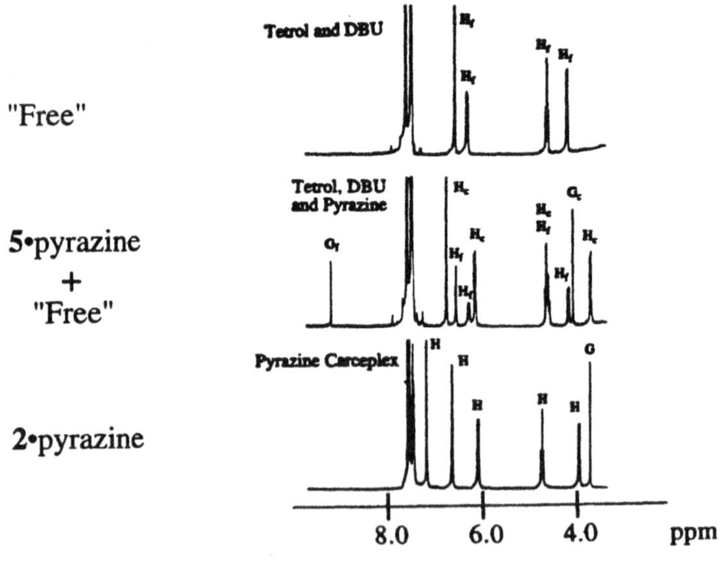

Figure 5. ^{1}H NMR spectra of "Free" species, **5**•pyrazine + "Free" and carceplex **2**•pyrazine. H = host; G = guest; f = free; c = complexed.

The interaction between the two tetrols was investigated by titration of tetrol **1** with base in the presence of pyrazine. The percent complex increased until two equivalents of base were added per tetrol. This indicates that the bowls form four charged hydrogen bonds as represented in structure **5** in *Figure 6*.

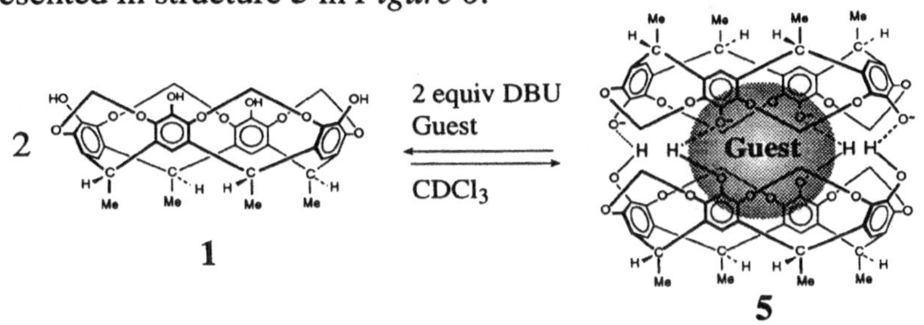

Figure 6. Formation of ternary complex **5**•guest.

To obtain association constants for the ternary complexes, we needed to determine the nature of the "free" species. We obtained ^1H NMR spectra in 1:1 $CDCl_3$ / $CHCl_3$ and suppressed the signal of free $CHCl_3$ at 7.24 ppm. At low temperatures, we observed a new signal at 4.6 ppm which we assigned to complexed chloroform. Thus the "free" species is a ternary complex of chloroform. The association constants for chloroform and six other guests are listed in Table 2 along with the corresponding carceplex template ratios. A plot of the log of the template ratios against the log of the associations constants gave a correlation of 0.97. This means that the forces at play in the transition state of the GDS of the carceplex reaction correlate with the forces at play in the association of the ternary complex. These forces are charged hydrogen bonds between the two bowls and van der Waals interactions between the host and guest.

TABLE 2. Association constants for ternary complexes **5**•guest

Guest	Association Constants (M^{-2}) for Ternary Complexes	Template Ratios (relative to benzene)
Pyrazine	110,000,000	420
Dioxane	4,100,000	120
DMSO	830,000	29
Pyridine	480,000	14
Acetone	61,000	5
Benzene	75,000	1

6. Conclusions

We have demonstrated that the carceplex and hemicarceplex reactions are driven by charged hydrogen bonds between the two bowls and by van der Waals interactions between the template molecules and the walls of the forming cavities. The organizational effect of the hydrogen bonds to form these selective, highly stable ternary complex explain the better than statistical yields obtained in the carceplex and hemicarceplex reactions. Also, these hydrogen bonded complexes represent a new family of switchable (by adjustment of pH) self-assembling structures. The molecular encapsulation in these self-assembly structures may have use as drug delivery devices.

Acknowledgments

I thank the University of British Columbia and the Natural Sciences and Engineering Research Council of Canada for financial support of this research. Acknowledgement is made to the donors of The Petroleum Research Fund, administered by the ACS, for partial support of this research.

References

1. Chapman, R.G., Chopra, N., Cochien, E.D., and Sherman J.C., (1994) Carceplex formation: scope of a remarkable efficient encapsulation reaction, *J. Am. Chem. Soc.* **116,** 369-370. For the original preparation of carceplex **2**, see Sherman, J.C., Knobler, C.B., and Cram, D.J. (1991) Synthesis and properties of soluble carceplexes, *J. Am. Chem. Soc.* **113,** 2194-2204.
2. Chopra, N. and Sherman, J.C. (in press) Templation effects on formation of a hemicarceplex, *Supramol. Chem.* For original preparation of hemicarceplex **4**, see Cram, D.J., Tanner, M.E., Knobler, C.B., (1991) Guest release and capture by hemicarcerands introduces the phenomenon of constrictive binding, *J. Am. Chem. Soc.* **113**, 7717-7727.

Cyclodextrin - Metallocene Inclusion Complexes

V.I.Sokolov

Institute of Organo-Element Compounds,
Russian Academy of Sciences, 28 Vavilov Street
Moscow, Russia

After a centenary history the stereochemistry of individual molecules is close to its principal completion. Now aggregates of molecules are becoming a central topic of modern stereochemistry. Therefore the great importance is gained by what can be called
'Supramolecular Stereochemistry'.

Host-guest chemistry does obviously match the area embraced by this term due to the importance of mutual steric requirements to be met during the formation of inclusion complexes. This is even more true if chiral aspect is involved when one component is optically active.

Among common host molecules natural cyclodextrins (CyD's) are available as optically pure substances capable of including a wide vari ety of organic molecules (1). Moreover, different kinds of organo-metallic molecules are able to be included into the internal cavity of CyD's. Existence of three cyclodextrin homologues composed of six (α) seven (β), and eight (γ) α-glucose residues permits to study the influence of the cavity diameter on the complex formation.

J. S. Siegel (ed.), Supramolecular Stereochemistry, 239–245.

The most convenient method to investigate the cyclodextrin complexes in solution is the spectroscopy of induced circular dicroism (ICD) because the absorption bands in UV-VIS spectra of the guest molecules exhibit Cotton effects only when embedded into the CyD cavity. Intensity of the Cotton effect is proportional to the concentration of a complexed part of the guest molecules in the equilibrium:

$$\text{CyD} + \text{Z} = [\text{CyD}\,.\,\text{Z}]$$

Ability of the first organometallic guest, ferrocene, to be complexed by β-CyD in solution had been reported in 1975 (2).Isolation of several solid inclusion complexes with ferrocenes and observation of the ICD spectra in ethylene glycol had been reportedin 1984 by Harada and Takahashi (3). Later on complexation of other kinds of organotransition metal π-complexes has been found (4-7): ruthenocene, osmocene, benchrotrene, cymantrene, $CpFe(CO)_2$ dimer, π-allylpalladium halides, π-benzeneruthenium dihalides, ferricinium and cobalticinium,(arene)cyclopentadienyliron salts,cyclopentadienylrhodium-(diene) etc.

Herein the results obtained by our group will be outlined previously published mostly in Russian journals.

We have studied the equilibria in ethylene glycol using CD spectroscopy and determined the association constants for a series of metallocenes including substituted ferrocenes and unsubstituted osmocene, ruthenocene, cymantrene, benchrotrene (Table 1). Besides β-CyD a covalent derivative (6-OTs-β-CyD), α- end γ-CyD's were used.

Introduction of substituents into the ferrocene molecule decreases the binding constant progressively probably due to steric reason. For ruthenocene and osmocene binding is even better than for ferrocene. The presence of electron-withdrawing CO ligands diminishes the constant value. However, the positive charge does not prevent to form inclusion complex (metallocenium salts). Subsequently, Klingert and Rihs have investigated the crystal structure of CpFe(arene) salts (9). X-ray structure of the neutral metallocene inclusion complex has been reported (10) for $Cp_2Fe.2\alpha$-CyD only.

Table 1

Equilibrium constants with β-cyclodextrin (ethylene glycol,25°C)

π-COMPLEX	λ_{max} (nm)	K_{ass} (M^{-1})
Cp_2Fe	460	510
Cp_2Ru	334	660
Cp_2Os	272	790
$CpMn(CO)_3$	327	8.8
$C_6H_6Cr(CO)_3$	312	33
$C_5H_4COCH_3$ Fe C_5H_5	495	250
$C_5H_4COCH_3$ Fe $C_5H_4COCH_3$	490	15
C_5H_4 C=O Fe CH_2 CH_2 C_5H_4	364	1.9

For four metallocene samples the temperature-dependent CD spectra have been studied in ethylene glycol in the 12-90° C range and thermodynamic parameters evaluated (11), Table 2. Only feeble tendency for increase of free energy has been observed on going from iron through ruthenium to osmium. On the contrary, the introduction of a single acetyl group in ferrocene causes the noticeable decrease of the free energy of binding.

Finally, we were interested in the covalent linkage of the organometallic fragments to a cyclodextrin core. Successful was the direct metallocarbonylation of per-benzoyl-CyD's which afforded arenechromiumtricarbonyl derivatives containing one organochromium moiety in the case of α–CyD and two in the case of β–CyD.Both compounds were characterized by conventional methods and exhibited in CD spectra Cotton effects corresponding to the benchrotrene chromophore (415 and 320 nm) as well as more intense Cotton effects for benzoyl chromophore (287 and 248 nm).

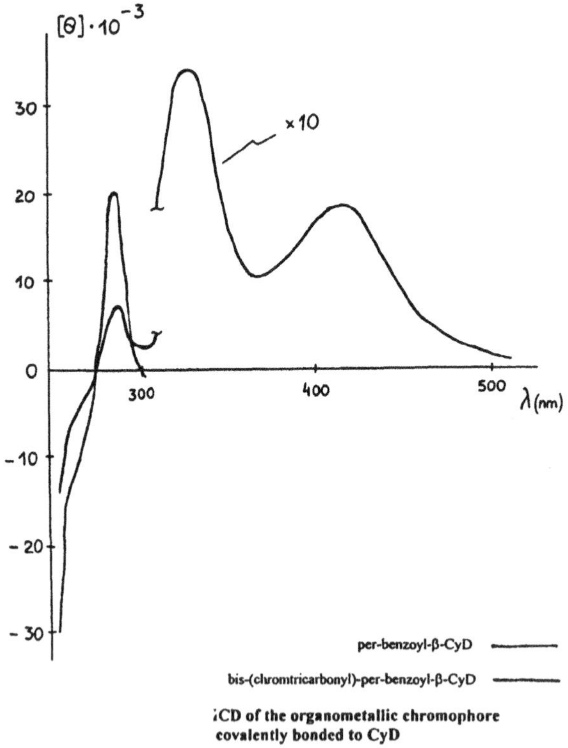

;CD of the organometallic chromophore covalently bonded to CyD

Table 2 **TEMPERATURE - DEPENDENT CD SPECTRA**

	AcC_5H_4FeCp	Cp_2Fe	Cp_2Ru	Cp_2Os
ΔG_{298}^{o}, kJ/mol	-12.4	-15.0	-16.1	-16.4
ΔH^{o}, kJ/mol	-49.6	-42.7	-42.7	-36.9
ΔS , J/mol K	-124.8	-92.9	-85.9	-68.8
t_0 , °C	88	114	144	155

Metalloorg.Khim. (Russian) ,1991, v.4 , N 3 , 697 -699

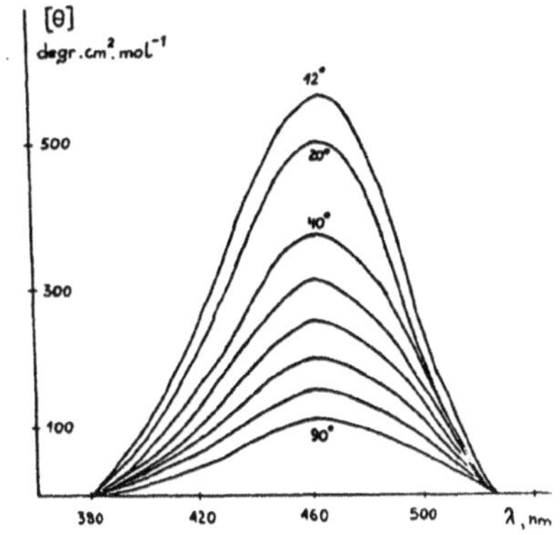

FcH-β-CyD (1·10^{-2} mol·L^{-1})

It seems possible to design the cyclodextrin derivatives which may be precursors for the synthesis of the complexes similar to 'carcerands'. The idea is to covalently bind a chain to a hydroxyl on the broad end of the conical cavity (2-OH or 3-OH) then put a guest molecule into the cavity and, finally to close a bridge across the broad end as shown on the scheme. The bridge will prevent the included molecule from going out of the cavity thus ensuring the formation of the species with topological bond. The bridge-closing reaction may be, for example, thermal or photochemical addition and can be done in the solid complex if an appropriate model will be found.

To summarize, it was demonstrated that metallocenes are able to form the inclusion complexes with cyclodextrins stable in solid state. The binding constants were determined in solution (ethylene glycol) for a series of cyclopentadienylmetals including substituted ferrocenes. Thermodynamic parameters were obtained for ferrocene, ruthenocene, osmocene, and acetylferrocene and found to be close to those of arene hydrocarbons that suggests the primary importance of hydrophobic interactions for the stability of the complexes.

Good perspectives can be anticipated for application in non-linear optics because inclusion into the optically active matrices of CyD's make the the crystals non-centrosymmetric. It seems that included compounds with a high second-order hyperpolarizability like, for instance, 1-ferrocenyl-2-p-nitrophenylethylene investigated by M.L.H.Green et al (13) can be used as NLO materials.

The experimental work had been carried out by Dr. V.Bondareva-Don, the circular dichroism studies by Drs. I.Golovaneva and G.Shustov.

References

1. M.L.Bender, M.Komiyama. Cyclodextrin Chemistry. Springer, Berlin, 1978.
2. B.Siegel, R.Breslow. J.Am.Chem.Soc., 90, 6869 (1975).
3. A.Harada, S.Takahashi. J.Chem.Soc., Chem.Commun., 645 (1984).
4. A.Harada, K.Saeki, S.Takahashi. Chem. Letters, 1157 (1985).
5. A.Harada, M.Takeuchi, S.Takahashi. Chem. Letters, 1893 (1986).
6. V.I.Sokolov, V.L.Bondareva et al. Izv. AN SSSR, ser.khim., 1689 (1987).
7. V.L.Bondareva-Don. Ph.D. Thesis, INEOS RAN, Moscow, 1992.
8. V.I.Sokolov, V.L.Bondareva, I.F.Golovaneva.
   (a) J.Organometal.Chem., 358 (1988) 401.
   (b) Metalloorg. Khim. (Russ.) 1, 716 (1988); 2, 1252 (1989).
9. B.Klingert, G.Rihs. J.Chem.Soc., Dalton, 2749 (1991).
10. Y.Odagaki, K.Hirotsu et al. J.Chem.Soc., Perkin I, 1230 (1990).
11. V.I.Sokolov, V.L.Bondareva, G.V.Shustov. Metalloorg. Khim. (Russ.) 4, 697 (1991).
12. V.I.Sokolov, V.L.Bondareva. Metalloorg. Khim. (Russ.) 1, 1195 (1988).
13. M.L.H.Green, J.Qin et al. Pure Appl. Chem., 61, 817 (1989).

DENDRITIC POLYPHOSPHATES: GLOBULAR POLYELECTROLYTES?

YITZHAK TOR
Department of Chemistry and Biochemistry
The University of California, San Diego
La Jolla, CA 92093
U.S.A.

ABSTRACT. The design and the synthetic approach to the automated solid-phase synthesis of dendritic polyphosphates are described. Some potential applications of these novel materials are discussed.

1. Introduction

Dendrimers are hyperbranched, highly-ordered oligomeric structures that emerge from a single polyfunctional core. The unique architecture possessed by these compounds is manifested in their coined names: cascade molecules, starburst polymers and arborols. Their size, shape and topology, as well as flexibility and surface functionalities, are controlled at the molecular level [1,2,3]. This entirely new family of molecules presents challenging problems in polymer physics and chemistry, and displays novel properties. Consequently, these dendritic compounds have been predicted to replace a range of existing materials in a variety of applications, including adhesives, drug delivery systems and catalyst carriers [4]. These hyperbranched oligomeric compounds--unique among polymeric compounds--have also been proposed as building blocks for nanostructured materials [5].

The "dendritic world" provides a unique opportunity to address fundamental questions regarding well defined oligomeric and polymeric compounds. Of particular interest to us is the opportunity to synthesize and study novel structures containing certain functional groups in an unusual chemical environment. In this account, I will describe our design and preliminary effort towards the solid-phase synthesis of dendritic polyphosphates. These novel compounds are especially attractive due to the following reasons.

Novel polyelectrolytes. Unlike natural cylindrical polyelectrolytes (*i.e.*, nucleic acids) or other "rod-like" polyions, dendritic polyphosphates possess an uncommon spherical distribution of negatively charged phosphate groups.

Metal binders. As polyanions of high symmetry such polymers are expected to have high affinity towards metal ions and can serve as "metal sponges".

Biodegradable materials. As phosphodiesters, these compounds are expected to be chemically rather stable but should be sensitive to certain phosphodiesterases.

J. S. Siegel (ed.), Supramolecular Stereochemistry, 247–253.

2. General Synthetic Strategy

Two fundamentally different approaches have been reported for the synthesis of dendritic molecules [2,3]. The "divergent-growth" strategy involves the initial reaction of a tri- or tetra-functional core with a multifunctional monomer. Growth proceeds radially outward with a rapid increase in the number of "surface" groups. The alternative convergent approach involves a covalent attachment of dendritic fragments to a polyfunctional central core where growth begins at the periphery and proceeds inward. Both of these synthetic approaches require elaborate chemistry and multistep procedures [2,3]. The main practical problems are slow reactions and reduced yields. Unlike other types of polymerization reactions, very high yields and high selectivities are required in order to generate well-defined, defect-free dendrimers.

In contrast to these approaches, we propose to use the well-established and optimized solid-phase phosphoramidite chemistry for the construction of negatively charged dendritic structures [6,7]. A general synthetic strategy is shown in Figure 1. The dendrimer's polyfunctional core is attached to a controlled pore glass (CPG) support using a succinate linker. In the first step of the synthetic cycle the 4,4'-dimethoxytrityl (DMT) groups are removed. The free hydroxyl groups are then condensed with the activated phosphoramidite. The trivalent phosphorous is oxidized to the pentavalent triester yielding the DMT-protected, first generation immobilized dendritic structure. The cycle of deprotection-condensation-oxidation is then repeated with the same phosphoramidite or with a different one, providing the second generation dendrimer. After the desired structures have been assembled on the solid support, treatment with concentrated ammonia cleaves the product from the support and hydrolyzes the phosphate triesters to the desired diesters.

The attractive features of the solid-phase approach are worth noting:

- Large excesses of reagents drive the reactions to completion; unreacted reagents are washed of the solid support, and no purification is needed after each monomer addition.
- The synthetic cycle is amenable to automation.
- Different phosphoramidites can be used in the condensation step under similar reaction conditions, thus affording easy access to virtually countless varieties of dendritic structures.
- The growing dendrimer is neutral and does not bear any charges until the final ammonia deprotection step. This eliminates any electrostatic repulsion that may lower the condensation efficiencies and inhibit dendritic growth.

3. Synthesis of the Building Blocks

Figure 2 shows the synthesis of two polyfunctional alcohols. The triol (**2**) was initially selected as the simplest building block. It was obtained by DIBAL reduction of the commercially available dimethyl-3-hydroxyglutarate (**1**). The primary alcohols were selectively protected as the 4,4'-dimethoxytrityl derivatives (DMT), followed by phosphitylation of the secondary OH to give phosphoramidite monomer **3**.

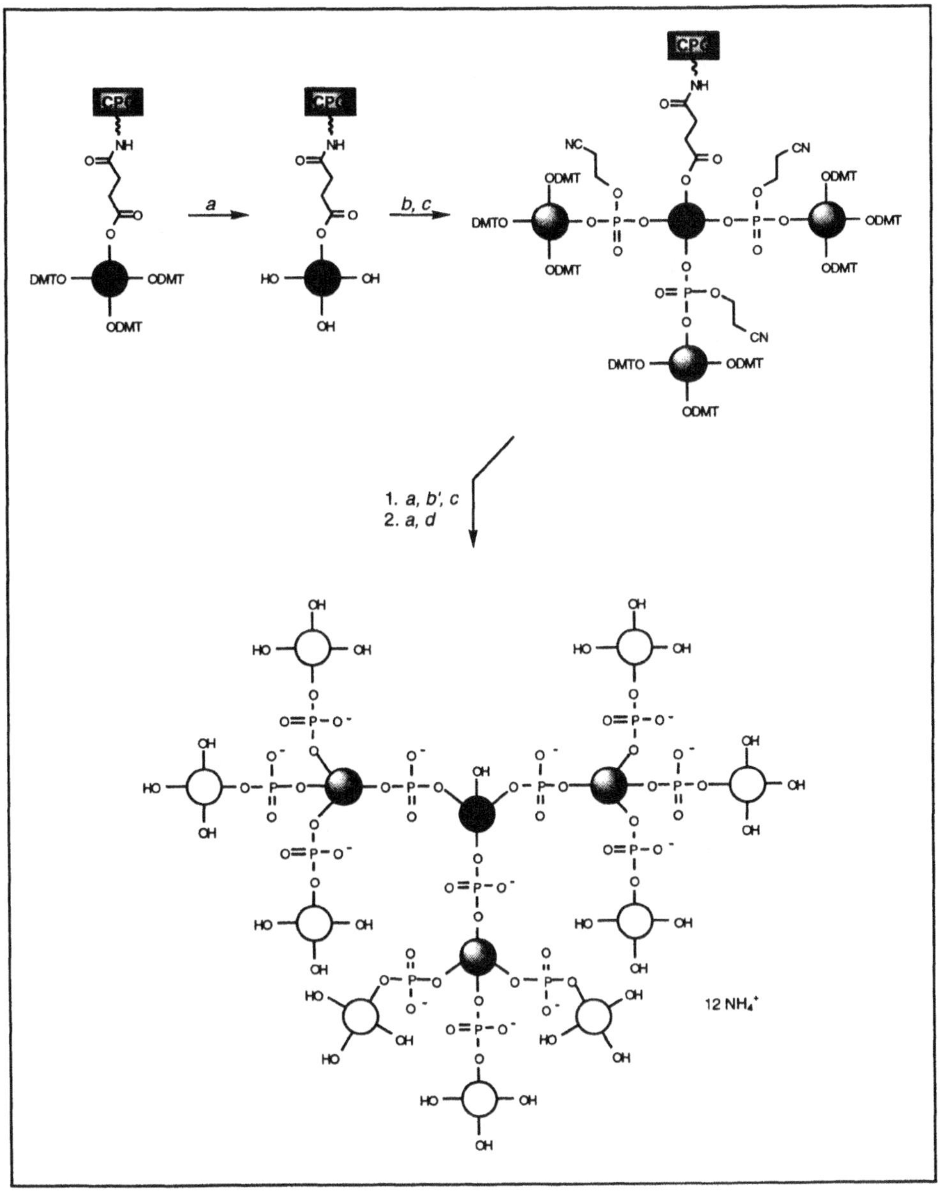

Figure 1. A general approach for the synthesis of polyphosphate dendrimers. Reagents: (a) Cl_3CCOOH, CH_2Cl_2; (b) $(DMTO)_3$●$OP(OCH_2CH_2CN)NiPr_2$, 1*H*-tetrazole; (b') $(DMTO)_3$○$OP(OCH_2CH_2CN)NiPr_2$, 1*H*-tetrazole; (c) I_2, THF-H_2O-2,6-lutidine; (d) NH_4OH, 55°C. ●, ◕ and ○ can be the same or different polyfunctional compounds.

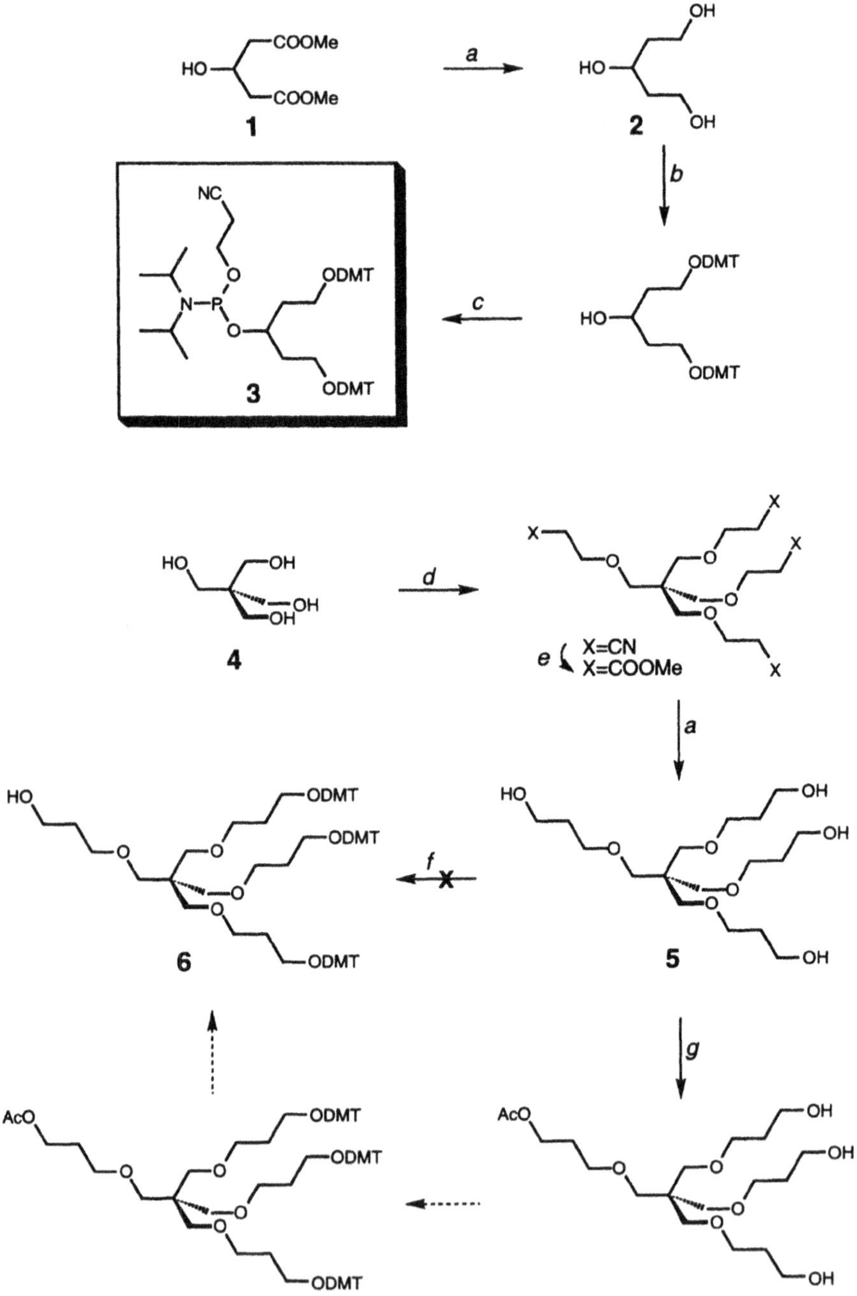

Figure 2. Synthesis of monomers. Reagents: (a) DIBAL, CH_2Cl_2; (b) DMT-Cl, pyridine; (c) $ClP(OCH_2CH_2CN)NiPr_2$, $EtNiPr_2$; (d) $H_2C{=}CHCN$, KOH; (e) HCl/MeOH; (f) 3.3 eq DMT-Cl; (g) Ac_2O, Et_3N, CH_2Cl_2.

Using pentaerythritol (**4**) as a starting material, extended tetrols can be made. Cyanoethylation of **4** followed by methanolysis and subsequent DIBAL reduction of the tetraester gave the desired tetrol **5**. Attempts to selectively protect three out of the four OH's resulted in non-statistical mixtures of products. It was found, though, that monoacetylation proceeds well. This monoacetate can then be transformed to the tris-DMT derivatives by per-tritylation. Mild basic hydrolysis is expected to give the desired tris-protected tetrol **6**.

A test coupling of phosphoramidite **3** and a polyfunctional alcohol was conducted in solution (Figure 3). A mixture of tetrol **5** and four equivalents of phosphoramidite **3** was condensed in the presence of 1*H*-tetrazole and subsequently oxidized. ^{31}P NMR showed a major signal at δ -1.9 ppm for crude **7**, about 150 ppm upfield of the phosphoramidite **3** (δ +148.2 ppm) and is in agreement with the presence of a pentavalent phosphotriester [8].

An important technical issue that remains to be solved is the selection of an appropriate solid-support. Controlled pore glass is available in various pore sizes, ranging from 500Å to 2000Å. Large pore sizes are usually preferred for the synthesis of long oligomers and presumably for the construction of dendritic structures. Yet, these solid supports usually suffer from a relatively low capacity (*i.e.*, low level of functionalizaion). A specific support might need to be developed for the solid-phase synthesis of dendritic polyphosphates.

4. Dendritic Polyphosphates--Potential Properties and Applications

Polyphosphate dendrimers represent a novel family of compounds which, due to their charged nature, are expected to show distinct properties. Unlike other dendritic structures which carry polar or ionizable groups on their periphery (and therefore have micellar characteristics), the phosphate groups are spread along the dendritic structure covering all generations. They therefore represent a unique family of polyelectrolytes that have not yet been studied experimentally, those with a spherical rather than linear distribution of negative charges[9]. The modular synthetic approach presented here allows us to introduce variations into their structures, thus offering a unique opportunity to study fundamental aspects of polyelectrolytes. Controlling the number of phosphate groups, the distance between them and their three dimensional distribution will allow examination of counter-ion condensation on different charge configurations--as well as the effect of ionic strength (and charge shielding) on their folding.

Having phosphate groups buried within the organic matrix of a dendritic structure introduces another intriguing aspect uniquely related to these compounds. Different phosphate groups, depending on the polarity of their microenvironment, are likely to display different pKa's. This will result in different degrees of dissociation and hence charges at different pH's. The need to neutralize high charge densities may generate binding cavities for certain metal ions. Distinct physical properties (*e.g.*, size, intrinsic viscosity) are expected with different metal ions, such as mono-, di- and trivalent, and also with polycationic species such as polyamines.

Metal ion binding by polyphosphate dendrimers may find some novel applications. Multiple binding sites for metal cations can lead to "oligomeric" chelates with relatively high binding constants due to the need to neutralize a large number of negative charges. These structures may function as "metal sponges" which may find applications as toxic metal scavengers or metal delivery systems. Dendritic lanthanide complexes carrying large numbers of metal ions (Gd^{3+} in particular) may find use as advanced Magnetic Resonance Imaging (MRI) contrast agents [10].

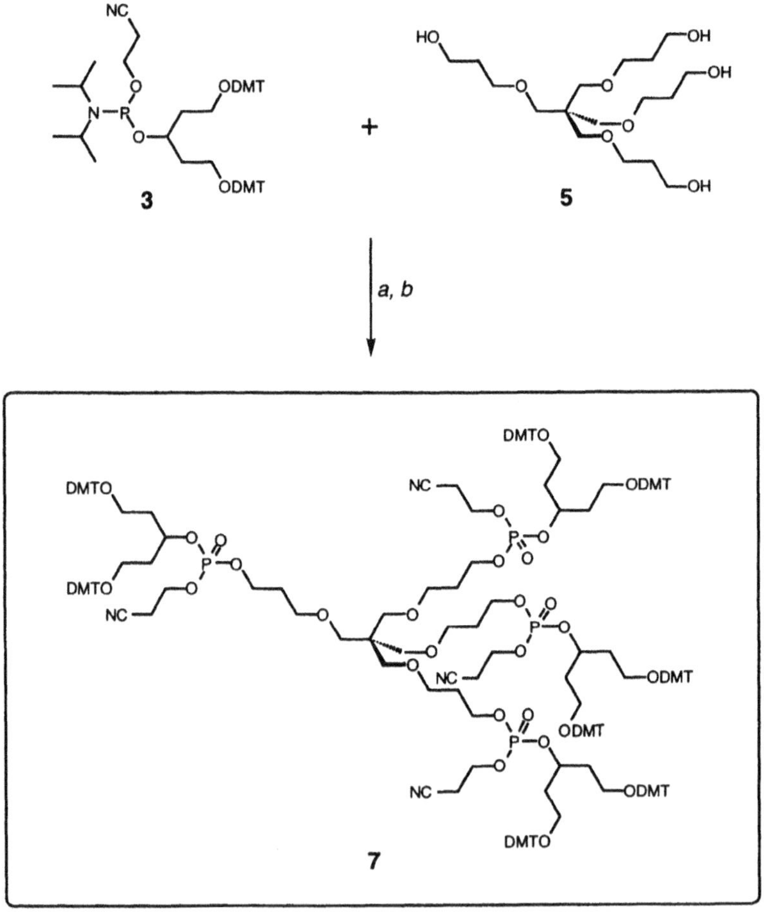

Figure 3. Solution coupling of phosphoramidite **3** and tetrol **5**. Reagents: (a) 1*H*-tetrazole, CH_3CN; (b) I_2, THF-H_2O-2,6-lutidine.

5. Acknowledgments.

I wish to thank Dr. Andrei A. Zlota for his contribution to this project.

6. References

[1] Tomalia, D.A., Naylor, A.M. and Goddard, W.A., III (1990) Starburst Dendrimers: Molecular-level Control of Size, Shape, Surface Chemistry, Topology, and Flexibility from Atoms to Macroscopic Matter, *Angew. Chem. Int. Ed. Engl.* **29**, 138-175.

[2] Mekelburger, H.-B., Jaworek, W. and Vögtle, F. (1992) Dendrimers, Arborols, and Cascade Molecules: Breakthrough into Generations of New Materials, *Angew. Chem. Int. Ed. Engl.* **31**, 1571-1576.

[3] Tomalia, D.A. and Durst, H.D. (1993) Genealogically Directed Synthesis: Starburst*/Cascade Dendrimers and Hyperbranched Structures, *Top. Curr. Chem.* **165**, 193-313.

[4] O'Sullivan, D.A. (1993) Dendrimers Nearing Availability for Commercial Evaluation, *Chem. & Eng. News* 20-23 (August 16).

[5] Miller, T.M.; Neenan, T.X.; Zayas, R.; Bair, H.E. (1992) Synthesis and Characterization of a Series of Monodisperse, 1,3,5-Phenylene-Based Hydrocarbon Dendrimers Including $C_{276}H_{186}$ and their Fluorinated Analogues, *J. Am. Chem. Soc.* **114**, 1018-1025.

[6] (a) Caruthers, M.H.; Barone, A.D.; Beaucage, S.L.; Dodds, D.R.; Fisher, E.F.; McBride, L.J.; Matteucci, M.; Staibinsky, Z.; Tang, J.-Y. (1987) Chemical Synthesis of Deoxyoligonucleotides by the Phosphoramidite Methods, *Meth. Enz.* **154**, 287-313. (b) Beaucage, S.L.; Iyer, R.P. (1992) Advances in the Synthesis of Oligonucleotides by the Phosphoramidite Approach, *Tetrahedron* **48**, 2223-2311. (d) Eckstein, F. (1991) *Oligonucleotides and Analogues, A Practical Approach*, Oxford University Press, New York.

[7] Ulrich, K.E., Boegeman, S., Fréchet, J.M.J. and Turner, S.R. (1991) The Solid-phase Synthesis of Dendritic Polyamides, *Polym. Bull.* **25**, 551-558.

[8] Schirmeister, H., Pfleiderer, W. (1994) Synthesis and Characterization of Modified 2'-5' Adenylate Trimers-Potential Antiviral Agents, *Helv. Chim. Acta* **77**, 10-22.

[9] Lifson, S. (1957) Improved Approximation for the Potential of Spherical Polyelectrolyte Molecules in Solution, *J. Chem. Phys.* **27**, 700-701.

[10] (a) Laufer, R.B. (1987) Paramagnetic Metal Complexes as Water Proton Relaxation Agents for NMR Imaging: Theory and Design, *Chem. Rev.* **87**, 901-927. (b) Wiener, E.C., Brechbiel, M.W., Brothers, H., Magin, R.L., Gansow, O.A., Tomalia, D.A., Lauterbur, P.C., (1994) Dendrimer-based Metal Chelates: A New Class of Magnetic Resonance Imaging Contrast Agents, *Magn. Reson. Med.* **31**, 1-8.

Author Index

(underline indicates participant)

Subject Index

GPSR Compliance

The European Union's (EU) General Product Safety Regulation (GPSR) is a set of rules that requires consumer products to be safe and our obligations to ensure this.

If you have any concerns about our products, you can contact us on ProductSafety@springernature.com

In case Publisher is established outside the EU, the EU authorized representative is:

Springer Nature Customer Service Center GmbH
Europaplatz 3
69115 Heidelberg, Germany

Batch number: 10761697

Printed by Printforce, the Netherlands